Smart Sensing Technology for Agriculture

Smart Sensing Technology for Agriculture

Editor: Elsie Kennedy

Callisto Reference
www.callistoreference.com

Callisto Reference,
118-35 Queens Blvd., Suite 400,
Forest Hills, NY 11375, USA

Visit us on the World Wide Web at:
www.callistoreference.com

ISBN: 978-1-64116-821-2 (Hardback)

Cataloging-in-Publication Data

Smart sensing technology for agriculture / edited by Elsie Kennedy.
p. cm.
Includes bibliographical references and index.
ISBN 978-1-64116-821-2
1. Agriculture--Remote sensing. 2. Wireless sensor networks. 3. Smart materials.
4. Agricultural innovations. I. Kennedy, Elsie.
S494.5.R4 S63 2023
577.82--dc23

Table of Contents

Preface

Every book is a source of knowledge and this one is no exception. The idea that led to the conceptualization of this book was the fact that the world is advancing rapidly; which makes it crucial to document the progress in every field. I am aware that a lot of data is already available, yet, there is a lot more to learn. Hence, I accepted the responsibility of editing this book and contributing my knowledge to the community.

Sensors are small devices that are used to measure or detect natural Earth conditions such as motion, heat, or light, and represent this in analog or digital form. In the agricultural sector, sensors are used in drones, weather stations and robots. Data collected by these sensors is used for monitoring and optimizing crops and protecting the crops from adverse impacts of the environment. Optical sensors are agricultural sensors that involve the use of various frequencies of light to measure the properties of soil. It can determine organic matter in the soil, soil moisture and clay content in soil. Electrochemical sensors are used to gather information about chemical data of soil for detecting the nutrients present in the soil. This book provides a detailed explanation of the applications of sensing technology in agriculture. A number of latest researches have been included to keep the readers updated with the global concepts in this area of study. The book aims to equip students and experts with the advanced topics and upcoming concepts in this area of agricultural technology.

While editing this book, I had multiple visions for it. Then I finally narrowed down to make every chapter a sole standing text explaining a particular topic, so that they can be used independently. However, the umbrella subject sinews them into a common theme. This makes the book a unique platform of knowledge.

I would like to give the major credit of this book to the experts from every corner of the world, who took the time to share their expertise with us. Also, I owe the completion of this book to the never-ending support of my family, who supported me throughout the project.

Editor

1

Precision Agriculture Application for Sustainable Nitrogen Management of *Justicia brandegeana* using Optical Sensor Technology

Ariel Freidenreich, Gabriel Barraza, Krishnaswamy Jayachandran and Amir Ali Khoddamzadeh *

Department of Earth and Environment, Florida International University, 11200 SW 8th St, Miami, FL 33199, USA; afrei006@fiu.edu (A.F.); gbarr077@fiu.edu (G.B.); jayachan@fiu.edu (K.J.)

* Correspondence: akhoddam@fiu.edu

Abstract: Over-fertilization is a common practice in ornamental nursery production. Oftentimes, visual analysis is used to determine plant nutrient levels, leading to less accurate estimates of fertilizer application. This study focused on exploring the suitability of two non-destructive sensors, Soil Plant Analysis Development (SPAD-502) and GreenSeeker^{TM}, for measuring plant tissue nutrient uptake. Florikan Top-Dress fertilizer 12N-6P-8K was applied to *Justicia brandegeana* in various increments (0, 10, 20, 30, 40, and 50 g) to simulate plants with deficient to excessive nitrogen rates. Various parameters were recorded including Normalized Difference Vegetation Index (NDVI) and SPAD readings, soil leachate analysis (nitrates and phosphate), and total leaf carbon:nitrogen (C:N). The NDVI and SPAD readings were recorded biweekly for three months after the initial controlled release fertilizer (CRF) treatments. Leaf C:N was analyzed through dry combustion while nitrates and phosphate were determined from soil leachate. Results suggest that the smaller amount (20 g) of CRF is as effective in providing N to *J. brandegeana* as larger amounts (30, 40, 50 g). Implementation of this fertilizer regimen will result in reduced agricultural nutrient runoff and overall negative environmental impacts. Application of optical sensor technology using SPAD and GreenSeeker^{TM} showed promising results in determining the fertilizer requirements of *J. brandegeana*. This method could serve as a guideline for nursery producers and landscape personnel as a fast and non-destructive tool for sustainable fertilizer management practices within the ornamental plant industry.

Keywords: specialty crop; fertilizer; runoff pollution; ornamental plant; SPAD; NDVI

---

1. Introduction

Throughout the United States, the nursery ornamental production industry has become an increasingly popular agricultural sector. The United States Department of Agriculture's 15 state program has generated a cumulative monetary worth greater than four billion dollars since 2006. California and Florida account for 49% of the 15 state total value, and Florida ranks second in the total wholesale of floriculture crops at 24% of the nationwide supply, making this an economically important industry for the state [1].

Large scale nursery production involves the use of containers to grow plants with the necessary medium, usually a light weight potting mix. Producers are responsible for providing appropriate nutrients for each plant with fertilizer application depending on the targeted species and age [2]. When establishing a fertigation regime, three macro nutrients are necessary for proper plant growth; nitrogen, phosphorus, and potassium (N, P, and K, respectively) [3]. A plant that lacks the required supplement of available N (NO_3^-, NH_4^+) will demonstrate various symptoms that denote feebleness such as slow or stunted growth, yellowing of foliage, leaf chlorosis, and leaf abscission [4]. In contrast, plants with excessive N show symptoms of delayed flowering patterns and overall diminished growth potential [5]. Nursery containers have openings at the bottom to permit aeration and drainage. This allows for water to drain, and in turn, leaching of nutrients [6]. Nutrient leaching has detrimental effects on the environment, particularly in creating eutrophic conditions in natural water bodies [7]. An acre of land in nursery production has the potential to house 40,000 to 300,000 containers, many of which receive excessive fertilizer application [8]. This is a primary driver in declining water quality throughout Florida, which ultimately contributes to algal blooms. Previous studies have shown that N found in fertilizer can be lost as runoff or leach through container [9,10]. These findings call for a reconsideration of recommended fertilization protocols in nursery production.

Justicia brandegeana (Brandegee) is an ornamental plant native to Mexico and naturalized in Florida, based on its tropical characteristic [11]. In its native semi-arid habitat, *J. brandegeana* grows as a shrub with attractive and colorful bracts that surround small white flowers. When observed together the bracts and flowers resemble a shrimp, hence its common name "shrimp plant" [12]. *Justicia brandegeana* has the potential to thrive in North America's hardiness zones 8–11 [13]. Due to its wide range of suitable habitat, *J. brandegeana* is a resilient candidate for landscaping projects. When grown in outdoor gardens, the nectar within its flowers entice native hummingbirds and other pollinators [14]. These factors make the shrimp plant attractive to homeowners for the purpose of landscaping [13]. This study focuses on demonstrating the use of non-destructive handheld sensors to estimate nutritional needs for *J. brandegeana*. GreenSeeker^{TM} (Trimble, Sunnyvale, CA, USA) gives a Normalized Difference Vegetative Index (NDVI) measurement that provides an indication of chlorophyll content within green plants material [15]. This allows for better understanding of a plant's overall health, vigor and nutrient content at a given time. GreenSeeker^{TM} handheld sensor contains two electroluminescent diodes that emit red (660 nm) and near infrared (NIR) light (770 nm) onto the canopy of the target plant [7,16,17]. Plants with green color exhibit a low reflectance of blue and red, and a high reflectance of NIR radiation [18]. The palisade layer within leaves contain the chlorophyll responsible for the reflectance of visible red light and scattering of NIR [19]. The reflectance of NIR light can be directly attributed to the mesophyll cells and cavities present between the cells [20]. As a result, NDVI is a ratio of NIR to red [21], with a range from 0 to 1. Values closer to 1 have a high reflectance of NIR representing a leaf with substantial chlorophyll content, while values closer to 0 represent low chlorophyll content [18].

As an addition to GreenSeeker^{TM} readings, Soil Plant Analysis Development (SPAD) meter (Konica Minolta, Ramsey, NJ, USA) was also used to estimate N. The SPAD-502 meter, similar to the GreenSeeker^{TM}, is a non-destructive handheld sensor that gives a numerical representation of the approximate amount of chlorophyll present in a given leaf [17]. The SPAD meter clips onto a single leaf and records light intensity of 650 nm and 940 nm, measuring the absorbance of light emitted [16]. The 650 nm wavelength is correlated to maximum chlorophyll activity while the 940 nm wavelength accounts for other factors including leaf thickness and moisture content [22]. The difference of the two absorbance values for the light emitted at a given wavelength gives the user a value ranging from −9.9 to 199.99 [23].

Past studies have been conducted for effectiveness of non-destructive handheld sensors on various crops and ornamentals in situ [17,24–27]. However, no studies have specifically focused on *J. brandegeana* and popular controlled release fertilizer (CRF) Florikan. Because *J. brandegeana* is an economically popular plant in South Florida, this trial will provide useful information for nursey producers in and out of the state. As such, the primary objective of this study is to determine effectiveness of non-destructive sensors SPAD and GreenSeeker^{TM} for sustainable fertilizer management in the ornamental production of *J. brandegeana*. Additionally, to investigate current industry fertilizer application compared to lesser rates of Florikan top-dressed fertilizer for sustainable production of this common landscaping plant. It is hypothesized that standard ornamental industry fertilization practices over-compensate with fertilizer application to prevent plant stress. Therefore, precision horticulture and non-destructive sensors can be employed to detect fertilization needs. The results of this study can provide indication for nursery producers to follow and manage adequate fertilizer amendments in appropriate intervals. Water resource contamination stemming from agricultural origins has consistently been a problem in the State of Florida [28]. Eutrophication of water bodies has resulted in various environmental issues including toxic algal blooms [29] and significant vegetation shifts and wildlife presence in Everglades ecosystems [30].

2. Materials and Methods

Two weeks prior to the first sampling date, *J. brandegeana* cuttings were obtained from Symphony Landscaping Nursery (Hollywood, FL, USA). The cuttings were received in 12 cm diameter containers with established roots and no prior fertilization. At Florida International University's greenhouse (Miami, FL, USA), media was prepared using a 2:1 ratio of Sun Gro Horticulture Professional Growing Mix (Sun Gro Horticulture Inc., Bellview, WA, USA) and perlite. This media mix contained no supplementary nutrients to ensure controlled fertilizer additions. Perlite was added to allow for maximum soil aeration in each pot for optimal plant growth [31]. Cuttings were transplanted into 16.5 cm diameter containers with the premixed media. Transplanting of cuttings assured that plants were not confined to small containers that would stunt their growth for the four-month experimental period. Containers were placed randomly onto greenhouse benches, equidistant from each other and exposed to natural photoperiods in a completely randomized design (CRD). Florikan Top-Dress fertilizer 12N-6P-8K (Florikan, Sarasota, Florida Top Dress 12-6-8 60 Series), a polymer coated CRF (2–3-month application rate) was applied in the same transplanting day to individually top-dressed each potted plant. The study included six treatments, each with five replications. Fertilizer treatments were distinguished by 10-gram increments and applied simultaneously onto pre-labeled potted plants (0, 10, 20, 30, 40, 50 g treatments/pot). This allowed the regulation of nutrient uptake for *J. brandegeana* at distinct fertilization rates. The range of fertilizer treatments applied was intended to simulate deficiencies in N at one end, and excessive quantities of N at the other end. All plants were irrigated with water three times per week until fully saturated (% leachate) with average greenhouse temperature of 23–25°C.

Beginning at two weeks after treatment (WAT), GreenSeeker^{TM} was used to obtain NDVI measurements on a biweekly basis. The plants green canopy was the intended target, all other colored surfaces contribute background noise when utilizing the GreenSeeker^{TM}. For this reason, each plant was placed on top of a black surface omitting any surrounding color that would otherwise have affected the readings. The GreenSeeker^{TM} was held at 45 cm above the sample plant for each reading [24]. This assured that the field of view for the sensor was ~11.85 cm in diameter [16]. In consideration of variation in foliage on the canopy top of each plant, three measurements were taken per replicate in each treatment. The mean of the three measurements were considered when analyzing data.

A Soil Plant Analytical Development meter was used simultaneously with the GreenSeeker^{TM} ensuring similar conditions for all plants during the sampling period. For SPAD-502 measurement, three mature leaves were selected at random from the middle to top canopy region for consistency of each plant measurement. The SPAD electrode dial was placed towards the central region of each leaf during measurement. Three mature leaves were sampled randomly with five replicates in each treatment. Initial baseline sampling took place prior to the introduction of any supplemental nutrients. Composite leaf tissue samples for each treatment were collected and analyzed for CN at the 4, 8, 12 WAT sampling periods.

Composite tissue samples of each treatment were analyzed using dry combustion via LECO Truspec Carbon and Nitrogen Analyzer (LECO Corporation, St. Joseph, MI, USA). Total carbon and N of composite leaf samples were quantified as a response of each treatment.

Additionally, the pour-through method [32] was used to obtain leachate samples from individual containers to determine nutrient runoff rates. Each plant was irrigated until a saturated state was reached. Once containers reached the water holding capacity stage, a saucer tray was placed underneath serving as a collection reservoir. Deionized water (350 ml) was applied allowing for the collection of 50 mL of leachate. Samples were stored in 50 mL conical tubes that were immediately refrigerated at 4 °C until laboratory analysis was performed. The pH of each leachate sample was measured. Nitrate and phosphate were quantified using a SEAL AQ2 Automated Discrete Analyzer (Mequon, WI, USA).

Statistical Analysis

Five treatments and a control were established with five replications for each treatment resulting in a total of 30 potted plants which were arranged in a CRD. Containers were placed at random onto greenhouse benches and were labeled by treatment and replicate number. Continuous response variables of NDVI, SPAD chlorophyll index, NO_3^- and P leachate were statistically analyzed using generalized linear mixed model methods. All statistical analysis was performed at $P < 0.05$ level using SPSS statistical software (Version 22.0., IBM Corp., Armonk, NY, USA) Data trends were compared using one-way ANOVA and correlation tests, significant differences between variables were found using post-hoc least significant differences (LSD) method.

3. Results

Normalized Difference Vegetation Index values increased from 4 WAT to 8 WAT in most of the fertilizer rates (Table 1). The inconsistent values associated with 2 WAT (initial reading) can be attributed to low number of leaves resulting in fertilizer background noise (white color). This led to higher recorded NDVI values for all the fertilizer treatments except the control (Table 1). The greatest NDVI values (0.82 and 0.81) were observed at 8 WAT with 30 and 40 g fertilizer rates; although, there was no significant difference between these two values (Table 1). In 8, 10, and 12 WAT, a decrease in NDVI values is evident. It is possible that this decrease in NDVI is due to the additional N requirement of *J. brandegeana* by the 10 WAT sampling period, a crucial time for fertilizer management in nursery production (Table 1).

Table 1. Normalized Difference Vegetation Index values for a plant in a single pot with different rates of fertilizer at different weeks after fertilizer treatment in *Justicia brandegeana*.

Fertilizer Rate (g)	2 WAT *	4 WAT	6 WAT	8 WAT	10 WAT	12 WAT
0	0.58 b	0.56 c	0.63 b	0.64 c	0.63 c	0.64 b
10	0.79 a	0.67 ab	0.73 a	0.76 b	0.72 b	0.68 ab
20	0.80 a	0.69 a	0.73 a	0.80 ab	0.76 ab	0.70 ab
30	0.76 a	0.71 a	0.74 a	0.81 a	0.79 a	0.70 ab
40	0.78 a	0.79 a	0.77 a	0.82 a	0.77 ab	0.73 a
50	0.72 a	0.73 a	0.73 a	0.80 a	0.79 a	0.68 ab

* WAT = Weeks after a top-dressed treatment of 12–6–8 Florikan CRF. Mean values ($n = 5$) followed by the same letter are not significant at $p \leq 0.05$ within WAT using least significant differences ($LSD_{0.05}$).

Furthermore, NDVI readings were significantly correlated with leaf N results at 4 and 8 WAT, although this was not the case at 12 WAT (Table 2).

Table 2. Bivariate correlation of all recorded parameters throughout the study at sampling times 4, 8, and 12 WAT.

	NDVI [a]	SPAD [b]	Total P/L [c]	Total NO_3^- /L	Leaf N	Leaf C
4 WAT [d]						
N Rate	0.549 ** [e]	0.393 *	0.367 *	0.838 ***	0.845 ***	0.646 ***
NDVI		0.333	0.501 ***	0.703 ***	0.604 ***	0.511 **
SPAD			0.129	0.463 **	0.726 ***	0.789 **
Total P/L				0.687 ***	0.427 *	0.047
Total NO_3^- /L					0.892 **	0.635 **
Leaf N						0.873 ***

Table 2. *Cont.*

	NDVI [a]	SPAD [b]	Total P/L [c]	Total NO_3^- /L	Leaf N	Leaf C
8 WAT						
N Rate	0.739 ***	0.283	0.893 **	0.690 ***	0.902 ***	0.607 ***
NDVI		0.683 ***	0.726 ***	0.354	0.880 ***	0.841 **
SPAD			0.298	0.050	0.530 **	0.716 ***
Total P/L				0.633 ***	0.857 ***	0.651 ***
Total NO_3^- /L					0.532 **	0.196
Leaf N						0.799 ***
12 WAT						
N Rate	0.320	0.824 ***	0.924 ***	0.515 **	0.931 ***	0.880 ***
NDVI		0.191	0.255	0.109	0.299	0.385 *
SPAD			0.829 ***	0.498 **	0.830 ***	0.779 ***
Total P/L				0.617 ***	0.893 ***	0.790 ***
Total NO_3^- /L					0.559 **	0.437 *
Leaf N						0.906 ***

[a] Normalized Difference Vegetation Index. [b] Soil Plant Analytical Development. [c] Leachate [d] WAT = days after initial treatment of 16–9–12 Florikan CRF. [e] Representing Pearson's correlation coefficient (r) significant at $p \leq 0.05$ (*), $p \leq 0.01$ (**), or $p \leq 0.001$ (***).

Soil Plant Analytical Development values varied from 20 to 50 throughout the various fertilizer rates. Similar to the NDVI values, the 50 g fertilizer rate did not show the highest chlorophyll index values and varied little throughout 4, 6, 8, and 10 WAT collection periods (Table 3).

Table 3. Soil Plant Analytical Development values for a plant in a single pot with different rates of fertilizer at different weeks after fertilizer treatment during in *Justicia brandegeana*.

Fertilizer Rate (g)	2 WAT *	4 WAT	6 WAT	8 WAT	10 WAT	12 WAT
0	21.1 c	22.0 b	21.34 c	30.60 c	21.81 d	20.43 c
10	41.5 ab	47.07 a	37.53 b	45.43 b	28.95 cd	22.97 c
20	41.71 ab	48.03 a	43.61 ab	50.82 a	38.76 b	31.22 ab
30	42.27 a	45.15 a	43.42 a	40.24 b	37.15 bc	29.60 b
40	42.69 a	45.34 a	45.67 a	43.98 b	46.65 a	33.92 ab
50	37.77 b	40.21 a	39.9 a	41.69 b	41.98 ab	35.66 a

* WAT = Weeks after a top-dressed treatment of 12–6–8 Florikan CRF. Mean values (n = 5) followed by the same letter are not significant at $p \leq 0.05$ within WAT using least significant differences ($LSD_{0.05}$).

There was no significant difference between 20, 30, and 40 g fertilizer rates with the exception of the 8 and 10 WAT sampling periods. The largest increase in chlorophyll index occurred in the transition of 2 to 4 WAT. The greatest SPAD value (50.82) was observed in 20 g fertilizer rate at 8 WAT (Table 3). The leaf total N showed varying results based on fertilizer treatments throughout 4 to 12 WAT (Table 4). Although not confirmed by statistical analysis, the greatest leaf total N content was observed in the 30 g fertilizer rate at 4 WAT, closely followed by the 50 g and 40 g fertilizers with 53.48, 52.23, and 52.91 mg/g dw, respectively (Table 4).

Table 4. Leaf total N concentration (mg^{-1} g dw) at different fertilizer rates at three sampling times 4, 8, and 12 WAT in *Justicia brandegeana*. Values are from combining five mature leaves and no petioles from five different pots per treatment for a single composite sample (n = 1).

Fertilizer Rate (g)	4 WAT *	8 WAT	12 WAT
0	17.99	11.45	14.32
10	39.68	22.89	19.93
20	47.99	30.56	29.66

Table 4. *Cont.*

Fertilizer Rate (g)	4 WAT *	8 WAT	12 WAT
30	53.48	32.05	30.74
40	52.23	40.34	29.33
50	52.91	35.25	37.49

* WAT = Weeks after a top-dressed treatment of 12–6–8 Florikan CRF.

At 8 WAT, the plants treated with 40 g fertilizer rate contained the highest leaf total N with 40.34 mg/g dw. The last reading, collected after three months after transplanting, showed the greatest leaf total N value (37.49 mg/g dw) with the 50 g fertilizer rate (Table 4). The leaf C:N ratio results reveal that at 4 WAT fertilizer treatments 20, 30, 40, and 50 g had similar values (Table 5). At 8 WAT C:N varied among treatments as the 10, 20, and 30 g, rates had higher ratios than the 40 g and 50 g treatments. At 12 WAT, C:N ratio values evened out and the 20, 30, and 40 g treatment showed similar values.

Table 5. Leaf total C:N ratio (mol:mol) at different fertilizer rates at three sampling times 4, 8, and 12 WAT in *Justicia brandegeana*. Values are from combining five mature leaves and no petioles from five different pots per treatment for a single composite sample ($n = 1$).

Fertilizer Rate (g)	4 WAT *	8 WAT	12 WAT
0	21.82	32.81	25.76
10	12.15	18.31	21.82
20	9.79	15.51	15.15
30	8.79	14.24	14.64
40	8.91	10.98	15.86
50	9.17	12.56	12.66

* WAT = Weeks after a top-dressed treatment of 12–6–8 Florikan CRF.

SPAD results were significantly correlated ($p \leq 0.05$ and $p \leq 0.01$) with leaf total N at 4, 8, and 12 WAT (Table 2). Nitrate leachate (mg NO_3^- /L) was collected from each treatment at various time intervals. The greatest NO_3^- leaching occurred 4 and 6 WAT as the fertilizer had been applied closer to those time periods (Table 6).

Table 6. Leachate (nitrate runoff) collected after irrigation (mg NO_3^- /L) collected for a plant in a single pot with different rates of CRF at different weeks after fertilizer treatment in *Justicia brandegeana*.

Fertilizer Rate (g)	4 WAT *	6 WAT	8 WAT	10 WAT	12 WAT
0	1.61 c	0.57 d	0.45 b	0.16 b	0.19 b
10	14.45 b	7.84 cd	0.72 b	0.38 b	0.36 b
20	41.44 a	16.01 bc	2.27 b	1.41 b	1.24 ab
30	48.00 a	24.69 b	3.39 b	4.40 ab	1.12 b
40	51.25 a	40.72 a	16.73 a	2.12 b	0.81 b
50	46.97 a	2.13 d	24.16 a	6.16 a	2.72 a

* WAT = Weeks after a top-dressed treatment of 12–6–8 CRF. Mean values ($n = 5$) followed by the same letter are not significant at $p \leq 0.05$ within WAT using least significant differences ($LSD_{0.05}$).

The newly transplanted red shrimp had not yet developed an expanded root and shoot system in comparison to the following weeks (6–8 WAT). Samples treated with 10, 20 and 30 g of fertilizer showed significantly less leaching when compared to the 40 and 50 g fertilizer rates. Towards the end of the experiment (10 and 12 WAT), samples treated with 10, 20, 30, and 40 g of fertilizer had significantly less NO_3^- leaching than the 50 g fertilizer rate. The NO_3^- leachate was significantly correlated ($p \leq 0.05$ and $p \leq 0.01$) with the N rate applied at 4, 8, and 12 WAT (Table 2).

4. Discussion

Timing the precise amount of fertilizer application is the most crucial task of horticultural nursery producers. The parameters measured were intended to determine ideal fertilization application rates for *J. brandegeana* as a guideline for nursery producers and landscape personnel. Furthermore, information collected can be used to determine if SPAD and/or GreenSeeker^{TM} are appropriate devices to estimate fertilizer application rates through leaf N measurements.

GreenSeeker^{TM} values revealed lack of significant difference between treatments at 2 WAT (Table 1). This can be attributed to insufficient differences in biomass between plants though all the fertilizer rates received adequate amounts of N [7,16]. At early developmental stages of *J. brandegeana*, the foliage forming the upper canopy is not completely established. Thus, vegetation does not fill the sensors field of view entirely and the soil reflectance properties can create background noise. This may lead to less accurate readings during earlier collection dates and little to no differentiation between treatments [17,33]. The greatest NDVI readings were observed at 8 WAT, although no significant difference between fertilizer rates occurred at this time (Table 1). There were no notable significant differences between fertilizer rates among NDVI readings throughout the sampling times. Correlations ($P \leq 0.01$) between fertilizer rates and NDVI readings at 4 and 8 WAT were significant. Therefore, it can be inferred that 20 g and 30 g of fertilizer are just as effective as 40 g and 50 g fertilizer rates according to NDVI results (Table 1, Table 2).

Readings obtained with the SPAD meter displayed the largest difference in fertilizer rate means between 2 and 4 WAT (Table 3). These results might be attributed to *J. brandegeana* coming out of a stress period during the transplanting procedure. Various metabolic processes in transplanted seedlings can be disturbed by the influence of changes in water content and root pruning [34]. Consequently, the growth and development of seedlings temporarily stagnates and could initially hinder overall plant health [35]. At the 4 WAT sampling period, *J. brandegeana* was responsive and assimilated N input as indicated by the two sensors chlorophyll index readings [7,16,36,37]. Dunn et al. [7] was also in line with the findings of this study, stating that SPAD reading values could differentiate fertilizer treatments clearly as early as 28 days after treatment (DAT). Similar to the NDVI results, SPAD readings were greatest at 8 WAT. The SPAD and NDVI values were only significantly correlated ($p \leq 0.05$ and $p \leq 0.01$) at the 8 WAT sampling date. The highest SPAD reading throughout the experiment was observed at 8 WAT via the 20 g fertilizer treatment. The 20 g fertilization regimen may be ideal for *J. brandegeana* as SPAD and NDVI readings were similar for this measurement time.

Both NDVI and chlorophyll concentration values reached a plateau at the 8 WAT mark. A sharp decrease followed at the 10 and 12 WAT collection period. This coincides with the manufactures claims that this specific fertilizer has the ability to supplement nutrients within a 2–3-month period before additional inputs are called for. Therefore, based on GreenSeeker^{TM} and SPAD meter values, *J. brandegeana* seems to need additional supplementation of N at around the 2-month (8 weeks) period when using Florikan Top-dress fertilizer (12N-6P-8K). Measurements from both optical sensors revealed that *J. brandegeana* was not able to effectively assimilate N when fertilizer rate was greatest (50 g). Rather, the 40 g treatment maintained large values (NDVI and SPAD) indicative of sufficient N presence throughout the experiment. The leaf total N was highest at 4 WAT, then began to taper off over time (Table 4). Leaf total N was correlated ($p \leq 0.05$ and $p \leq 0.01$) with NDVI and SPAD throughout the experiment at 4, 8, and 12 WAT, with the exception of NDVI at the last sampling time (Table 2). These results imply that generally, SPAD and NDVI via GreenSeeker^{TM} can be used as reliable tools to determine leaf N in *J. brandegeana*. The non-destructive optical sensors are most effective in estimating leaf total N concentration at 4 weeks after fertilizer treatment ($p \leq 0.01$) (Table 2). At 12 WAT, the leaf total N results were nearly the same between the 20 g, 30 g, and 40 g fertilizer treatments indicating that any of these fertilization rates are suitable for the successful growth of *J. brandegeana*. The C:N ratio results (Table 5) coincide with the leaf N and SPAD/NDVI results.

These results specify that nursery growers can significantly reduce their fertilizer input without altering plant growth and salable characteristics. This is beneficial economically and environmentally, as not all N applied to crops is taken up by plants. Nitrogen is often lost by leaching into water resources and volatilization into the atmosphere [38]. Nutrient pollution issues have arisen within the last 70 years and will continue if sustainable practices are not properly executed [39]. The application of optical sensor technology in large production systems can allow for spot treatment of plants in need of fertilizer, rather than applying unnecessary amounts to each individual pot. Moreover, NO_3^- leaching occurred heavily at the beginning of the experiment, soon after the fertilizer was applied (Table 6). At 4 WAT, the 10 g fertilizer rate had leached significantly less NO_3^- ($p \leq 0.05$ and $p \leq 0.01$) than the other treatments and continuously performed well, displaying the least leached NO_3^- over time. While the 10 g treatment did show the least amount of leaching, the 20 g and 30 g fertilizer rates were not significantly different from the 10 g treatment except for the first sampling time. The 20 g and 30 g treatments did prove to be more effective than the 10 g treatment regarding leaf total N concentration. All results considered, the 20 and 30 g fertilizer rates were similar and, in some cases, showed greater sensor readings than the 40 and 50 g fertilizer rates. The NO_3^- leachate analysis from the 20 g and 30 g treatments rates were significantly ($p \leq 0.05$ and $p \leq 0.01$) less than the higher fertilizer rates in some cases at 8, 10, and 12 WAT. Total P leachate was also correlated to N application rate and leaf tissue total N at all three sampling times, and NDVI readings for 4 and 8 WAT sampling times (Table 2). This implies that GreenSeeker™ may be used as a tool to infer relationships between total N and P. Additionally, optical sensors can be recommended to determine appropriate times for fertilizer supplements. GreenSeeker™ and SPAD meter both showed positive correlation ($p \leq 0.05$ and $p \leq 0.01$) with N rates. Both optical sensors can be used as guidelines when establishing a fertilization regime, though further analysis of tissue samples and leachate are necessary to differentiate treatments in greater detail.

5. Conclusions

It can be concluded that the 20 g fertilizer treatment is the ideal recommendation for ornamental nursery producers that are making effort to reduce their fertilizer cost. The 20 g fertilizer treatment showed no significant difference in sensor reading results as compared to the higher fertilizer rates. Additionally, the leaf tissue total N content and C:N ratio did not show much quantitative difference compared to 30, 40 and 50 g treatments. The 20 g treatment showed significantly less NO_3^- leaching than higher application rates as well. These precision horticulture recommendations can be helpful for nursery producers to save money and lessen overall environmental impact. These findings are especially important for South Florida, as natural areas such as the Everglades [40], Biscayne Bay [41], and off shore reefs [42] have been severely impacted by nutrient loading since the popularization of synthetic fertilizers. With respect to environmental concerns, coupled with economic efficiency, precision agricultural applications using non-destructive optical sensor technology can offer beneficial outcomes for ornamental nursery producers and environmental stakeholders alike.

Author Contributions: Conceptualization, A.A.K.; Methodology, A.F., G.B., A.A.K.; Validation, all authors.; Formal Analysis, A.F.; Investigation, A.F. and G.B.; Resources, K.J., A.A.K.; Data Curation, A.F., G.B., A.A.K; Writing – Original Draft Preparation, A.F. and G.B., Writing – Review & Editing, A.F., K.J., A.A.K; Visualization, all authors; Supervision, K.J. and A.A.K.; Project Administration, A.A.K.; Funding Acquisition, K.J.

Acknowledgments: The authors would like to thank David Riera, Scott Zona and Eric Betancourt for assisting in the project, providing greenhouse space and for guidance in presentation preparation, respectively.

Abbreviations

ANOVA	Analysis of variance
C	Carbon
CRF	Controlled release fertilizer

CRD	Completely randomized design
DW	Dry weight
K	Potassium
LSD	Least significant difference
N	Nitrogen
NDVI	Normalized Difference Vegetation Index
NIR	Near infrared
P	Phosphorus
SPAD	Soil Plant Analysis Development
WAT	Weeks after treatment

References

1. U.S. Department of Agriculture. Floriculture Crops 2014 Summary. Available online: http://usda.mannlib.cornell.edu (accessed on 6 February 2019).
2. Demotes-Mainard, S.; Boumaza, R.; Meyer, S.; Cerovic, Z.G. Indicators of nitrogen status for ornamental woody plants based on optical measurements of leaf epidermal polyphenol and chlorophyll contents. *Sci. Hortic.* **2008**, *115*, 377–385. [CrossRef]
3. Fageria, N.; Filho, M.B.; Moreira, A.; Guimarães, C. Foliar fertilization of crop plants. *J. Plant Nutr.* **2009**, *32*, 1044–1064. [CrossRef]
4. Uchida, R. Essential nutrients for plant growth: Nutrient functions and deficiency symptoms. In *Plant Nutrient Management in Hawaii's Soils*; Silva, J.A., Uchida, R., Eds.; College of Tropical Agriculture and Human Resources, University of Hawaii: Honolulu, HI, USA, 2000; pp. 31–55.
5. Swiader, J.; Chyan, Y.; Freiji, F. Genotypic differences in nitrate uptake and utilization efficiency in pumpkin hybrids. *J. Plant Nutr.* **1994**, *17*, 1687–1699. [CrossRef]
6. Ristvey, A.G.; Lea-Cox, J.D.; Ross, D.S. Nutrient uptake, partitioning and leaching losses from container-nursery production systems. *Acta Hort.* **2004**, *630*, 321–328. [CrossRef]
7. Dunn, B.L.; Goad, C. Effect of foliar nitrogen and optical sensor sampling method and location for determining ornamental cabbage fertility status. *HortScience* **2015**, *50*, 74–77. [CrossRef]
8. Barile, P.J. Evidence of anthropogenic nitrogen enrichment of the littoral waters of east central Florida. *J. Coast. Res.* **2004**, *20*, 1237–1245. [CrossRef]
9. Cabrera, R.I.; Evans, R.Y.; Paul, J. Leaching losses of N from container-grown roses. *Sci. Hortic.* **1993**, *53*, 333–345. [CrossRef]
10. Evans, R.Y.; Dodge, L.; Newman, J. Nutrient Management in Nursery and Floriculture. Available online: https://escholarship.org/uc/item/3cr3c5k5 (accessed on 6 February 2019).
11. Wunderlin, R.; Hansen, B.; Hansen, B. *Atlas of Florida Vascular Plants [SM Landry and KN Campbell (Application Development), Florida Center for Community Design and Research]*; Institute for Systematic Botany, University of South Florida: Tampa, FL, USA, 2008.
12. Scheper, J. *Floridata: Justicia brandegeana*; Floridata: Tallahassee, FL, USA, 2007; Available online: http://www.floridata.com/ref/j/just_bra.cfm (accessed on 3 November 2009).
13. O'Neill, C.S. Anatomy of the shrimp plant, *Justicia brandegeana* (Acanthaceae). *SURG J.* **2010**, *3*, 41–47.
14. Calvino-Cancela, M. Time-activity budgets and behaviour of the Amazilia hummingbird, *Amazilia amazilia* (Apodiformes: Trochilidae) in an urban environment. *Rev. Biol. Trop.* **2006**, *54*, 873–878. [CrossRef] [PubMed]
15. Wu, C.; Wang, L.; Niu, Z.; Gao, S.; Wu, M. Nondestructive estimation of canopy chlorophyll content using Hyperion and Landsat/TM images. *Int. J. Remote Sens.* **2010**, *31*, 2159–2167. [CrossRef]
16. Dunn, B.L.; Shrestha, A.; Goad, C.; Khoddamzadeh, A.A. Use of optical sensors to monitor *Gaillardia* Foug. nitrogen status. *J. Appl. Hortic.* **2015**, *17*, 181–185.
17. Khoddamzadeh, A.A.; Dunn, B.L. Application of optical sensors for nitrogen management in chrysanthemum. *HortScience* **2016**, *51*, 915–920. [CrossRef]
18. Govaerts, B.; Verhulst, N. *The Normalized Difference Vegetation Index (NDVI) GreenseekerTM Handheld Sensor: Toward the Integrated Evaluation of Crop Management Part A: Concepts and Case Studies*; International Maize and Wheat Improvement Center (CIMMYT): Texcoco, Mexico, 2010.

19. Horler, D.; Dockary, M.; Barber, J. The red edge of plant leaf reflectance. *Int. J. Remote Sens.* **1983**, *4*, 273–288. [CrossRef]
20. Shaver, T.; Khosla, R.; Westfall, D. Evaluation of two crop canopy sensors for nitrogen variability determination in irrigated maize. *Precis. Agric.* **2011**, *12*, 892–904. [CrossRef]
21. Jackson, R.D.; Huete, A.R. Interpreting vegetation indices. *Prev. Vet. Med.* **1991**, *11*, 185–200. [CrossRef]
22. Blackmer, T.; Schepers, J. Use of a chlorophyll meter to monitor nitrogen status and schedule fertigation for corn. *J. Prod. Agric.* **1995**, *8*, 56–60. [CrossRef]
23. Basyouni, R.; Dunn, B.L. *Use of Optical Sensors to Monitor Plant Nitrogen Status in Horticultural Plants (HLA-6719-4)*; Oklahoma Cooperative Extension Service: Stillwater, OK, USA, 2013.
24. Basyouni, R. Use of Nondestructive Optical Sensors in Assessing Nitrogen Status in Six Popular Greenhouse Plants. Master's Thesis, Oklahoma State University, Stillwater, OK, USA, 2014.
25. Cerovic, Z.G.; Masdoumier, G.; Ghozlen, N.B.; Latouche, G. A new optical leaf-clip meter for simultaneous non-destructive assessment of leaf chlorophyll and epidermal flavonoids. *Physiol. Plant.* **2012**, *146*, 251–260. [CrossRef] [PubMed]
26. Muñoz-Huerta, R.; Guevara-Gonzalez, R.; Contreras-Medina, L.; Torres-Pacheco, I.; Prado-Olivarez, J.; Ocampo-Velazquez, R. A review of methods for sensing the nitrogen status in plants: Advantages, disadvantages and recent advances. *Sensors* **2013**, *13*, 10823–10843. [CrossRef]
27. Tomkiewicz, D.; Piskier, T. A plant-based sensing method for nutrition stress monitoring. *Precis. Agric.* **2012**, *13*, 370–383. [CrossRef]
28. Badruzzaman, M.; Pinzon, J.; Oppenheimer, J.; Jacangelo, J.G. Review: Sources of nutrients impacting surface waters in Florida: A review. *J. Environ. Manag.* **2012**, *109*, 80–92. [CrossRef]
29. Smith, V.H.; Tilman, G.D.; Nekola, J.C. Eutrophication: Impacts of excess nutrient inputs on freshwater, marine, and terrestrial ecosystems. *Environ. Pollut.* **1999**, *100*, 179–196. [CrossRef]
30. Sklar, F.; McVoy, C.; VanZee, R.; Gawlik, D.E.; Tarboton, K.; Rudnick, D.; Miao, S.; Armentano, T. The effects of altered hydrology on the ecology of the Everglades. In *The Everglades, Florida Bay, and Coral Reefs of the Florida Keys*; CRC Press: Boca Raton, FL, USA, 2002; pp. 39–82.
31. Grable, A.R. Soil Aeration and Plant Growth. In *Advances in Agronomy*; Norman, A.G., Ed.; Academia Press: Cambridge MA, USA, 1966; pp. 57–106.
32. Wright, R.D. The pour-through nutrient extraction procedure. *HortScience* **1986**, *21*, 227–229.
33. Gao, B. NDWI-A normalized difference water index for remote sensing of vegetation liquid water from space. *Remote Sens. Environ.* **1996**, *58*, 257–266. [CrossRef]
34. Sasaki, R.; Gotoh, K. Characteristics of rooting and early growth of transplanted rice [*Oryza sativa*] nursling seedlings with several plant ages in leaf number. *Jpn. J. Crop Sci. (Japan)* **1999**, *68*, 194–198. [CrossRef]
35. Richner, W.; Soldati, A.; Stamp, P. Shoot-to-root relations in field-grown maize seedlings. *Agron. J.* **1996**, *88*, 56–61. [CrossRef]
36. Grossnickle, S.C. Importance of root growth in overcoming planting stress. *New For.* **2005**, *30*, 273–294. [CrossRef]
37. Basyouni, R.; Dunn, B.L.; Goad, C. The use of nondestructive sensors to assess nitrogen status in potted dianthus (Dianthus chinensis L.) production. *Can. J. Plant Sci.* **2016**, *97*, 1–9. [CrossRef]
38. Terman, G.L. Volatilization losses of nitrogen as ammonia from surface-applied fertilizers, organic amendments, and crop residues. *Adv. Agron.* **1979**, *31*, 189–223.
39. Di, H.J. Nitrate leaching in temperate agroecosystems: Sources, factors and mitigating strategies. *Nutr. Cycl. Agroecosyst.* **2002**, *64*, 237–256. [CrossRef]
40. Belanger, T.V.; Scheidt, D.J.; Platko, J.R. Effects of nutrient enrichment on the Florida Everglades. *Lake Reserv. Manag.* **1989**, *5*, 101–111. [CrossRef]
41. Caccia, V.G.; Boyer, J.N. A nutrient loading budget for Biscayne Bay, Florida. *Mar. Pollut. Bull.* **2007**, *54*, 994–1008. [CrossRef] [PubMed]
42. Lapointe, B.E.; Barile, P.J.; Matzie, W.R. Anthropogenic nutrient enrichment of seagrass and coral reef communities in the Lower Florida Keys: Discrimination of local versus regional nitrogen sources. *J. Exp. Mar. Biol. Ecol.* **2004**, *308*, 23–58. [CrossRef]

2

Using Optical Sensors to Identify Water Deprivation, Nitrogen Shortage, Weed Presence and Fungal Infection in Wheat

Gerassimos G. Peteinatos [1,*], Audun Korsaeth [2], Therese W. Berge [2] and Roland Gerhards [1]

1 Institute of Phytomedicine, University Hohenheim, Otto-Sander-Straße 5, Stuttgart 70599, Germany; roland.gerhards@uni-hohenheim.de

2 Norwegian Institute of Bioeconomy Research (NIBIO), P.O. Box 115, N-1431 Ås, Norway; audun.korsaeth@nibio.no (A.K.); therese.berge@nibio.no (T.W.B.)

* Correspondence: g.peteinatos@uni-hohenheim.de

Abstract: The success of precision agriculture relies largely on our ability to identify how the plants' growth limiting factors vary in time and space. In the field, several stress factors may occur simultaneously, and it is thus crucial to be able to identify the key limitation, in order to decide upon the correct contra-action, e.g., herbicide application. We performed a pot experiment, in which spring wheat was exposed to water shortage, nitrogen deficiency, weed competition (*Sinapis alba* L.) and fungal infection (*Blumeria graminis* f. sp. *tritici*) in a complete, factorial design. A range of sensor measurements were taken every third day from the two-leaf stage until booting of the wheat (BBCH 12 to 40). Already during the first 10 days after stress induction (DAS), both fluorescence measurements and spectral vegetation indices were able to differentiate between non-stressed and stressed wheat plants exposed to water shortage, weed competition or fungal infection. This meant that water shortage and fungal infection could be detected prior to visible symptoms. Nitrogen shortage was detected on the 11–20 DAS. Differentiation of more than one stress factors with the same index was difficult.

Keywords: fluorescence; multi-stress; precision agriculture; site-specific crop management; spectral indices; stress symptoms

---

1. Introduction

Abiotic stress, as excess or shortage of water, inadequate nutrients, and biotic stressors such as insects or fungi have the potential to reduce crop production significantly [1]. Important biotic stressors are pest organisms like weeds and plant pathogenic fungi. The potential global loss in wheat due to weeds, fungal diseases and bacteria has been estimated to 38% [2]. In conventional farming, fertilizers, herbicides and fungicides are applied to avoid nitrogen deficiency, weed competition and fungal diseases, respectively. However, the application of a uniform rate of these external inputs is rarely the best approach since the nitrogen demand and weed presence are generally heterogeneously distributed within fields (e.g., [3,4]). Hence, some parts will receive excess levels at the expense of others. Site-specific crop management, or precision agriculture, seeks to adjust the rates of external inputs to this spatial heterogeneity. To be cost-effective, precision farming requires sensors to measure this heterogeneity.

Sensing techniques for crop management have been suggested since the early 1980s but only a few have reached the market. The precision agriculture procedure that has been widely adopted, is precision fertilization. The development of a tailor-made sensor to estimate the nitrogen demand of the crop on-the-go has been a decisive factor. A series of optical sensors applicable for outdoor use exist, including spectrometers, fluorometers and optoelectronic sensors. Compared to imaging techniques, these can almost instantaneously provide simple measures like single bands, ratios and indices based on spectral reflectance. Hence, such measures are particularly suitable for automatic on-the-go field mapping for precision farming implements.

The applicability of spectrometers and fluorometers has been investigated for single stress factors like water shortage, nitrogen deficiency and weed identification in a series of crops especially wheat [5–14]. **Nitrogen deficiency** can result in slower growth rate, smaller plants and reduced yield. Normalized Difference Vegetation Index (NDVI), which is one of the most widely used vegetation indices, has been used to determine nitrogen status, vegetation vigor or crop density in cereals [8,9]. Red Edge Inflection Point (REIP), NDVI, Modified Chlorophyll Absorption in Reflectance Index (MCARI), Greenness Index (G), Optimized Soil Adjusted Vegetation Index (OSAVI), an index from Zarco-Tejada and Miller (ZM), Photochemical Reflectance Index (PRI) and Normalized Phaeophytinization Index (NPQI) have been shown to correlate with nitrogen stress, including nitrogen deficiency, in maize and wheat [10,15]. Similar results are obtained by fluorescence indices (Table 2) [13].

Drought stress in crops has been detected by various spectral bands and indices like REIP, NDVI, OSAVI, MCARI, G, ZM, PRI, a simple ratio proposed by Vogelmann *et al.* [16] (VOG1), and Plant Pigment Ratio (PPR) [7,9,17–20]. Fluorescence measurements with sensors like the Multiplex® and Dualex® sensor have also been correlated with water stress in wheat [14].

Weed density is correlated with Leaf Area Index (LAI) and biomass. Different weed species may be discriminated based on their spectral reflectance curves [21,22]. Based on UV-induced fluorescence, Longchamps *et al.* [23] successfully classified maize and weeds into three plant groups (four maize hybrids, four dicotyledonous weed species and four monocotyledonous weed species). Tyystjärvi *et al.* [24] used chlorophyll fluorescence to classify six weed species, maize and barley into weeds and crop with a high correct classification rate (86.7%–96.1%).

Plant pathogenic fungi depend on their host plants for nutrients and carbon assimilates, therefore disturbing the plant growth. Rumpf *et al.* [25] demonstrated the potential of pre-symptomatic detection of plant diseases in sugar beet by use of spectral indices obtained by hyperspectral reflectance. Several indices like REIP, NDVI, MCARI, G, ZM, PRI, OSAVI, Red Edge Vegetation Stress Index (RVSI), Renormalized Difference Vegetation Index (RDVI), along with fluorescence indices have been correlated with fungal infection in various crops [11,26–28].

Since multiple stressors often occur simultaneously in a field, it is of interest to develop a sensor-based method that is able to identify the type of stressors. Previous studies addressing concurrent, multiple stressors have only explored imaging technologies. For example, Karimi *et al.* [29] used hyperspectral imagery to identify combinations of various nitrogen application rates and weediness in maize. Backoulou *et al.* [12] used multispectral imagery to separate stress by a pest aphid from other concurrent stressors in wheat. Imaging technologies can be helpful in identifying weeds but not the rest of the stressors used in the current study. Our approach is to explore single spectral bands, simple ratios and indices measured by on-the-shelf non-imagery sensors. For this purpose we implemented both spectral and fluorescence parameters since the technologies by themselves are not able to indicate the nature of the stress factor. To our knowledge, such measures have not been tested to identify concurrent biotic and abiotic stressors in wheat until now.

The aim of this study was to determine whether simple sensor-based measures like single bands, ratios and indices obtained by on-the-shelf optical sensors can be used to detect abiotic and biotic stressors in spring wheat. The four stressors tested were water deficiency, nitrogen shortage, weed competition (*Sinapis alba* L.) and fungal infection (powdery mildew). We hypothesized that the selected sensors can be used to detect single stressors, even in the co-existence of other stress factors.

2. Experimental Section

2.1. Experimental Design

An outdoor pot experiment was performed twice during 2013 at the University of Hohenheim, Stuttgart, Germany. In the experiments, spring wheat (*Triticum aestivum* L. *cv.* Toras) was exposed to four stress factors: (a) water shortage; (b) nitrogen deficiency; (c) weeds (*Sinapis alba* L.) and (d) fungal infection (*B. graminis* f. sp. *tritici*). For each stressor, a non-stressed treatment was included, and using a randomized complete block design, all 16 treatment combinations were tested with four replicate blocks (n = 64 in each experiment). Table 1 presents all different combinations used in this experiment. Mitscherlich pots were each filled with 6 kg of soil consisting of composted soil, loamy soil and coarse sand in the ratio 2:2:1, as measured by volume. These pots were filled until 90% of their volume and are ideal for cereal test plants [30]. The soil had previously been sieved through a net with a mesh size of 8 mm. The soil surface of filled pots was approximately 300 cm^2. An ample amount of spring wheat seeds were planted per pot. At growth stage BBCH 12 [31], some plants were removed, so that each pot had 12 remaining plants with an even distribution corresponding to a crop density of about 40 plants m^{-2}. The first experiment was performed from 10 March–17 May . Unfortunately, the fungal infection was not successful in this first experiment, and the data related to the fungi infection in the first experiment were thus excluded from further analyses. The experiment was repeated in the 30 June–17 August period.

The pots were weighted and irrigated daily to maintain a soil moisture content, corresponding to a water holding capacity (WHC) of 50% until the wheat plants had reached BBCH 12. Then water stress (+) was induced by reducing the soil moisture level to a maintenance level of 30% WHC. The soil moisture in pots without water stress (−) was adjusted up to a level of 70% WHC.

A basic fertilization of 73 mg nitrogen was given as calcium ammonium nitrate to all pots. The pots with nitrogen stress (+) did not receive any additional fertilizer, whereas the non–stressed pots (−) were fertilized at BBCH 21 and BBCH 32 as well, totaling 189 mg nitrogen per pot.

Weed competition (+) was induced when the wheat plants had reached BBCH 12, by planting 10 plants of *Sinapis alba* L. evenly into each pot, corresponding to a density of 33 plants m^{-2}. In the pots without weed competition (−), volunteer weeds were removed during the experiment. Fungal infection was realized by the inoculation of *Blumeria graminis* f. sp. *tritici* at the same time as the weed planting. Spores of *B. graminis* f. sp. *tritici* were suspended from highly infested wheat leaves into a water solution. Inoculation was performed by vaporizing the spore suspension. The infected pots were kept under high humidity conditions for 24 h, in order to facilitate the infection. During the last measurement day (BBCH 40), the plants were harvested and dried for 48 h at 80 °C to calculate the total biomass and the root to shoot ratio. Species were not separated in the weedy pots.

Table 1. All 16 treatment combinations, with "+" indicating presence of stress, while "−" represents absence of stress.

Treatment	1	2	3	4	5	6	7	8	9	10	11	12	13	14	15	16
Fungi	+	+	+	+	+	+	+	+	−	−	−	−	−	−	−	−
Water	+	+	+	+	−	−	−	−	+	+	+	+	−	−	−	−
Nitrogen	+	+	−	−	+	+	−	−	+	+	−	−	+	+	−	−
Weeds	+	−	+	−	+	−	+	−	+	−	+	−	+	−	+	−

2.2. Sensors

Three different sensors were selected to provide data for stress detection, a passive spectrometer, an active spectrometer and a fluorometer.

2.2.1. HandySpec

A hand-held passive spectrometer device with a spectral range of 360–1000 nm with 10 nm bandwidth (HandySpec Field, Tec5, Oberursel, Germany) (Figure 1a) was used to measure the spectral reflectance. Before each measuring period, the device was calibrated with a white standard ($BaSO_4$). This spectrometer consists of two independent sensors, one pointing upwards to measure the ambient light through a cosine diffuser, and the other pointing downwards to measure the ground reflection. Based on the spectral data 16 spectral indices were calculated (Table 2).

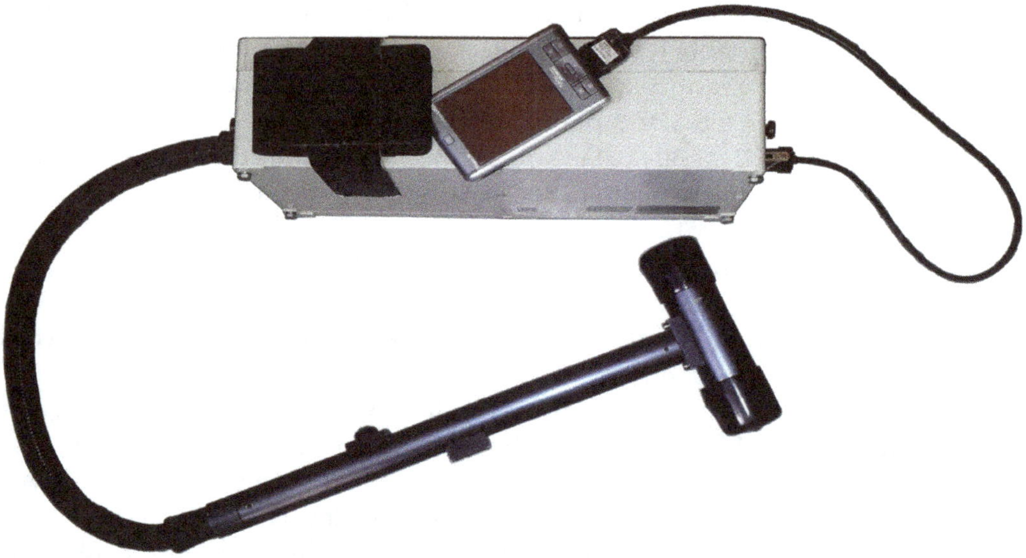

(a) HandySpec®

(b) Isaria®

(c) Multiplex®

Figure 1. Sensors used in the current experiment.

Table 2. Overview of the spectral indices and fluorescence parameters measured with the 3 sensors (HandySpec® Field sensor, Multiplex®, Isaria®) used in this study.

Index Reference	Explanation	Formula
	Structural indices	(HandySpec®)
NDVI [32]	Normalized Difference Vegetation Index	$\frac{R_{780} - R_{670}}{R_{780} + R_{670}}$
OSAVI [33]	Optimized Soil-Adjusted Vegetation Index	$(1 + 0.16)\frac{R_{800} - R_{670}}{R_{800} + R_{670} + 0.16}$
RDVI [34]	Renormalized Difference Vegetation Index	$\frac{R_{800} - R_{670}}{R_{800} + R_{670}}$
	Red Edge Inflection Point	(HandySpec® & Isaria®)
REIP [35]	Red Edge Inflection Point	$700 + 40\frac{(\frac{R_{670} + R_{780}}{2}) - R_{700}}{R_{740} - R_{700}}$
	Chlorophyll indices	(HandySpec®)
G [36]	Greenness Index	$\frac{R_{554}}{R_{677}}$
MCARI [37]	Modified Chlorophyll Absorption in Reflectance Index	$((R_{700} - R_{670}) - 0.2(R_{700} - R_{550}))(\frac{R_{700}}{R_{670}})$
NPQI [38]	Normalized Phaeophytinization Index	$\frac{R_{415} - R_{435}}{R_{415} + R_{435}}$
PPR [20]	Plant Pigment Ratio	$\frac{R_{550} - R_{450}}{R_{550} + R_{450}}$
PVR [20]	Photosynthetic Vigor Ratio	$\frac{R_{550} - R_{650}}{R_{550} + R_{650}}$
VOG1 [16]	Simple Ratio 740/720	$\frac{R_{740}}{R_{720}}$
GM1 [39]	Simple Ratio 750/550	$\frac{R_{750}}{R_{550}}$
LIC1 [40]	Lichtenthaler Index 1	$\frac{R_{800} - R_{680}}{R_{800} + R_{680}}$
ZM [41]	Zarco-Tejada & Miller	$\frac{R_{750}}{R_{710}}$
	Stress–Pigment indices	(HandySpec®)
PRI [42]	Photochemical Reflectance Index	$\frac{R_{531} \quad R_{570}}{R_{531} + R_{570}}$
CTR1 [43]	Simple Ratio 695/420	$\frac{R_{695}}{R_{420}}$
RVSI [44]	Red-edge Vegetation Stress Index	$\frac{R_{714} + R_{752}}{2} - R_{733}$
	Fluorescence indices	(Multiplex®)
ANTH [45]	Anthocyanins	$log(FER_{RG})$

Table 2. *Cont.*

Index Reference	Explanation	Formula
RF_R [45]	Red Fluorescence (Red Excitation)	—
FRF_{UV} [45]	Infra-red Fluorescence (UV Excitation)	—
BGF_G [45]	Blue Green Fluorescence (Green Excitation)	—
BGF_{UV} [45]	Blue Green Fluorescence (UV Excitation)	—
FER_{RUV} [45]	Fluorescence Excitation Ratio (Red & UV Excitation)	$\frac{FRF_R}{FRF_{UV}}$
FER_{RG} [45]	Fluorescence Excitation Ratio (Red & Green Excitation)	$\frac{FRF_R}{FRF_G}$
$FLAV$ [45]	Flavonoids	$log(FER_{RUV})$
NBI_G [45]	Nitrogen Balance Index	$\frac{FRF_{UV}}{RF_G}$
NBI_R [45]	Nitrogen Balance Index	$\frac{FRF_{UV}}{RF_R}$
SFR_G [45]	Simple Fluorescence Ratio (Green Excitation)	$\frac{FRF_G}{RF_G}$
SFR_R [45]	Simple Fluorescence Ratio (Red Excitation)	$\frac{FRF_R}{RF_R}$

2.2.2. Isaria

Isaria® sensor, which is an multi-spectral sensor, was also used (ISARIA®, Fritzmeier Umwelttechnik, Großhelfendorf, Germany) (Figure 1b). The sensor has four illumination sources in the range of 660–780 nm and receives their reflectance with a detector integrated in the sensor head. The REIP and Isaria Biomass Index (IBI) were determined for each pot. IBI is a proprietary sensor index used in the sensor to estimate plant density.

2.2.3. Multiplex

The Multiplex® (Force-A, Centre-Universitaire Paris Sud, Cedex, France) sensor (Figure 1c) is an optical, non-contact, active sensor. It is a fluorometer which measures the fluorescence of molecules inside plant tissues. The sensor has three silicon photodiodes as light sources, which generate light at UV, blue, green and red wavelengths. Three detectors are measuring the emitted fluorescence at the wavelengths of blue, red and far-red light. Ghozlen *et al.* [46] and Cerovic *et al.* [47] provide a more detailed description of the sensor. Table 2 gives an overview of the different ratios measured by the Multiplex® sensor. Table 3 summarizes the characteristics of the implemented sensors.

Table 3. Technical characteristics of the senors used in this study.

	HandySpec®	Isaria®	Multiplex®
Type	Spectrometer	Spectrometer	Fluorometer
Has Illumination	No	Yes	Yes
Needs Calibration	Yes	No	No
Plant-Sensor Distance (cm)	55 ± 5	60	10
Field of View (cm^2)	200	700	50

2.3. Data Processing and Statistical Analyses

Sensor measurements were acquired every third day from when the wheat had two leaves until booting (BBCH 12 to 40). At each measuring day, five measurements were taken per pot and sensor. Average values of the spectral indices and fluorescence ratios were derived from these five measurements. First, all data excluding the fungal infection data were pooled from both experiments (that is treatment combinations # 9-16, as presented in Table 1). The outcome of the study could be estimated using following the linear mixed effect model:

$$\begin{aligned} y_{ijklmn} = {} & \mu + \alpha_i + \beta_j + \gamma_k + (\alpha\beta)_{ij} + (\beta\gamma)_{jk} + (\alpha\gamma)_{ik} + \\ & (\alpha\beta\gamma)_{ijk} + r_l + \delta_m + (r\delta)_{lm} + \zeta_n + \theta + e_{ijklmn}, \end{aligned} \quad (1)$$

where y_{ijklmn} is the spectral index or fluorescence ratio, with subscript i indicating water treatment, j nitrogen treatment, k weed treatment, l block replicate at the m day of the n experiment. Further, μ is the general mean, whereas α, β, γ are the main effects of water stress, nitrogen stress and weed stress, respectively, with two-way interactions, $(\alpha\beta)$, $(\beta\gamma)$ and $(\alpha\gamma)$ and with $(\alpha\beta\gamma)_{ijk}$ as the three-way water-nitrogen-weed stress interaction. The block effect is represented by r, δ is the day effect and $(r\delta)$ is their interaction, ζ is the experiment effect, θ the day covariance matrix and e the residual. Then, all data from the second experiment were tested separately with the following mixed effects model:

$$\begin{aligned} y_{ijklmno} = {} & \mu + \alpha_i + \beta_j + \gamma_k + \eta_o + (\alpha\beta)_{ij} + (\beta\gamma)_{jk} + (\alpha\gamma)_{ik} + \\ & (\alpha\eta)_{io} + (\beta\eta)_{jo} + (\gamma\eta)_{ko} + (\alpha\beta\gamma)_{ijk} + (\alpha\beta\eta)_{ijo} + \\ & (\beta\gamma\eta)_{jko} + (\alpha\beta\gamma\eta)_{ijko} + r_l + \delta_m + (r\delta)_{lm} + \theta + e_{ijklmn}, \end{aligned} \quad (2)$$

with subscript o indicating fungal treatment, η the main effect of the existence or not of fungal stress, $(\alpha\eta)$, $(\beta\eta)$ and $(\gamma\eta)$ are the additional two-way interactions, $(\alpha\beta\eta)$ and $(\beta\gamma\eta)$ are the additional three-way interactions, and $(\alpha\beta\gamma\eta)$ the four-way water-nitrogen-weed-fungi stress interaction. Results for stress caused by water shortage, nitrogen deficiency, and weed presence were derived from the model shown in Equation (1). Results for stress caused by fungal infection were derived from the model in Equation (2). Spectral indices calculations and data analysis was conducted with R 3.2.0 RC [48]. The means of every stress factor were compared with Tukey's HSD (honest significant difference) test. In order to study the effects of time in more detail, the original time series of sensor data per stress factor were pooled into three discrete measuring time periods: 0–10, 11–20 and 21–30 days after stress induction (DAS). Our goal was to present the differences of the sensor signals as a function of the development of the crop at different growth stages, but on the same time to smoothen effects due to temperature and illumination variations during the measurements.

3. Results

The four stressors affected the growth of the wheat plants significantly, either in terms of biomass or the root to shoot ratio (Figure 2). Even in the nitrogen stressed pots and the fungi infected pots the total biomass of the stressed plants was lower than the non-stressed ones, but the result was not statistically significant. Pots with weeds showed higher total biomass than the weed free pots, due to the accumulated measurement of plant and weed biomass. The differences even though statistically significant did not present a high contrast. The severity of the disease was small. On average, wheat plants with fungal disease stress treatment had an estimated infestation with *B. graminis* f. sp. *tritici* of 4.4% of the leaf area on the first examination two weeks after inoculation and 6.5% on the last examination four weeks later. In addition, slight symptoms of wilting were observed on wheat plants with water stress treatment, especially in the afternoon and on hot days, compared to the non stressed ones. HandySpec® (Section 3.1.1) and the Multiplex® (Section 3.1.2) sensors were able to differentiate between non-stressed and stressed plants for all four stress factors. Isaria® could only discriminate stress due to weed competition (Section 3.1.3).

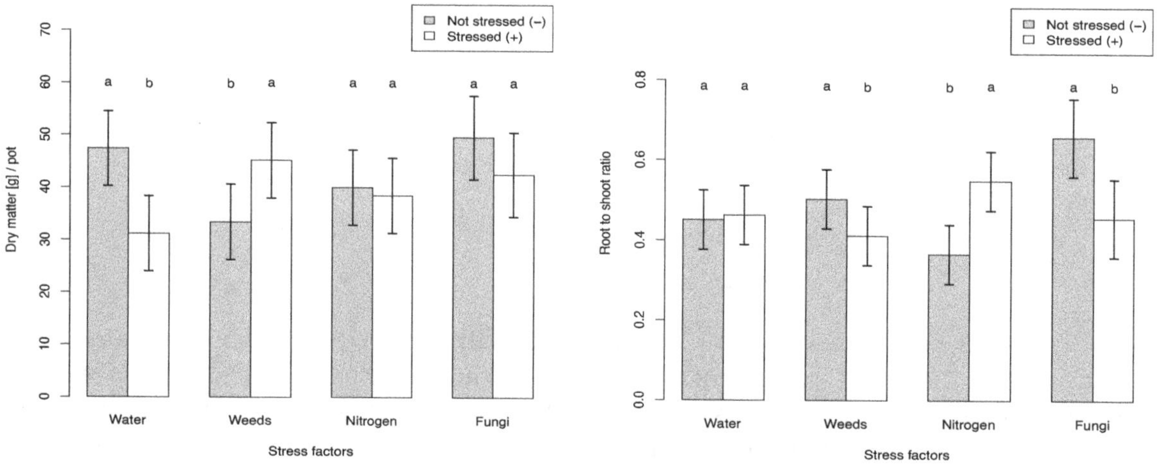

Figure 2. Mean values with their upper and lower confidence intervals of the total (including weeds) below and aboveground biomass and the root to shoot ratio of spring wheat at the end of the experiment (BBCH 40 = 30 days after onset of stress). Different letters indicate significant differences within each category (Tukey's HSD test).

3.1. *Single Stressors*

3.1.1. HandySpec

Several spectral indices could be used to differentiate between **nitrogen** deprived plants and non-nitrogen-stressed plants (Table 4). Using REIP, we were able to differentiate between the two *N*-treatments from the second time period (11–20 days), whereas ZM, VOG1, GM1 and NPQI could only be used to differentiate on the last time period (Figure 3a).

All tested indicators except NPQI contained sufficient information to enable a differentiation between plants that were under **water deprivation** and non–water-stressed plants (Table 4). The G index along with the PVR could clearly separate between stressed and non–stressed plants throughout the entire experimental period (Figure 3b). PPR and MCARI could also be used to differentiate between the two groups from the second time period (11–20 days) and onwards. To a lesser extent, LIC1 could also detect water stress. For all the above indices, water stressed plants had lower values. G, PPR and MCARI had a relatively stable result from the 10th day after treatment onwards, while LIC1 showed the typical increase in both groups as days passed.

Weediness was detected by all tested indices except CTR1 and VOG1 (Table 4). In all cases, apart from RVSI, the index values for weedy pots were higher than those for weed free pots. Most of the indices increased their values as time passed (Figure 3c). G and MCARI provided relatively similar values for the second and third 10-day period, but their values also increased slightly.

Several indices could be used to discriminate **fungal** infected and non-infected pots. Both structural, chlorophyll and stress related indices showed statistically significant differences between the two groups (Table 4). RVSI and NDVI differentiated between infected and non-infected plants already at the beginning of the experiment until the second 10-day period, but were not able to differentiate afterwards (Figure 3d). On the other hand, indices like RDVI, NPQI, PRI, and CTR1 showed statistically significant differences between the two groups in the second and third period.

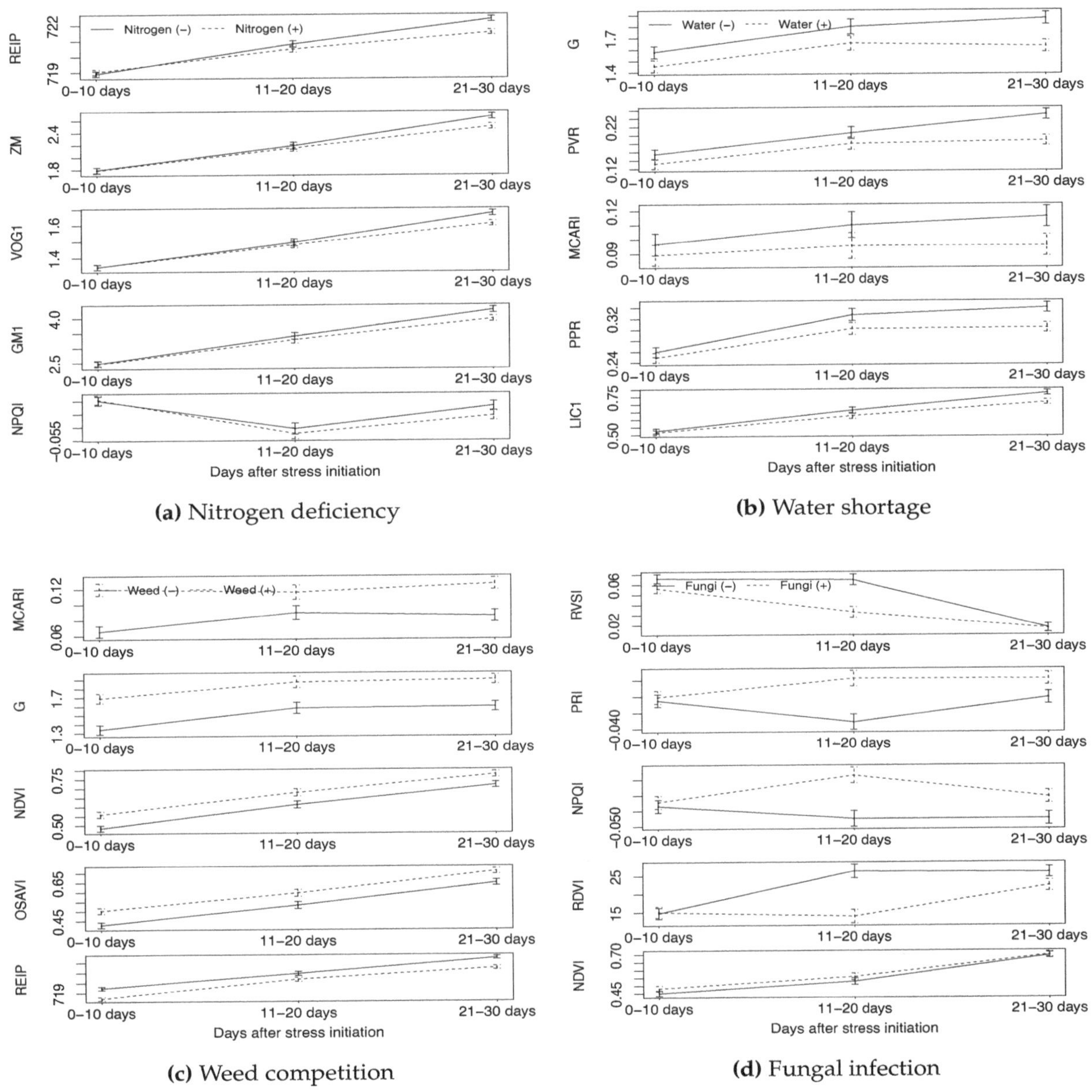

Figure 3. Ten day mean values of spectral indices derived by the HandySpec® sensor. The error bars demonstrate their upper and lower 95% confidence intervals (Tukey's HSD). The most important indices per stress are shown, separated into three periods: 0–10, 11–20 and 21–30 days after stress induction. '−' indicates absence of stress, '+' indicates presence of stress.

Table 4. Mean values of spectral indices for the main effects derived by the HandySpec® sensor. '−' indicates the absence of stress and '+' indicates its presence. The significance level is indicated as *** for $p < 0.001$; ** for $p < 0.01$; * for $p < 0.05$, and NS for non-significant differences.

Index	Nitrogen Deficiency			Water Shortage			Weed Competition			Fungal Infection		
	−	+	Sign	−	+	Sign	−	+	Sign	−	+	Sign
REIP	721	720	***	720	721	*	721	720	***	722	721	***
NDVI	0.64	0.63	NS	0.65	0.62	***	0.60	0.67	***	0.56	0.58	**
PVR	0.19	0.18	NS	0.20	0.17	***	0.15	0.22	***	0.10	0.12	***
OSAVI	0.58	0.56	NS	0.58	0.56	**	0.54	0.60	***	0.52	0.51	NS
MCARI	0.11	0.11	NS	0.12	0.10	***	0.08	0.14	***	0.06	0.05	***
RVSI	0.030	0.032	NS	0.028	0.034	**	0.034	0.028	***	0.047	0.036	***
RDVI	23.5	22.8	NS	24.0	22.3	*	20.6	25.7	***	21.9	17.5	***
G	1.68	1.65	NS	1.75	1.58	***	1.50	1.83	***	1.26	1.36	***
ZM	2.23	2.15	***	2.23	2.15	**	2.15	2.23	**	2.07	2.09	NS
NPQI	−0.043	−0.045	**	−0.045	−0.044	NS	−0.046	−0.042	***	−0.046	−0.039	***
PRI	−0.021	−0.024	*	−0.02	−0.025	***	−0.026	−0.019	***	−0.033	−0.027	***
CTR1	1.52	1.56	*	1.52	1.55	*	1.53	1.55	NS	1.51	1.40	***
LIC1	0.64	0.63	NS	0.65	0.61	***	0.60	0.66	***	0.56	0.58	**
VOG1	1.51	1.48	**	1.50	1.48	**	1.48	1.50	NS	1.46	1.47	NS
GM1	3.40	3.24	***	3.38	3.26	*	3.22	3.42	***	2.93	2.88	NS
PPR	0.29	0.30	NS	0.30	0.28	***	0.26	0.32	***	0.22	0.22	NS

3.1.2. Isaria

The two indices obtained by the Isaria-sensor, REIP and its proprietary biomass index (IBI), could differentiate between the two weed treatments only (Table 5). Weed free pots resulted in higher values for IBI and lower for REIP compared with the corresponding values for pots with weeds. The differences were significant throughout the entire experiment (Figure 4).

Table 5. Mean values of spectral indices for the main effects derived by the Isaria® sensor. '−' indicates the absence of stress and '+' indicates its presence. The significance level is indicated as *** for $p < 0.001$; ** for $p < 0.01$; * for $p < 0.05$, and NS for non-significant differences.

Index	Nitrogen Deficiency			Water Shortage			Weed Competition			Fungal Infection		
	−	+	Sign	−	+	Sign	−	+	Sign	−	+	Sign
REIP	724	723	NS	723	724	NS	725	722	***	726	726	NS
IBI	79.6	76.3	NS	80.8	75.1	NS	64.3	91.6	***	63.2	64.5	NS

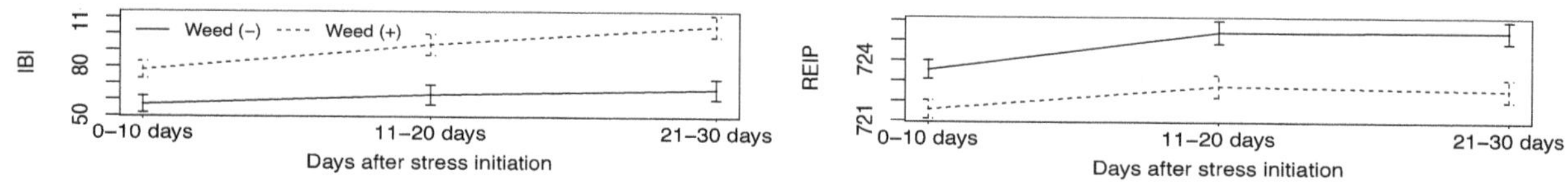

Figure 4. Ten day mean values of spectral indices derived by the Isaria® sensor. The error bars demonstrate their upper and lower 95% confidence intervals (Tukey HSD $p = 0.05$). The most important indices per stress are shown, separated into three periods: 0–10, 11–20 and 21–30 days after stress induction. '−' indicates absence of stress, '+' indicates presence of stress.

3.1.3. Multiplex

Nitrogen Balance Index excited by green or red light, NBI_G and NBI_R, measured by the Multiplex sensor could clearly differentiate between **nitrogen** stressed and non-stressed plants, giving lower values for the stressed plants (Table 6). An index correlated with flavonoids (FLAV) also correlated with nitrogen content along with red and far red fluorescence under ultraviolet excitation (RF_{UV} and FRF_{UV}). Multiplex® measurements differentiated between stressed and non-stressed plants already at the second 10-day period (Figure 5a). NBI_G gave the best results followed by NBI_R and FLAV. Multiplex could only differentiate **water stress** indirectly also through the chlorophyll and the flavonoid content. The distinction between water deprived and non water deprived plants is statistically significant with NBI_G, flavonoids (FLAV), NBI_R, and FER_{RUV}. The value NBI_G and NBI_R is higher on the stressed plants than the non–stressed ones. The opposite applies for FLAV and FER_{RUV}. All of the above indices could differentiate between the two groups in the first days, and apart from NBI_R, also in the last days after the water stress treatment (Figure 5b).

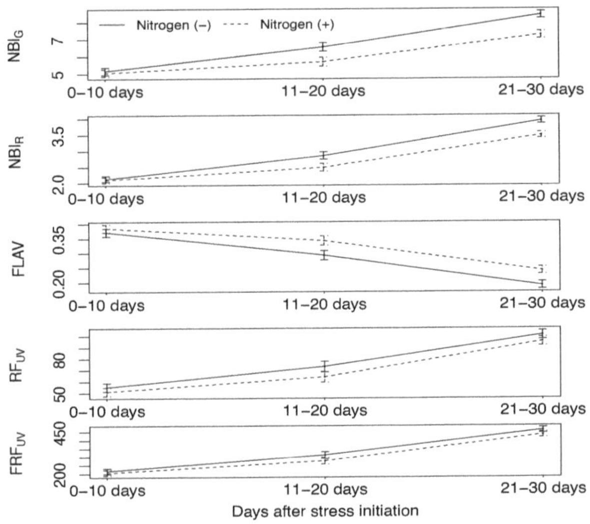

(a) Nitrogen deficiency

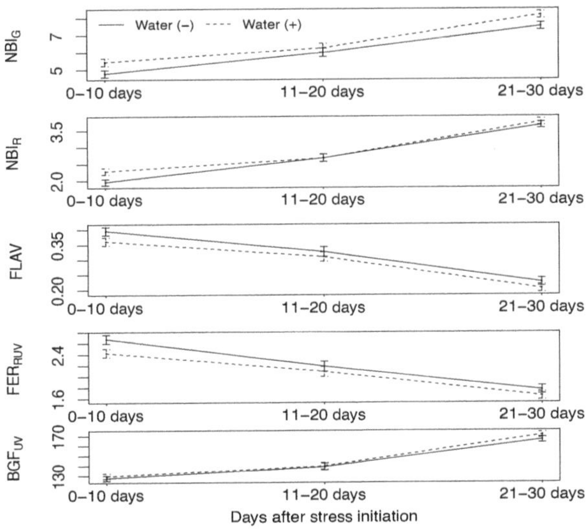

(b) Water shortage

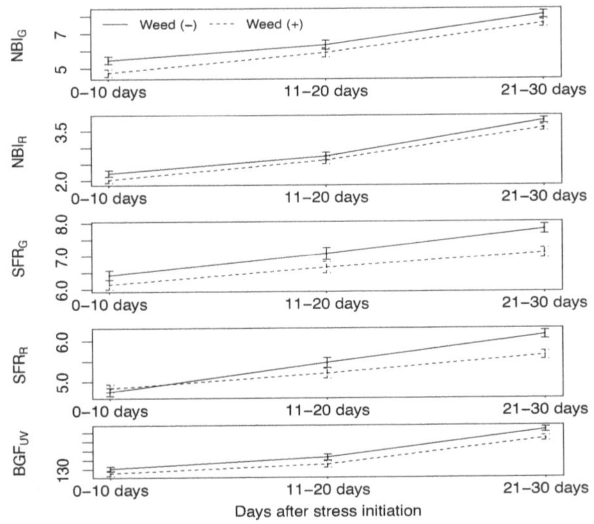

(c) Weed competition

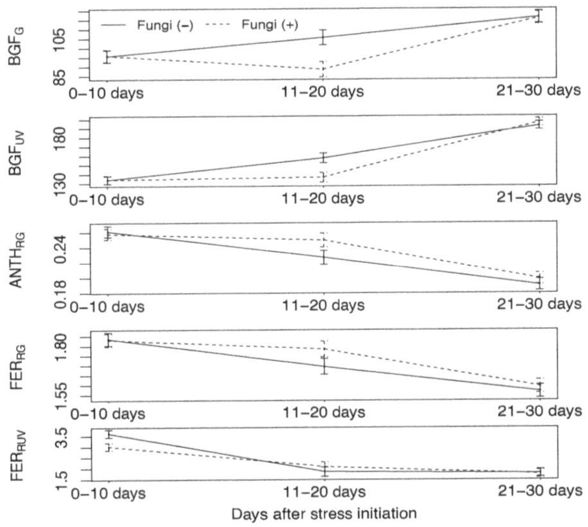

(d) Fungal infection

Figure 5. Ten day mean values of spectral indices derived by the Multiplex® sensor. The error bars demonstrate their upper and lower 95% confidence intervals (Tukey's HSD). The most important indices per stress are shown, separated into three periods: 0–10, 11–20 and 21–30 days after stress induction. '−' indicates absence of stress, '+' indicates presence of stress.

According to the Multiplex®sensor, identification of **weed presence** could be detected with the Nitrogen Balance Indices and the Simple Fluorescence Ratios both under green and red excitation (Table 6). In all cases, the values of the weedy pots were higher than the weed free pots. Differentiation was performed in all three periods with NBI_G, NBI_R, and SFR_G, yet the typical increase in their values between the periods occurred (Figure 5c). For SFR_R, significant differences were visible from the second period. Multiplex® did not provide many different indices that could differentiate fungal infection. Blue green fluorescence under green and ultraviolet excitation (BGF_G, BGF_{UV}), FER_{RUV}, FER_{RG} and $ANTH_{RG}$ were able to identify the **fungi** stress (Table 6). Fungal infected plants showed higher values than the non–infected ones on FER_{RG} and $ANTH_{RG}$ and the opposite on the rest. FER_{RUV} was able to differentiate between the fungal infected and the healthy plants the first ten days after inoculation (Figure 5d). The remaining indices were able to differentiate between the two groups in the second period. FER_{RG} and $ANTH_{RG}$ showed correlation only with fungi stress.

Table 6. Mean values of spectral indices for the main effects derived by the Multiplex® sensor. '−' indicates the absence of stress and '+' indicates its presence. The significance level is indicated as *** for $p < 0.001$; ** for $p < 0.01$; * for $p < 0.05$, and NS for non-significant differences.

Index	Nitrogen Deficiency			Water Shortage			Weed Competition			Fungal Infection		
	−	+	Sign	−	+	Sign	−	+	Sign	−	+	Sign
BGF_{UV}	147	146	NS	146	148	**	150	144	***	162	158	**
RF_{UV}	80.1	73.5	***	76.7	76.8	NS	74.7	78.8	*	101	100	NS
FRF_{UV}	347	315	***	327	335	NS	332	331	NS	462	465	NS
BGF_G	88.8	90.1	NS	88.9	90	NS	90.6	88.3	**	106	102	***
SFR_G	7.15	7.03	NS	7.00	7.19	*	7.35	6.83	***	7.80	7.75	NS
SFR_R	5.48	5.38	*	5.37	5.48	*	5.54	5.32	***	6.35	6.41	NS
FER_{RUV}	2.08	2.33	***	2.29	2.12	**	2.16	2.25	NS	2.49	2.31	*
$FLAV$	0.29	0.33	***	0.32	0.29	***	0.30	0.31	**	0.35	0.34	NS
FER_{RG}	1.80	1.83	NS	1.83	1.80	NS	1.83	1.80	NS	1.70	1.73	*
$ANTH_{RG}$	0.25	0.25	NS	0.25	0.25	NS	0.25	0.25	NS	0.23	0.23	*
NBI_G	6.83	6.15	***	6.22	6.76	***	6.74	6.24	***	6.34	6.47	NS
NBI_R	3.04	2.72	***	2.80	2.97	***	2.97	2.79	***	3.14	3.17	NS

3.2. Combinations of Stressors

3.2.1. HandySpec

More than one stress factor can be described with the same HandySpec® index as presented in Figure 6. These results are more complicated to be explained and less robust. If the result increases or decreases simultaneously for both stressors, then the control and the combined stressed treatment occupy the edge values, while the values from only one stress combination are between the two aforementioned values. Examples of the above situation can be REIP for nitrogen deficiency and weed presence, VOG1 for nitrogen deficiency and water shortage, RDVI for water shortage and fungal infection and G for weed presence and fungal infection. In cases where the values for the stressed plants increase for one of the two stressors and decrease for the the other one, then the effect of one stressor counteracts with the effect of the second stressor. We notice that the highest and the lowest values belong to the pots containing only one stress value. The control along with the combination of stressors are in between. We can see this outcome in indices like PPR, NDVI and MCARI for water shortage and weed presence and PVR for water shortage and fungal infection.

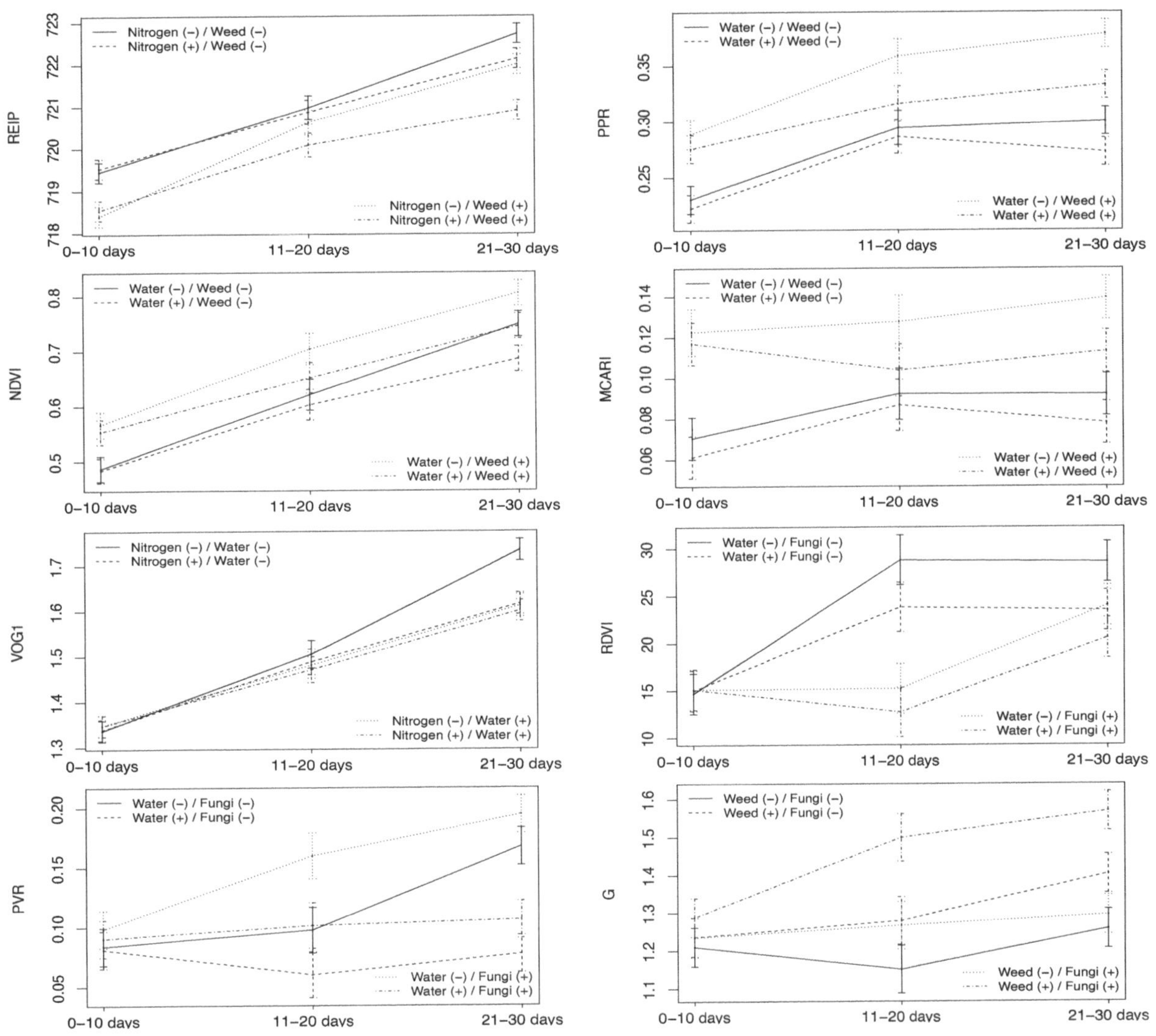

Figure 6. Ten day mean values of spectral indices derived by the HandySpec® sensor. The error bars demonstrate their upper and lower 95% confidence intervals (Tukey HSD). The most important interactions per combination of two stressors are showed, separated into three periods: 0–10, 11–20 and 21–30 days after stress induction. '−' indicates absence of stress, '+' indicates presence of stress.

3.2.2. Multiplex

For Multiplex®, apart from indices like FRF_{UV} that can only identify nitrogen deficiency, FER_{RG} and $ANTH_{RG}$ that can only pinpoint fungi stress, most of the indices showed statistically significant results for more than one stress factor. As described in HandySpec®, we have two distinct cases: (i) if the values increase in the appearance of both stressors; or (ii) in the appearance of one of the two, the index value increases and for the other the value decreases. If the values shift simultaneously for both stressors, then the highest and lowest values belong to the control and the combination of both stressors. Treatments with only one stress factor provide values between the two aforementioned edges. In Multiplex® we can clearly see these trends in SFR_R and NBI_G for identifying the combination of nitrogen deficiency and weed presence and in BGF_{UV} for identification of weed presence and fungal infection (Figure 7). If the index increases for one stressor and decreases for the other, then the values for treatments with only one stressor are the highest/lowest and the values of the control and the combination of stressors are between those values. For the Multiplex® data, we can see the results in $FLAV$ for the identification of nitrogen deficiency and water infection and BGF_G for the identification of water shortage and fungal infection (Figure 7).

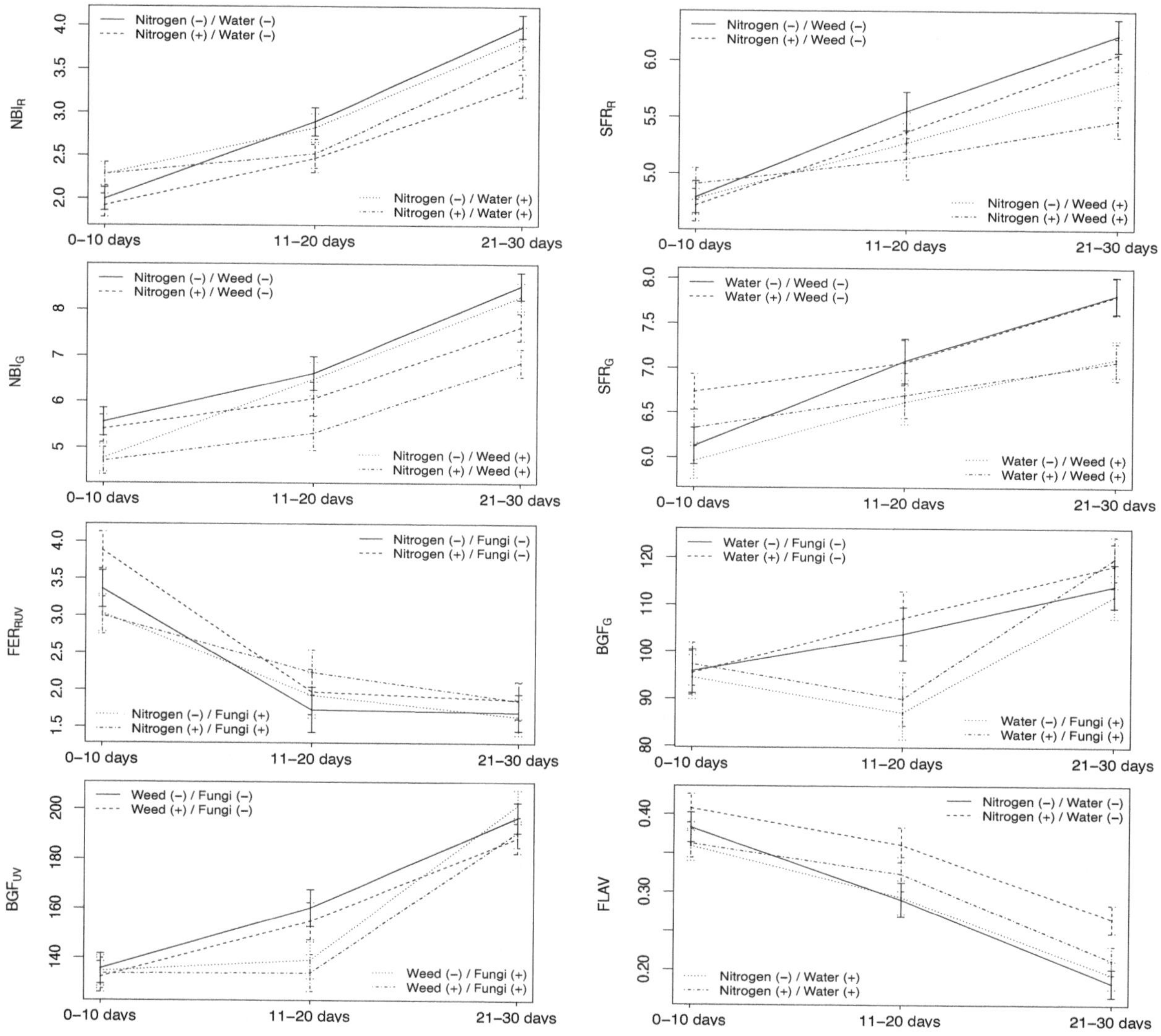

Figure 7. Ten day mean values of spectral indices derived by the Multiplex® sensor. The error bars demonstrate their upper and lower 95% confidence intervals (Tukey's HSD). The most important interactions per combination of two stressors are showed, separated into three periods: 0–10, 11–20 and 21–30 days after stress induction. '—' indicates absence of stress, '+' indicates presence of stress.

4. Discussion

The biomass and the shoot-to-root ratio measurements showed that the stressors affected plant development. Even though the stress factors were there, they were not causing extreme differences between stressed and non-stressed plants. The largest effect was measured for water content on dry matter production, where the dry matter yields in the water stressed pots averaged 29.5% less than that obtained in the non-water-stressed pots. Hence, the data set should be a good starting point for stress recognition.

4.1. Could Nitrogen Deficiency Stress Be Detected by the Sensors?

When using the HandySpec indices REIP, ZM, VOG1, and GM1, we were able to discriminate between spring wheat grown with or without sufficient nitrogen supply. REIP was the only index enabling the detection of nitrogen deficiency as early as 11–20 days after onset of stress. It should be noted that all plants had the same nutrient conditions until BBCH 12 (ample resources). The delay in the differentiation between N-stressed and fertilized plants may be attributed to the early growth stages of the plants at time of fertilization, and their relatively limited N demand at this developing stage. As expected, REIP was lower for the stressed than the non-stressed plants. This agrees with the well documented red shift of the REIP due to higher concentrations of chlorophyll, which normally follows increased plant N availability. [16,49–53]. The indices VOG1, ZM and GM1 have also been proven to correlate with chlorophyll content [16,54,55]. Taking into account that VOG and ZM are simple ratios centered around 730 nm (red), both can be considered as gross estimators of the REIP. In contrast, GM1 also involves reflection from the green area (dividing reflectance at 750 nm with that at 550 nm), and has been reported to correlate well with total chlorophyll along with nutrition and fertilization level [8,52,53,56].

Five of the twelve indices measured with the Multiplex® sensor could be used to detect nitrogen deficiency as early as the 11–20 day period. The indices NBI_G and NBI_R clearly differentiated between N-stressed and non-stressed plants, giving lower index values for the stressed than the non-stressed plants. Similar findings have also been reported for bermudagras and turfgrasses [57,58]. Longchamps and Khosla [59] were able to distinguish between all their four N-levels in an experiment with maize, using NBI_G and NBI_R. In our experiment, we also found out that an index related to flavonoids, FLAV, also contained information usable for nitrogen stress detection. In contrast with NBI_G and NBI_R, FLAV had higher, not lower values for the stressed plants. This agrees well with results from Agati *et al.* [58]. Several authors have reported that FLAV correlates with the flavonoid levels in fruits [13,60–62]. Cartelat *et al.* [13] found a negative correlation between flavonoid levels and chlorophyll content in wheat. All in all concerning nitrogen deficiency, both the spectrometer and the fluorometer were able to detect nitrogen deficiency. REIP, VOG1, ZM, NBI_G, NBI_R and FLAV can be used for this identification. On the other hand, the Isaria®sensor did not provide significant data.

4.2. Could Water Stress Be Detected by the Sensors?

Among the five indices presented for HandySpec® capable in detecting shortage in water, PVR and G could detect it already 0–10 DAS. This was well before any visual symptoms of water deprivation occurred. PVR, MCARI and G showed relatively stable values from the 10th DAS and onwards. All the above indices utilize the reflection of 550 nm (green) in their formula. Lin *et al.* [17] pinpointed a shift of the region around 535–540 nm with the water content of *Cinnamomum camphora* (Linn.) Seib. Kusnierek and Korsaeth [63] identified the region 560–610 nm as one of three spectral regions in the range 400–950 nm containing significant information related to water status in spring wheat. Thenkabail *et al.* [8] associates the wavebands around 550 nm with total chlorophyll and biomass, therefore water stress measurements also derive indirectly from chlorophyll measurements. Wang *et al.* [64] also showed a robust correlation between PPR and chlorophyll concentration. In the current study, NDVI, SAVI, OSAVI, VOG1, LIC1, GM1 and ZM could also be used to differentiate between water deprived and non stressed plants 20–30 DAS. These indices have also previously been associated with water content in plants [9,17–20]. The correlations between the indices NDVI, SAVI, OSAVI, VOG1, LIC1, GM1 and ZM at one side and plant water status at the other were probably an indirect result, as water deprivation affects other parameters of plant growth, like chlorophyll content and leaf area index, which these indices correlate better with.

Using the Multiplex® device, four indices (FER_{RUV}, FLAV, NBI_G, and NBI_R) enabled early detection of water shortage. Two of the indices, FLAV and NBI, have previously been associated with water tolerance in wheat [14]. Concerning water stress for both the spectrometer and the fluorometer, the presented approach appears promising. It can be argued that the spectrometer performed better than the fluorometer in this task, since the basic recognition indices in the fluorometer are the same as for nitrogen. PVR and G from the spectrometer can be used for water identification. However, the robustness of all indices related with water should be confirmed by further studies.

4.3. Could Weed Competition Be Detected by the Sensors?

As expected, weed competition was correlated with many indices calculated from the HandySpec® data. As for water shortage, G and other indices correlating with the Leaf Area Index

(LAI), like NDVI and OSAVI, could be used to identify weediness from the first days of the experiment and onwards. The stress factor weeds, differed markedly from the other stressors we imposed, as the weed plants were physically present from the first day of measurement. Since the pots occupy a predefined space, the plant biomass and the LAI were thus already higher at the beginning of the measurement period, compared with pots without any weeds planted.

Vegetation Indices like NDVI, REIP, OSAVI, ZM, RDVI and LIC1, contained not only the information needed to separate between pots with or without weeds, but they also showed an increase with time, reflecting the growth in biomass during the experiment (both treatments). Measurements were performed at the early growing stages of the plants (BBCH 12–40). Therefore, crop plants were also rapidly growing, resulting in a continuous increase in e.g., LAI, commonly shown to be positively correlated with NDVI. Indices like PVR and PPR could be used to identify weed presence from days 11–20 and beyond. Since the pot size and volume are finite, this result might be an indirect result from water stress, due to assumedly higher transpiration from pots with weed and wheat plants than from pots with wheat plants alone.

Weed treatments could be separated by means of the indices REIP and IBI as obtained by the active spectrometer Isaria®. The REIP as calculated from the Isaria sensor gave higher values than the REIP calculated by the Handyspec®-sensor. This agrees with the findings of Peteinatos *et al.* [65]. The Multiplex provided useful data, as four of its output-indices (BGF_{UV}, SFR_G, SFR_R and NBI_G) could be used to separate the weed treatments.

All three sensors provided information for weed identification. HandySpec®and Isaria®provided bigger differences faster than the fluorometer. It should be noted that the classical method of detecting weeds is by combining high-density RGB-images and image analysis. Such an approach is, however, less useful for detecting the other stressors of interest in the current study. Moreover, image-based weed detection is normally performed with the weed plants at a very early development stage, mainly to reflect the timing of herbicide application in practice. In this study, we focused on the combined effect of more stressors, and thus selected sensors, which had a potential for identifying more than one stressor. Therefore, from our perspective, indices MCARI, G, NDVI and OSAVI were able to perform this task along with SFR_G and SFR_R.

4.4. Could Fungal Infection Be Detected by Sensors?

Two indices calculated from data obtained by the HandySpec® could be used to identify infection by powdery mildew (*B. graminis*) as soon as 0–10 DAS, NDVI and RVSI, the latter also 11–20 DAS. This clearly suggests that RVSI is suitable for early, pre-visual, detection of powdery mildew in wheat. Bauriegel *et al.* [11] and Bauriegel and Herppich [27] identified the spectra around 550–560 nm, and 665–675 nm as important for identifying *Fusarium* infection in wheat, but they could not find any significant results with NDVI, G and LIC1. Two of the above indices, MCARI and RDVI, use at least one of the above two spectral ranges in their calculations. Time-wise, the results showed an interesting point. A series of four indices could discriminate between infested and non-infested in the second time period. These could also provide indices which enabled a differentiation in the third period, but differences became smaller. Zhang *et al.* [26] made a similar observation when investigating yellow rust infection over 17 days (four measurements at 216 till 233 days after sowing) in wheat with NDVI, PRI, RVSI and MCARI. Their results also showed the highest correlations at the second measuring date (225 days after sowing), whereas five days later only PRI was correlated with yellow rust infection. We cannot explain this phenomenon based on the current data, but this should be investigated in more detail, since this may be a potential important issue related to pre-symptomatic fungi detection. Indices like PRI and MCARI that identified water stress and NDVI and MCARI that identified weed presence, were also able to identify fungi infection. Zhang *et al.* [26] reported similar results for PRI. The fact that all types of spectral indices correlated with presence of fungi can be attributed to the result that fungi infected plants had a reduced growth dynamic, a lower total biomass and LAI.

Multiplex® could also differentiate between fungi infected and non-infected plants. Blue-Green fluorescence under green or ultraviolet excitation (BGF_G and BGF_{UV}) and indices relative to Ferodoxines and Anthocyanins could be used to identify the fungal infection. FER_{RG} and $ANTH_{RG}$ seemed sensitive only on fungal infection. This can make them an indicator for *B. graminis* . Latouche *et al.* [66] used Multiplex® to successfully identify *Plasmopara viticola* in vineyards, yet the sensor setup was modified. Time-wise, BGF_G, BGF_{UV}, FER_{RG} and $ANTH_{RG}$ were able to identify the fungal infection in the second 10-day period. The result seems similar to the result of the HandySpec® sensor. Only FER_{RUV} was able to identify the fungal infection in the first 10-day period. However, it identified it only in the first 10-day period and the results, even if it is not statistically significant, were reversed in the second 10-day period. Therefore, more experimentation is needed to see if this index can be used as a tool to identify *B. graminis* . Concerning fungal infection, the small time windows that it can be measured in, makes it challenging to be identified by both spectral and fluorescence data.

4.5. Could Combinations of Stressors Be Detected by Sensors?

A lot of indices both from the HandySpec® and the Multiplex® sensor showed correlations with more than one stress factor. In the combination of stress factors, if we take into account the different way that each index reacts to each stressor, the increase through time for most of the indexes and the different rate of increase per stressor, identifying combinations of stressors with the same index can be quite challenging. For example, REIP as presented in Figure 3c in the first 10-day period can clearly differentiate weedy from weed free pots, but could not differentiate per weed group the nitrogen stress. As days passed by, the non-nitrogen stressed plants increased in value at a higher rate than the non-stressed plants. In the third 10-day period, there was a clear distinction, between the control and the stressed pots. The pots containing only one of the two stressors could not be differentiated from each other. Similar results can be noted for RDVI where, in the second 10-day period, it can clearly differentiate between all four water shortage and fungal infection combinations. On 21–30 DAS, only three groups are clearly distinguishable (control, combinations, one stressor). For weed presence and fungal infection, G presents the aforementioned three groups on the 10–20 DAS, but in the third 10-day period, fungal infection cannot be differentiated on the weed free pots. Weedy pots could clearly be differentiated from weed free and in their case fungal infection could also be pinpointed. On the other hand, VOG1 can only differentiate the control from the other three treatments in the third 10-day period. Zhang *et al.* [26] also pinpointed the difficulty of identifying more than one stress factor with the same index.

In cases where the results of the two stressors are heading in the opposite direction, identifying the existence of one stressor or the other can be easy. We can see this in PPR, NDVI and MCARI for separating water shortage and weed presence and in PVR for the identification of weed presence and fungal infection from 10–20 DAS onwards. On the other hand, identification between the control and the combinations of the two stressors is harder. The ability to identify the control from the combination of two stressors relies only on the condition, if the results on the index for one stressor are higher than the other. NDVI is not able to perform this, while the rest of the aforementioned indices can perform this separation in the third 10-day period.

Combination of more than one stressor with the aid of the same index seems harder for the Multiplex® sensor. NBI_R can separate nitrogen deficiency in the combination with water shortage from the second 10-day period onwards. NBI_G can do the same for the combination of nitrogen deficiency and weed presence. For both indices, only the pots having nitrogen deficiency could differentiate the second stressor (water deficiency and weed presence) in the third 10-day period. BGF_G and BGF_{UV} could differentiate fungal infection in the second 10-day period in the combination of water shortage and weed presence, respectively. In the third 10-day period, BGF_G could separate only water shortage and BGF_{UV} only weeds. Identification of both stressors was not performed simultaneously but in different time periods. That pinpoints the constraints of creating a robust stress identification based on sensor values. Depending on the circumstances, a similar result could be attributed to one or more different stressors. A similar example can be SFR_G which can differentiate water stress in weed free pots in the first 10-day period. However, from the 11–20 DAS onwards, the differentiation performed is only weedy and weed free pots regardless of the water stress. Concerning interaction of more than two factors, no significant results were identified.

5. Conclusions

Water deficiency and fungal infection could be detected pre-symptomatic, *i.e.*, 0–10 days after stress induction (DAS), with one or more of the sensors tested. The presence or absence of weeds could also be identified using the same sensors (present during the entire experiment).

Nitrogen shortage could be detected 10–20 DAS by the index REIP (Red Edge Inflection Point) measured with HandySpec® and several indices based on Multiplex. The lack in detection of nitrogen shortage earlier could be attributed to the early growth stage (BBCH 12–14) and hence the limited nitrogen needs. Water deprivation in spring wheat could be detected as early as 0–10 days after stress induction (DAS) by the Photosynthetic Vigor Ratio (PVR; 550 nm − 650 nm/550 nm + 650 nm), the Plant Pigment Ratio (PPR; 550 nm − 450 nm/550 nm + 450 nm) and Greenness (G; 554 nm/677 nm) from the HandySpec® spectrometer and several indices based on the Multiplex® fluorometer. Weediness (*Sinapis alba*) could be detected from 0–10 DAS and onwards by indices measured by all three sensors, including REIP and biomass index (IBI) from the Isaria sensor. Fungal infection in spring wheat (*Blumeria graminis* f. sp. tritici) could be detected from 0–10 DAS by Red-edge Vegetation Stress Index (RVSI; (714 nm + 752 nm/2) − 733), Photochemical Reflectance Index (PRI; 531 nm − 570 nm/531 nm + 570 nm) and Normalized Phaeophytinization Index (NPQI; 415 nm − 435 nm/415 nm + 435 nm) measured with HandySpec® and by the ratio FER_{RG} (Fluorescence Excitation Ratio; Red/Green excitiation) measured with Multiplex®. One of the main findings of this study was that the results both from spectral and from fluorescence data varied more due to the plant development rather than the presence or absence of a stress factor. In a lot of cases, where values were attributed to stressed plants at the early stages, ten or twenty days later the same values are then attributed to the non stressed plants. The study shows that it is possible to distinguish between a stressed plant and an unstressed control plant at the same growth stage, but we cannot pinpoint a general threshold value to use for discrimination regardless of the plant development stage. This hinders the stress identification process. More work needs to be done in order to model how the spectral indices change with the growth stage of the plants. In some cases, interactions of more than one stress factor could be viewed with the aid of the same index, yet these results are more complicated to interpret. Interactions with more than two stress factors could not present any significant results. None of the indices showed a significant three-way interaction.

Acknowledgments: The authors would like to thank the Alexander S. Onassis Public Benefit Foundation, the Carl-Zeiss foundation, and The Research Council of Norway (program BIONÆR, project MULTISENS, project number 207829) for funding. We would like to thank Victor Rueda-Ayala for his valuable ideas to set up the experiment and analyze the data. We would also like to thank Christoph Kunz, Manuel Geisser, Julian Börner and Yuxuan Wang for their help in the measurements.

Author Contributions: All authors contributed extensively to this work. Gerassimos Peteinatos organized the execution of the experimental setup, analysed the data, evaluated the results and drafted the manuscript. Audun Korsaeth helped in the design of the experimental setup, data analysis and evaluation and aided the drafting of the manuscript. Therese W. Berge participated in the draft of the manuscript and its revisions. Roland Gerhards helped in the design of the experiment, the drafting and revision of the paper.

References

1. Farooq, M.; Hussain, M.; Wahid, A.; Siddique, K.H.M. Drought Stress in Plants: An Overview. In *Plant Responses to Drought Stress*; Springer: Berlin, Germany; Heidelberg, Germany, 2012; pp. 1–33.

2. Oerke, E. Crop losses to pests. *J. Agric. Sci.* **2006**, *144*, 31–43.
3. Zhao, G.; Miao, Y.; Wang, H.; Su, M.; Fan, M.; Zhang, F.; Jiang, R.; Zhang, Z.; Liu, C.; Liu, P.; *et al.* A preliminary precision rice management system for increasing both grain yield and nitrogen use efficiency. *Field Crops Res.* **2013**, *154*, 23–30.
4. San Martín, C.; Andújar, D.; Fernández-Quintanilla, C.; Dorado, J. Spatial distribution patterns of weed communities in corn fields of Central Spain. *Weed Sci.* **2015**, *63*, 936–945.
5. Tremblay, N.; Wang, Z.; Cerovic, Z.G. Sensing crop nitrogen status with fluorescence indicators. A review. *Agron. Sustain. Dev.* **2011**, *32*, 451–464.
6. Peteinatos, G.G.; Weis, M.; Andújar, D.; Rueda Ayala, V.; Gerhards, R. Potential use of ground-based sensor technologies for weed detection. *Pest Manag. Sci.* **2014**, *70*, 190–199.
7. Govender, M.; Govender, P.J.; Weiersbye, I.M.; Witkowski, E.T.F.; Ahmed, F. Review of commonly used remote sensing and ground-based technologies to measure plant water stress. *Water SA* **2009**, *35*, 741–752.
8. Thenkabail, P.S.; Lyon, J.G.; Huete, A. (Eds.) *Hyperspectral Remote Sensing of Vegetation*; Taylor & Francis, CRC Press: Boca Raton, FL, USA, 2011.
9. Tilling, A.K.; O'Leary, G.J.; Ferwerda, J.G.; Jones, S.D.; Fitzgerald, G.J.; Rodriguez, D.; Belford, R. Remote sensing of nitrogen and water stress in wheat. *Field Crops Res.* **2007**, *104*, 77–85..
10. Li, F.; Miao, Y.; Hennig, S.D.; Gnyp, M.L.; Chen, X.; Jia, L.; Bareth, G. Evaluating hyperspectral vegetation indices for estimating nitrogen concentration of winter wheat at different growth stages. *Precis. Agric.* **2010**, *11*, 335–357.
11. Bauriegel, E.; Giebel, A.; Herppich, W.B. Hyperspectral and chlorophyll fluorescence imaging to analyze the impact of Fusarium culmorum on the photosynthetic integrity of infected wheat ears. *Sensors* **2011**, *11*, 3765–3779.
12. Backoulou, G.F.; Elliott, N.C.; Giles, K.L.; Mirik, M. Processed multispectral imagery differentiates wheat crop stress caused by greenbug from other causes. *Comput. Electron. Agric.* **2015**, *115*, 34–39.
13. Cartelat, A.; Cerovic, Z.; Goulas, Y.; Meyer, S.; Lelarge, C.; Prioul, J.L.; Barbottin, A.; Jeuffroy, M.H.; Gate, P.; Agati, G.; *et al.* Optically assessed contents of leaf polyphenolics and chlorophyll as indicators of nitrogen deficiency in wheat (*Triticum aestivum* L.). *Field Crops Res.* **2005**, *91*, 35–49.
14. Bürling, K.Z.C.; Cornic, G.; Ducruet, J.; Noga, G.; Hunsche, M. Fluorescence-based sensing of drought-induced stress in the vegetative phase of four contrasting wheat genotypes. *Environ. Exp. Bot.* **2013**, *89*, 51–59.
15. Strachan, I.B.; Pattey, E.; Boisvert, J.B. Impact of nitrogen and environmental conditions on corn as detected by hyperspectral reflectance. *Remote Sens. Environ.* **2002**, *80*, 213–224.
16. Vogelmann, J.E.; Rock, B.N.; Moss, D.M. Red edge spectral measurements from sugar maple leaves. *Int. J. Remote Sens.* **1993**, *14*, 1563–1575.
17. Lin, C.; Popescu, S.C.; Huang, S.C.; Chang, P.T.; Wen, H.L. A novel reflectance-based model for evaluating chlorophyll concentrations of fresh and water-stressed leaves. *Biogeosciences* **2015**, *12*, 49–66.
18. Suárez, L.; Zarco-Tejada, P.; Sepulcre-Cantó, G.; Pérez-Priego, O.; Miller, J.; Jiménez-Muñoz, J.; Sobrino, J. Assessing canopy PRI for water stress detection with diurnal airborne imagery. *Remote Sens. Environ.* **2008**, *112*, 560–575.
19. Zarco-Tejada, P.; González-Dugo, V.; Berni, J. Fluorescence, temperature and narrow-band indices acquired from a UAV platform for water stress detection using a micro-hyperspectral imager and a thermal camera. *Remote Sens. Environ.* **2012**, 117, 322–337.
20. Metternicht, G. Vegetation indices derived from high-resolution airborne videography for precision crop management. *Int. J. Remote Sens.* **2003**, *24*, 2855–2877.
21. Vrindts, E.; Baerdemaeker, J.D.; Ramon, H. Weed Detection using canopy reflection. *Precis. Agric.* **2002**, *3*, 63–80.
22. Haboudane, D.; Miller, J.R.; Pattey, E.; Zarco-Tejada, P.J.; Strachan, I.B. Hyperspectral vegetation indices and novel algorithms for predicting green LAI of crop canopies: Modeling and validation in the context of precision agriculture. *Remote Sens. Environ.* **2004**, *90*, 337–352.
23. Longchamps, L.; Panneton, B.; Samson, G.; Leroux, G.; Thériault, R. Discrimination of corn, grasses and dicot weeds by their UV-induced fluorescence spectral signature. *Precis. Agric.* **2010**, *11*, 181–197.
24. Tyystjärvi, E.; Norremark, M.; Mattila, H.; Keranen, M.; Hakala-Yatkin, M.; Ottosen, C.; Rosenqvist, E. Automatic identification of crop and weed species with chlorophyll fluorescence induction curves. *Precis. Agric.* **2011**, *12*, 546–563.
25. Rumpf, T.; Mahlein, A.K.; Steiner, U.; Oerke, E.C.; Dehne, H.W.; Plümer, L. Early detection and classification of plant diseases with Support Vector Machines based on hyperspectral reflectance. *Comput. Electron. Agric.* **2010**, *74*, 91–99.

26. Zhang, J.; Pu, R.; Huang, W.; Yuan, L.; Luo, J.; Wang, J. Using in-situ hyperspectral data for detecting and discriminating yellow rust disease from nutrient stresses. *Field Crops Res.* **2012**, doi:10.1016/j.fcr.2012.05.011.
27. Bauriegel, E.; Herppich, W.B. Hyperspectral and chlorophyll fluorescence imaging for early detection of plant diseases, with special reference to Fusarium spec. infections on wheat. *Agriculture* **2014**, *4*, 32–57.
28. Calderón, R.; Navas-Cortés, J.A.; Zarco-Tejada, P.J. Early detection and quantification of verticillium wilt in olive using hyperspectral and thermal imagery over large areas. *Remote Sens.* **2015**, *7*, 5584–5610.
29. Karimi, Y.; Prasher, O.S.; Patel, M.R.; Kim, H.S. Application of support vector machine technology for weed and nitrogen stress detection in corn. *Comput. Electron. Agric.* **2006**, *51*, 99–109.
30. Mitscherlich, E.A. Das Gesetz des Miniraums und das Gesetz des abhnehlmenden Bodenertrags. *Land. Jahrb* **1909**, *38*, 5371–5552.
31. Hess, M.; Barralis, G.; Bleiholder, H.; Buhr, L.; Eggers, T.; Hack, H.; Stauss, R. Use of the extended BBCH scale—General for the descriptions of the growth stages of mono and dicotyledonous weed species. *Weed Res.* **1997**, *37*, 433–441.
32. Rouse, J.W.; Haas, R.H.; Schell, J.A.; Deering, D.W.; Harlan, J.C. *Monitoring the Vernal Advancement and Retrogradation (Greenwave Effect) of Natural Vegetation*; Texas A & M University, Remote Sensing Center: College Station, TX, USA, 1974.
33. Rondeaux, G.; Steven, M.; Baret, F. Optimization of soil-adjusted vegetation indices. *Remote Sens. Environ.* **1996**, *55*, 95–107.
34. Roujean, J.L.; Breon, F.M. Estimation PAR absorbed by vegetation from bi-directional reflectance measurements. *Remote Sens. Environ.* **1995**, *51*, 375–384.
35. Guyot, G.; Baret, F.; Major, D.J. High spectral resolution: Determination of spectral shifts between the red and infrared. *Int. Arch. Photogram. Remote Sens.* **1988**, *11*, 750–760.
36. Zarco-Tejada, P.; Berjón, A.; López-Lozano, R.; Miller, J.; Martín, P.; Cachorro, V.; González, M.; de Frutos, A. Assessing vineyard condition with hyperspectral indices: Leaf and canopy reflectance simulation in a row-structured discontinuous canopy. *Remote Sens. Environ.* **2005**, *99*, 271–287.
37. Daughtry, C.S.; Walthall, C.L.; . Kim, M.S.; Brown de Colstoun, E.; McMurtrey, J.E. Estimating corn leaf chlorophyll concentration from leaf and canopy reflectance. *Remote Sens. Environ.* **2000**, *74*, 229–239.
38. Barnes, J.; Balaguer, L.; Manrique, E.; Elvira, S.; Davison, A. A reappraisal of the use of DMSO for the extraction and determination of chlorophylls a and b in lichens and higher plants. *Environ. Exp. Bot.* **1992**, *32*, 85–100.
39. Gitelson, A.A.; Merzlyak, M.N. Remote estimation of chlorophyll content in higher plant leaves. *Int. J. Remote Sens.* **1997**, *18*, 2691–2697.
40. Lichtenthaler, H.K. Vegetation stress: An introduction to the stress concept in plants. *J. Plant Physiol.* **1996**, *148*, 4–14.
41. Zarco-Tejada, P.J.; Miller, J.R.; Noland, T.L.; Mohammed, G.H.; Sampson, P.H. Scaling-up and model inversion methods with narrow-band optical indices for chlorophyll content estimation in closed forest canopies with hyperspectral data. *IEEE Trans. Geosci. Remote Sen.* **2001**, *39*, 1491–1507.
42. Gamon, J.A.; Peñuelas, J.; Field, C.B. A narrow-waveband spectral index that tracks diurnal changes in photosynthetic efficiency. *Remote Sens. Environ.* **1992**, *41*, 35–44.
43. Carter, G.A. Ratios of leaf reflectances in narrow wavebands as indicators of plant stress. *Int. J. Remote Sens.* **1994**, *15*, 697–703.
44. Merton, R. *Monitoring Community Hysteresis Using Spectral Shift Analysis and the Red-Edge Vegetation Stress Index*; Jet Propulsion Laboratory: Pasadena, CA, USA, 1998.
45. FORCE-A. *Users Guide, Multiplex 3, Hand-Held Multi-Parameter Optical Sensor*; FORCE-A, Centre Universitaire Paris-Sud ORSAY: Cedex, France, 2010.
46. Ghozlen, N.B.; Cerovic, Z.G.; Germain, C.; Toutain, S.; Latouche, G. Non-destructive optical monitoring of grape maturation by proximal sensing. *Sensors* **2010**, *10*, 10040–10068.
47. Cerovic, Z.; Moise, N.; Agati, G.; Latouche, G.; Ghozlen, N.B.; Meyer, S. New portable optical sensors for the assessment of winegrape phenolic maturity based on berry fluorescence. *J. Food Compos. Anal.* **2008**, *21*, 650–654.
48. R Development Core Team. *R: A Language and Environment for Statistical Computing*; R Foundation for Statistical Computing: Vienna, Austria, 2008.
49. Boochs, F.; Kupfer, G.; Dockter, K.; Kühbauch, W. Shape of the red edge as vitality indicator for plants. *Int. J. Remote Sens.* **1990**, *11*, 1741–1753.

50. Miller, J.R.; Hare, E.W.; Wu, J. Quantitative characterization of the vegetation red edge reflectance 1. An inverted-Gaussian reflectance model. *Int. J. Remote Sens.* **1990**, *11*, 1755–1773.
51. Belanger, M.; Miller, J.; Boyer, M. Comparative relationships between some red edge parameters and seasonal leaf chlorophyll concentrations. *Can. J. Remote Sens.* **1995**, *21*, 16–21.
52. Zhao, D.; Reddy, K.R.; Kakani, V.G.; Reddy, V. Nitrogen deficiency effects on plant growth, leaf photosynthesis, and hyperspectral reflectance properties of sorghum. *Eur. J. Agron.* **2005**, *22*, 399–403.
53. Corp, L.A.; Middleton, E.M.; Campbell, P.E.; Huemmrich, K.F.; Daughtry, C.S.; Russ, A.; Cheng, Y.B. Spectral indices to monitor nitrogen-driven carbon uptake in field corn. *J. Appl. Remote Sens.* **2010**, *4*, doi:10.1117/1.3518455.
54. Zarco-Tejada, P.J.; Berjón, A.; Miller, J.R. Stress detection in crops with hyperspectral remote sensing and physical simulation models. In Proceedings of the Airborne Imaging Spectroscopy Workshop, Bruges, Belgium, 8 October 2004.
55. Zarco-Tejada, P.J.; Miller, J.R.; Harron, J.; Hu, B.; Noland, T.L.; Goel, N.; Mohammed, G.H.; Sampson, P. Needle chlorophyll content estimation through model inversion using hyperspectral data from boreal conifer forest canopies. *Remote Sens. Environ.* **2004**, *89*, 189–199.
56. Campbell, P.E.; Middleton, E.; Corp, L.; Kim, M. Contribution of chlorophyll fluorescence to the apparent vegetation reflectance. *Sci. Total Environ.* **2008**, *404*, 433–439.
57. Agati, G.; Foschi, L.; Grossi, N.; Volterrani, M. In field non-invasive sensing of the nitrogen status in hybrid bermudagrass (*Cynodon dactylon* × *C. transvaalensis* Burtt Davy) by a fluorescence-based method. *Eur. J. Agron.* **2015**, *63*, 89–96.
58. Agati, G.; Foschi, L.; Grossi, N.; Guglielminetti, L.; Cerovic, Z.G.; Volterrani, M. Fluorescence-based *versus* reflectance proximal sensing of nitrogen content in Paspalum vaginatum and Zoysia matrella turfgrasses. *Eur. J. Agron.* **2013**, *45*, 39–51.
59. Longchamps, L.; Khosla, R. Early detection of nitrogen variability in maize using fluorescence. *Agron. J.* **2014**, *106*, 511–518.
60. Cerovic, Z.; Ounis, A.; Cartelat, A.; Latouche, G.; Goulas, Y.; Meyer, S.; Moya, I. The use of chlorophyll fluorescence excitation spectra for the nondestructive *in situ* assessment of UV-absorbing compounds in leaves. *Plant Cell Environ.* **2002**, *25*, 1663–1676.
61. Bilger, W.; Veit, M.; Schreiber, L.; Schreiber, U. Measurement of leaf epidermal transmittance of UV radiation by chlorophyll fluorescence. *Physiol. Plant* **1997**, *101*, 754–763.
62. Bilger, W.; Johnsen, T.; Schreiber, U. UV-excited chlorophyll fluorescence as a tool for the assessment of UV-protection by the epidermis of plants. *J. Exp. Bot.* **2001**, *52*, 2007–2014.
63. Kusnierek, K.; Korsaeth, A. Simultaneous identification of spring wheat nitrogen and water status using visible and near infrared spectra and powered partial least squares regression. *Comput. Electron. Agric.* **2015**, *117*, 200–213.
64. Wang, C.; Qi, J.; Moran, S.; Marsett, R. Soil moisture estimation in a semiarid rangeland using ERS-2 and TM imagery. *Remote Sens. Environ.* **2004**, *90*, 178–189.
65. Peteinatos, G.G.; Keller, M.; Weis, M.; Gerhards, R. *Comparison of Isaria Sensor with a Typical Spectrometer in a Series of Diverse Conditions*; Edicions de la Universitat de Lleida: Lleida, Spain, 2013; pp. 32–33.
66. Latouche, G.; Debord, C.; Raynal, M.; Milhade, C.; Cerovic, Z.G. First detection of the presence of naturally occurring grapevine downy mildew in the field by a fluorescence-based method. *Photochem. Photobiol. Sci.* **2015**, *14*, 1807–1813.

3

Infrared Thermography Applied to Tree Health Assessment

Daniele Vidal and Rui Pitarma *

Unit for Inland Development, Polytechnic of Guarda, Avenida Doutor Francisco Sá Carneiro N○50, 6300-559 Guarda, Portugal

* Correspondence: rpitarma@ipg.pt

Abstract: The tree is a fundamental living being. It contributes to nature and climate behaviour, as well to urban greening. It is also a source of wealth and employment. Most tree health inspection techniques are invasive or even destructive. Infrared thermography (IRT) is not invasive, and it has shown advantages when applied for inspection to trees and wood to detect deterioration or voids that could compromise its structure, stability, and durability. This study reviews the literature about IRT applied to a tree health inspection. It is framed in the context of the importance of trees for the balance of ecosystems, and the different techniques to detect tree deterioration. It highlights the difference when applied to wood or trees and the main factors that have been proven to cause disturbances in the thermal pattern of trees. The IRT, as other non-destructive methods, does not distinguish what type of damage it is, nor its causative agent. However, it enables identifying healthy and deteriorated tissues. The technology is very promising since it reveals that is efficient, fast, economical, and sustainable.

Keywords: infrared thermography; IRT; tree inspection; inspection techniques; tree monitoring

---

1. Introduction

The tree is a fundamental living being of ecosystems [1–3]. It contributes to regulating nature, climate, and urban greening [1]. It is also a source of wealth and employment [4]. Some trees are classified as natural and cultural heritage [5]. Trees provide enormous benefits, such as the ecosystems balance, prevention of desertification, and global warming, as well as the well-being of human populations and urban ecosystems [1]. However, trees are also prone to the risk of damage to people and assets associated with the falling of branches/trees, above all in urban spaces [6,7]. Risk increases when they have defects because it compromises their health and stability. Consequently, the safety of people and goods are compromised. To support decision making about trees, it is essential to monitor their health condition to understand their biological viability, cost-effectiveness, and associated risks.

The first discovery of the infrared spectrum was made by Herschel in 1800 during his study about optical filters to reduce the brightness of the sun in telescopes [8]. In his findings, Herschel also observed that these new rays had similar behaviour to the visible ones, i.e., they were absorbed, transmitted, refracted, and reflected [9]. The first thermal image (thermograph) was obtained by Herschel in 1840, by experiments with differential evaporation of a film of oil submitted to a heat pattern [8]. However, the crucial mark in the sensitivity of infrared detection was achieved by Langley in 1880 through the invention of the bolometer. The first infrared camera was developed by Tihanyi in 1929 and was applied for anti-aircraft defense by the British army [10]. The first uses of Infrared thermography (IRT) in trees was the assessment by infrared imaging of the crown condition, using monochrome and false colour films [11]. IRT is a non-contact technology that allows detection of damaged tissue in real-time. It is based on the detection of thermal differences between healthy and damaged zones.

This review paper concerns identifying the IRT effectiveness for inspection of the health of trees. For that purpose, this review was structured as follows. It starts with the research methods and then the importance of trees and their classification. The mechanism that trees use to protect themselves, as well as risks associated with them, and the way they should be managed, is referred to. Some methods and inspection techniques are also mentioned. The importance of sustainable techniques of inspection is highlighted. Principles of IRT, usefulness, and relevance are mentioned. A description of the IRT application for inspection of trees is done, and several experiments are described. The main differences between wood and tree properties are reviewed. Finally, the advantages and limitations associated with the use of IRT are described.

2. Materials and Methods

A database query research paper was carried out in B-on, DOAJ, Google Scholar, Microsoft Academic, ProQuest, SciELO, ScienceDirect, Scopus, and Web of Science. The review was done in Portuguese, English and Spanish. It was carried out between October 2018 and February 2019. The keywords were: Infrared thermography, inspection, inspection techniques, non-destructive, non-invasive, sustainable, monitoring, classification, deterioration detection, relevance, benefits, risks, management, tree, tree heritage, and timber. The keywords were combined with each other. Articles from the reference list of reviewed papers were also taken into account. Whenever combinations of keywords were found, the abstracts and keywords were read. When the abstract proved to be insufficient for classification, the full article was read. The selected articles were classified according to sections, subsections, and theoretical references. The gathered data was organized by type of item, title, authors, abstract, keywords, and classification. These procedures facilitated the identification and analysis of the main references used by the authors, as well as citation frequency. All possible contributions to the application of IRT for health tree monitoring were considered. Duplicate references were withdrawn. A total of 239 papers were collected, of which only 81 papers were selected.

3. Tree Relevance and Risks

In 1950, the world's urban population was 30%; nowadays, the urban population reached 55%, and by 2050, 68% of the world population is to probjected to live in urban areas [12]. That is, the urbanization in the world population is increasing. Trees and green spaces play important roles in integrating the inhabitants into their urban communities. Trees provide well-being and quality of life; they establish the connection of inhabitants with the natural world and contribute to the mitigation of climate change. Urban trees provide numerous benefits for their inhabitants, but they are also prone to risk [6,13,14].

Trees interact with the surrounding environment, making it more pleasant. The basic function of trees is the improvement of air quality. Trees reduce the rainwater flow, allowing the infiltration into underground aquifer environments. Trees increase biodiversity by providing shelter for wildlife. Trees provide wind protection to people and animals since their leaves alter the direction and the speed of the wind. The tree roots hold the soil, and the tree canopy protects the soil from falling rain. Trees reduce urban noises and promote a sense of calm and rest. Trees create spaces for recreation and tourism. Consequently, trees favour the reduction of stress and even crime, since public open space with trees tends to be used much more than space without trees and this increases surveillance [15]. They influence on the decrease of energy consumption in the summer season, providing shade and a low incidence of solar radiation [16]. Trees have influence in the increasing the property value, if they are in or even outside the property. Trees produce aesthetic benefits because they decorate the urban landscape and enhance the season' landscape [14,17–20].

Trees interact in the ecosystem. They are exposed to aggression caused by birds, insects, animals, fire, atmospheric conditions, or even human beings during their activities, and these interactions could result in wounds, which act as entrance doors for pathogenic microorganisms. Trees have developed a system of compartmentalisation known as CODIT (Compartmentalization of Decay in Trees), described by Shigo and Marx [21]. In this system, chemical and physical barriers are created; cells undergo changes to form four types of walls around the wound; the walls are organized in axial, radial, and tangential directions. They act against the proliferation of pathogens by isolating the wounded tissues and thus impeding their dissemination to prevent/mitigate tree deterioration [22]. Wood deterioration and trees ageing are natural processes, but they carry on risks to the tree itself, as well as people and property [23].

On the other hand, some trees are of special interest because of age, size, physical characteristics, as well as cultural and historical connections [5]. Some examples of classification are:

- Heritage: trees that are linked to history and culture; trees that are rare and/or botanically relevant (see for instance Reference [24]).
- Notable: trees that have reached maturity stand out from their surroundings because they are larger than the trees around them (see for instance Reference [25]).
- Ancient: trees that have overpassed the maturity and despite the hollow trunk they continue to be healthy. Their value is intrinsically linked to their age (see for instance Reference [26]).
- Veterans: trees that survived wounds and deterioration (decay); young trees that developed old trees characteristics (see for instance Reference [27]).

Nevertheless, trees are also a risk factor for people and property. Large numbers of imposing trees characterize most urban gardens and avenues. These trees are subjected to undergoing stress hostile environment characteristic of urban centres, and very often under poor management. As a consequence, many are unhealthy, and a damaged tree poses a threat to people safety [11]. Three people die each year as a result of falling trees in public spaces in the United Kingdom. It is a low risk when considered that one in 10 million die as a result of it [6,7]. While there is a low rate of tree incidents, it ends up altering the risk perception. The risk perception of the falling tree has led tree managers (owners and managers of urban green spaces) to knock down trees at the slightest sign of danger. Tree felling is being aggravated, as who is responsible for tree management bear the costs and legal duties related to trees, meanwhile the public benefits of trees without knowing about their management issues [6]. In order to continue providing benefits, it is crucial to managing trees in such a way as to minimize risks and conflicts they may cause [20]. However, the safety of people and goods are not the only factor to take into account. Environmental and aesthetic factors are also considered when assessing costs and benefits [6].

As described above, trees represent enormous environmental and social benefits, but they are also a source of risk. The aim of tree health monitoring is, consequently, multifunctional. There are techniques to tree health inspection that support decision making related to trees management, in which the IRT presents several specific advantages [11].

4. Methods and Techniques of Inspection

VTA (Visual Tree Assessment) is an inspection method for tree diagnosis. It was developed by Mattheck and Breloer (1994), and according to VTA, interprets the body language of trees and provides the expert with failure criteria [28]. The method consists of three steps: a) External visual inspection in search of evidence of internal damage; b) the use of diagnostic tools for confirmation and measurement of damage; and c) assessment of damage extent and failure risk.

4.1. Instruments for Detecting Deterioration

Visual inspection alone does not provide enough evidence of internal damage extent. Then, additional diagnostic tools are required. According to several authors, those tools are classified as:

- Invasive and non-invasive: invasive instruments require drilling for deep penetration in the sapwood through one or more holes; the sapwood is living wood. Non-invasive instruments do not need contact, or they penetrate superficially in the sapwood. It is noted that any type of wound is an entry for pathogens into the tree [29].

- Destructive and non-destructive: Unlike the destructive instruments, non-destructive ones allow the identification of damage presence or not in the trees without causing harm [30]. Then, non-contact instruments are considered as non-destructive. Table 1 shows only non-contact instruments since they were considered non-destructive.
- Screening, diagnostic and evaluation: Screening tools allow a quick assessment to identify healthy and non-healthy trees. Diagnostic instruments allow a more accurate assessment, but require more time to identify the extent and type of damage in the tree. An intermediate method is an evaluation; it is a combination of screening speed and diagnostic accuracy methods [31].

Table 1. Classification of instruments for detecting deterioration in trees (adapted from Reference [31]).

	Unidentified Method in [31]	Screening	Evaluation	Diagnostic
Invasive	Increment borer, Boroscope	Resistograph	Shigometer, Fractometer	
Non-Invasive	-	Stress wave velocity	Electrical resistance	Stress wave tomography, Electromagnetic tomography
Non-contact	Nuclear magnetic resonance (NMR)	IRT, Electronic nose	-	Gamma-ray computed Tomography

Table 1 summarizes the main instruments for deterioration detection in trees, allowing to relate the purposes of the instruments to the invasive or destructive nature.

4.2. *Sustainable Techniques Relevance*

Besides IRT, the equipment performance evaluation and its advantages/disadvantages for detecting tree deterioration are briefly described in Table 2, and more details can be found in References [29,31]. It is noted that invasive techniques can increase tree damage [22], and radiation exposure is life detrimental for any living thing [11]. Inspection equipment must be safe for users and people around. It is therefore relevant to choose sustainable equipment. This type of equipment maintains a good relationship between portability, use speed, and they are not invasive or ionising. Besides, they have low running costs due to low energy consumption and low use of consumables [32].

Table 2. A short summary of the techniques for detecting trees deterioration.

Technique	Principle and Brief Description	Key Highlights
Increment borer [29]	Visual inspection: a sample of tree core is extracted for visual inspection.	Measures tree growth rate, age and soundness. Requires experience of decay potential causes. Invasive method (it may itself be a decay factor).
Boroscope [29]	Remote visual inspection: the tree trunk is drilled and it is used a small video camera to observe inside.	Enables visual confirmation from inside. Same disadvantages of the technique "Increment borer".
Resistograph [33]	Penetration resistance: a small drill/needle is inserted into the tree; drilling resistance is measured and registered.	Fast and easy to execute as well as interpret the graph. Does not detect early to intermediate decay stages; Requires comparison with known patterns (samples without decomposition).

Table 2. *Cont.*

Technique	Principle and Brief Description	Key Highlights
Shigometer [34,35]	Electrical resistivity: a pulsed direct electric current generated by electrodes is applied into a drilled hole; it goes into the wood or the tree bark; electrical resistance is measured and registered.	Detects deterioration in early stages. The information is limited by the probe length.
Fractometer [28,35]	Strength and stiffness: it measures radial bending fracture strength and stiffness value.	Small device and easy to carry. Portable compression meter has depth limited.
Stress wave velocity [35,36]	Single-path acoustic wave velocity: detects cavities by measuring the acoustic wave velocity as it passes through the tree stem.	Quickly performed; defines the location and extent of internal decay. Difficult to determine early stages of decay.
Electrical resistance [29,37,38]	Electrical resistivity: resistivity is determined from the voltage difference between electrodes when the electricity is injected.	Effective to detect advanced stages of tree decay.
Stress wave tomography [38]	Multiple path acoustic wave velocity: the tomogram is obtained by measuring and registering the sound waves travel time generated by acoustic transducers positioned around the circumference of the tree stem.	Detects internal decay; accurately locates the anomalies; sensitive to early stages of decay. High cost and difficult to operate.
Electromagnetic tomography [31,38,39]	Electromagnetic wave permittivity: the tomogram is obtained by variations in the return signal. The variation is reflected by the receiving antenna when the electromagnetic wave emitted by the transmitting antenna encounters a boundary with a different dielectric constant.	Higher frequencies provide better resolution but penetration depth decreases. High cost and difficult to operate.
Nuclear magnetic resonance (NMR) [29,40]	Magnetic properties: uses the magnetic properties of certain atomic nuclei (as the hydrogen nucleus). They are aligned and oscillated using a strong magnetic field in the scanner.	Non-ionising radiation; delivers very detailed images that facilitate the analysis of the structure and function of the tissues. High cost and difficult to operate.
Electronic nose [41]	Odour: it distinguishes healthy and decayed wood through changes in the volatile organic compounds released by wood decay fungi.	Provides high levels of accuracy and reliability. Difficult to determine early stages of decay.
Gamma-ray computed Tomography [31,42]	Gamma-ray transmissivity: the tomogram is obtained from radiation absorption after directing gamma rays in multiple directions on a tree trunk thin slice.	Reliable and non-invasive; evaluates fungal decay and its extension. Ionizing radiation; high cost and difficult to operate; difficult to carry.

5. Infrared Thermography Applied to Trees

5.1. Principles of IRT

Infrared thermography (IRT) is a non-contact, non-destructive, and non-invasive technique that allows the detection of radiated heat energy from objects and bodies in the infrared range of the electromagnetic spectrum (wavelength range between 0.8–14 μm [43]). The conversion of infrared energy into visible imaging is possible through instruments that produce false-colour images, such as an infrared camera [44]. The principle underlying IRT is a method or equipment, which detects IR energy emitted from a surface, converts it to temperature, and displays the image of temperature field [43,45–50]. The fundamental concept behind it is that all bodies (alive and non-alive) have temperatures above absolute zero degrees (0 K) and emit infrared radiation that is captured by equipment capable of transforming that energy into pictures. The internal structure of the body shows different thermal behaviour depending on the health conditions of its parts. The differences in thermal behaviour result in differences in the colour pattern of the body surface image.

Figure 1 illustrates the heat transfer processes involved in a thermographic measurement. When the camera views the object, it receives radiation from the object, from the ambient between the camera and the object's surface, and from the atmosphere itself [8].

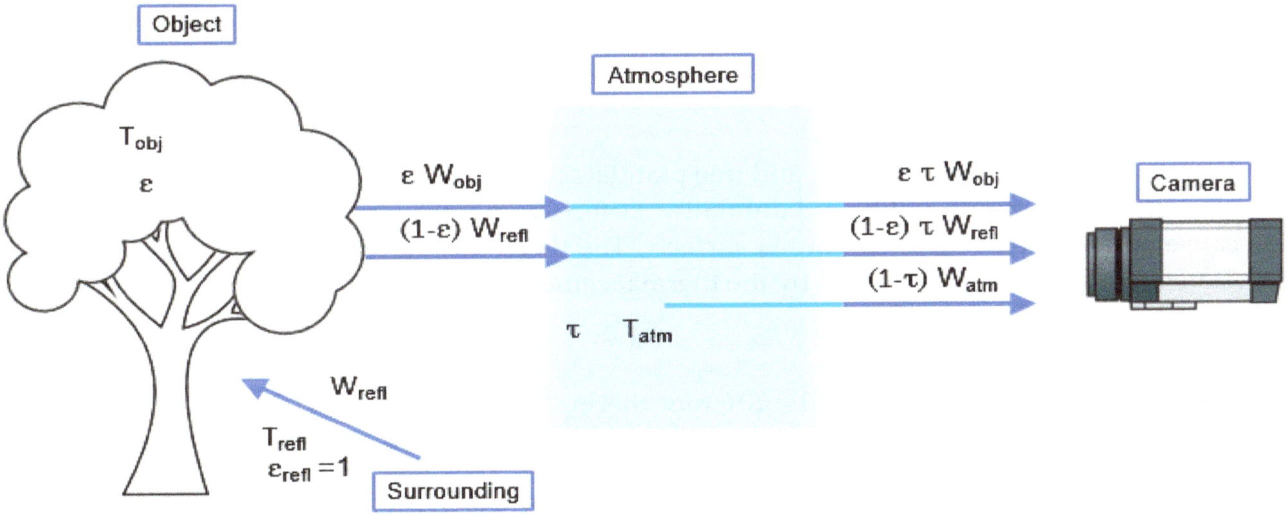

Figure 1. Schematic representation of a generic thermographic measurement (adapted from [8]).

The total received radiation power that the camera receives is equal to the sum of the emission of the object ($\varepsilon\ \tau\ W_{obj}$), the reflected emission of environmental sources ($(1-\varepsilon)\ \tau\ W_{refl}$), and the emission from the atmosphere ($(1-\tau)\ W_{atm}$). This is represented in the following equation (Equation (1)):

$$W_{tot} = \varepsilon\ \tau\ W_{obj} + (1-\varepsilon)\ \tau\ W_{refl} + (1-\tau)\ W_{atm} \tag{1}$$

where:

ε	Object emissivity
τ	Air transmittance
W_{obj}	Emission from the object
W_{refl}	Reflected emission from ambient sources
W_{atm}	Emission from the atmosphere

The appropriate infrared camera selection depends on the characteristics of the object to be inspected, the atmospheric conditions, and the observation distance. Depending on the wavelength in the infrared range of the electromagnetic spectrum, much of the radiation is absorbed by the atmosphere. Lower absorption allows more radiation to reach the camera sensor. It happens at bands:

Mid-wavelength infrared (MWIR) between 2 μm and 5 μm, and long-wavelength infrared (LWIR) between 8 μm and 14 μm. Limitations: Atmospheric humidity influences MWIR cameras, and LWIR cameras may capture optical and electronic noise [43].

The object emissivity depends on several factors such as wavelength. The emissivity is also highly sensitive to the nature and type of the material surface. As the emissivity increases, the influence of parasitic reflections diminishes. Metallic materials have low emissivity as they are highly reflective, whereas nonmetallic materials are high emissive and low reflection, leading to a better measurement [43,51]. Handling and interpretation of IRT information need cameras equipped with screens to visualise the energy emitted in the image format. The recording is an accessory that allows comparison between thermograms to follow the evolution of cases/pathologies [11]. It is also advantageous when the camera is equipped to capture thermograms and photographs simultaneously. It is preferably an integrated illumination accessory in the case of low illumination; this feature facilitates data interpretation [52].

There are two main thermographic procedures: Passive and active modes. In the active procedure, an external energy source is used to obtain the thermal contrast into the object of study. Different processes can trigger the heat flow, for example, thermal sources like lamps or heaters. The defects and damages on or near the object surface cause thermal discontinuities producing thermal contrast. It is detected during the thermographic inspection. In the passive procedure, the thermal contrast is generated by natural sources such as sunlight [44,50]. In most of the thermographic application scenarios, it is necessary to introduce an emissivity factor to calibrate the camera temperature measurement. The determination of the correct emissivity value of a specific wood remains a relevant issue [8].

IRT can be applied to both wood and tree plantlets to identify deterioration and voids that may compromise the structure, stability, and durability. Defects within the object disrupt the flow of energy and cause temperature differences on its surface. The thermal contrast results from differences in radiation emission that are registered by the thermal camera [43].

5.2. Wood and Trees

Wood and trees have significantly different thermal characteristics. That is, they differ in the variables that determine the temperature, density, humidity, thermal conductivity and thermal diffusivity, specific heat, convective coefficient, and emissivity [53–55].

After determining the density and the remaining properties of a species of wood, it is possible to predict its thermal behaviour. Therefore, when biodegradation changes the density and humidity of the wood, it also changes the thermal behaviour and temperature distribution along the surface [55–58]. Figure 2 shows the conventional photograph of a wood sample surface and its thermogram obtained in the active IRT procedure. By the active mode, this type of nodes and cracks show colour heterogeneity in relation to the general pattern denoting higher temperature. As shown in Figure 2, the cracks are barely visible in the photograph, but they are well seen in the thermogram [50].

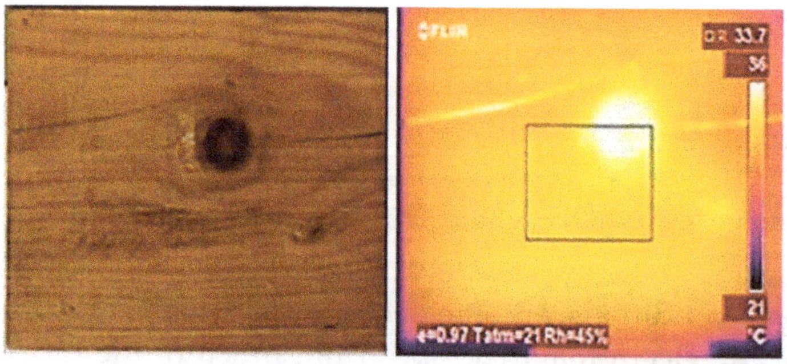

(a) (b)

Figure 2. Pine sample: (**a**) photograph and (**b**) IRT thermogram (active mode) [50].

The sap circulation causes the temperature gradient to vary along the trunk, and trees with more available water have better sap circulation [59]. This characteristic allows the differentiation between functional and dysfunctional tissue since the transport is done through the functional tissue. Therefore this feature allows that health and vitality can be verified, however, does not allow extrapolation of its thermal behaviour to the other trees, even of the same species, in order to identify damages. In fact, even for the same type of pathology, trees must be analyzed on a case-by-case basis [52,60]. Actually, two damaged trees of similar species, even when they have the same pathology, can generate different thermal patterns, because the availability of water to which the tree is subject is different and varies the temperature gradient along the trunk/tree. Thus, the temperature pattern, which allows the identification of functional or dysfunctional tissues, is unique for each tree.

Figure 3 shows the difference in surface temperature between a tree and a wooden stake [2]. The tree analysed was a specimen *Prunus domestica* L. (common name plum-tree) that was observed about 4 h after the sunset in the summer. The authors used an emissivity of 0.97 and a Rainbow colour pallet. During the observation, the atmospheric temperature was 22.5 °C, and the relative humidity was 55%. Most of the wooden stake presented a lower temperature than the tree, except its lower part. The lower part of the wood stake showed a slightly higher temperature compared to the lower trunk because the tree was irrigated before observation. The tree keeps a balanced relationship with the environment temperature, so tree temperatures are usually lower than the atmospheric temperature when the sun heating effect is over [61]. Therefore, the tree in Figure 3 showed higher temperatures in the healthier parts and lower temperatures in the deteriorated parts, as well as in parts where the tree was recently water wet [46].

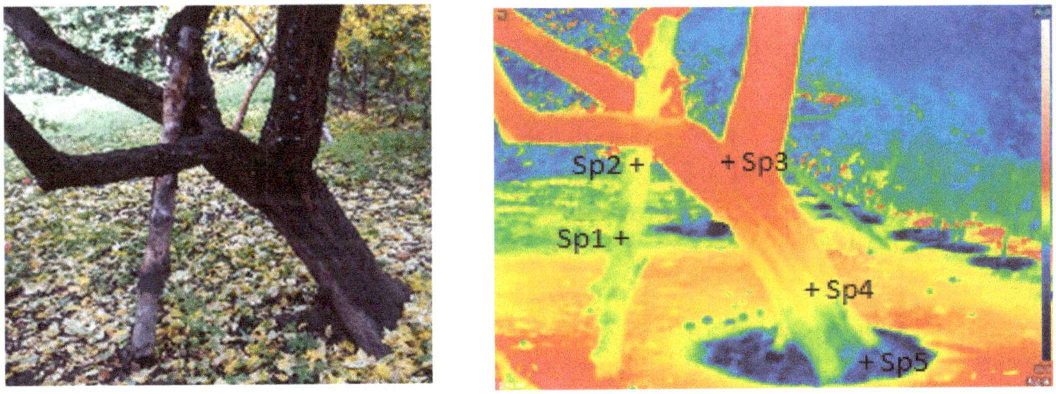

(a) (b)

Figure 3. A tree (*Prunus domestica*) and a wooden stake that supports it: (**a**) photograph; (**b**) IRT thermogram (passive mode). Temperature values (°C): Sp1 = 20.0; Sp2 = 20.5; Sp3 = 21.5; Sp4 = 20.5; and Sp5 = 19.5. Adapted from Reference [2].

IRT is commonly applied to wood in laboratory ambient. In fact, the lab is the first selected scenario because it is possible to control several environmental parameters (humidity, temperature, and luminosity), the thermal energy of the material, as well as the easy manipulation. Some studies in the lab have induced damage in the samples to verify the ability of IRT to detect wood degradation [55,56,62]. In the case of trees, most thermographic studies are performed on trees that are already programmed to be felled or when it is possible to observe the anomalies without causing damage [60,61].

The application of IRT to wood has been used for diverse purposes. Inspection of wood pieces quality in a production line is presented by Reference [63]. The results showed that TIV can be used to mark the debonded areas or to completely remove materials from the assembly line. IRT associated with ultrasonic equipment in the inspection of historic buildings is given by Reference [64]. Oratorio San Felipe Neri in Cadiz, Spain, was analyzed, and the authors conclude that the union of the ultrasound technique and the thermography represents a good tool for wooden structure inspection.

IRT associated with ultrasonic equipment, supported by laboratory analyses of timber samples in the inspection of the Aslanhane Mosque in Ankara, Turkey, which can be found in Reference [65]. The combined analysis of these techniques was done to evaluate the condition of the structural elements of wood in terms of their state of preservation, moisture problems, and recent incompatible repairs that affect them. This combination of techniques was reported as useful in assessing the reliability of timber, increasing the accuracy and effectiveness of the survey, facilitating and distinguishing the work of urgent intervention from long-term conservation programs. The detection of termite pests in wood or wood-clad structures is analyzed by Reference [66]. The author points out that TIV enables finding two basic elements that an inspector looks to identify the presence of pest infestation: Areas exhibiting anomalous elements that could be associated with the presence of moisture or hot spots and the presence of subsurface defects. Additionally, in large wooden structures such as a railway bridge where, in addition to other techniques, the IRT was used to identify a rapidly significant amount of structural damage that was not identified by other techniques [67].

5.3. *Analysis of Tree Health*

The thermal contrast captured by the thermal camera only represents the temperature on the surface of the tree bark. However, the tree bark near deteriorated tissue and voids show a lower temperature than the area around it. As in wood, tree deteriorated tissue and voids undergo a change in their thermal properties. Whenever there are significant differences in the thermal properties of the wood and tree bark, they can be detected by IRT [60,68]. IRT applied to trees and offers a means to differentiate damaged and deteriorated tissue from healthy tissues. However, due to the specific characteristics of each species, whenever a species is evaluated for the first time, it is advisable to observe the representation of the thermal pattern of the tree bark in the thermal image for better results interpretation [11].

The first use of IRT in trees was for aerial surveillance of the canopies to detect the distribution and spread of forest damage [60]. Later, IRT was applied to tree bark (trunks and branches), and it is now possible to evaluate some types of damage in the lower trunk and to deduce the cause that affects roots (root system) [60,69].

Several factors affect tree health, and consequently, can change the thermal properties of the tree trunk and branches. Some of the conditions that IRT can identify are described below, namely, diseases, pest attack, water stress, and formation of new functional tissue.

Regarding diseases, the application of IRT in the detection of tree bark lesions such as hemorrhagic cancers caused by *Phytophthora* spp., even before its consequences can be detected with the naked eye [60,70]. Fungi such as *Inonotus hispidus* (Bull.) P. Karst., *Phellinus punctatus* (Fr.) Pilát, *Coriolus pubescens* (Schumach.) and *Corticium* sp. were found in samples taken from the woody material of platan previously analyzed by IRT [71].

IRT can also be used for detection of insect pests on trees. Insects attack both leaves for feeding as well as trunks for their habitat, feeding, and egg-laying. While trees have a certain tolerance to pest attack on their leaves, it repeated attacks and factors such as poor irrigation, flooding, and inadequate trimming generate greater stress on the tree and consequently compromise its ability to regenerate or resist. On the other hand, larvae and insects create wounds, and these wounds make the trees susceptible to other pathogens. Some larvae can dig deep into the trunk creating tunnels that ultimately affect the transportation of nutrients and water causing rapid tree deterioration. Some of the insects recently detected by IRT were: White Pine Cone Beetle, Australian 'fire-beetle', Citrus Long-horned Beetle, Woodworm, and Red Palm Weevil (RPW) [72]. However, the IRT was not able to detect larval stage insects in some young tree species, as in the case of goat moth larvae. A hypothesis was the thermal adaptation of the larvae to the environment, that even with the elevation of temperature due to the effect of the activity of the larvae, it was not sufficient to detect them [73].

IRT is also applied for trees water stress assessment. The temperature in the canopy is an indicator of the amount of water available in the soil because the leaves transpiration is a mechanism to dissipate excess energy (heat). The tree uses surplus energy to transform water into water vapour through the leaves to cool it. However, when the soil is in water shortage, the tree perspires less, and as a consequence, there is an increase in leaf temperature when compared to trees in soil adequately wet. Water stress is a relevant indicator because the monitoring provides information for the adequate water supply of each tree species [74–77].

IRT detects the formations of new functional tissue. It was confirmed after dissection that the thermogram showed a band of higher wood formation in the area of the thermogram. This happens as a consequence of the adaptive growth of the change, because as Shigo (1984) has defined, the vascular cambium is the cell generator, and the new cells have different thermal properties from the other tissues [22]. This formation of new cells has different thermal properties than the other tissues. This tree had emptiness along the trunk that could also be observed in the thermogram. Catena and Catena [60] suggest that an adequate prognosis of the actual state of the tree requires the identification of the cause for tissue forming, that is, as result of damage and/or mechanical stress.

5.4. Advantages and Limitations of IRT Applicated to Trees

It is an advantage that IRT allows the tree to be observed as a whole, then the damage is early identified, even in trees that do not yet have external visual signs [52]. IRT requires little time to perform the inspection and is relatively easy to interpret the results. It does not require contact with the tree, and observations can be made at a distance of up to 25 m (depending on the camera and lens) safely. It is a safe technology for both the trees and the examiner as it does not emit harmful radiation. It is not necessary to use ladders or lifting platforms to observe the high parts of the tree; it reduces time and costs [60,68,78]. IRT also allows real-time assessment of root damage by observation of the lower trunk. It is a tool for stability and safety evaluation [69,79]. It eases to monitor the health status of trees over time, following up the evolution of pathologies previously identified as well as the identification of new patterns that indicate the development of new pathologies [69]. However, the major advantage of IRT compared to other inspection methods is the ability to differentiate functional tissue from dysfunctional tissue. In fact, IRT provides information for analysing the vitality and health status of a tree in a non-destructive, rapid, and cost-effective manner [60]. In essence, agriculture has been mechanized by the green revolution and now, like any other industry, agriculture is being digitalized [80]. According to this study, most farmers are ready to accept technology if it is profitable, less complex, and makes their life easier. IRT may be one of the relevant components of smart farming and agriculture, as can better manage and help to carry out real-time events.

As in other non-invasive methods, the main limitation is that IRT does not identify whether the damage detected is a void or a deteriorated tissue, nor its causative agent. Nor can it give precise indications of the magnitude of the damage. However, the fact that IRT identifies damaged areas early, it reduces the tests time and thus the damage progression. The acquired data leads to a more precise investigation of the pathologies, and identify the locations where invasive techniques are required [11,81]. Interferences are observed in thermograms when the surface of the tree is obscured by mosses or other vegetation, and when the tree is wet, or when IRT has been carried out directly to sunlight. Some of these limitations can be overpassed by avoiding inspections after rainy days or carry them out before the trees are watered. In the case of sunlight, the tests should preferably be performed at night. Another commonly noted limitation is the cost of the equipment when compared to simpler equipment. This is not the case when compared to more sophisticated equipment, or when multiple applications beyond the inspection of tree health are taken into account [60,68,78].

A brief search result summary of IRT applied to tree health analysis is presented in Table 3.

Table 3. A short summary of infrared thermography applied to tree health analysis.

Study	Main Focus	Findings
Al-doski et al. [72]	Pest detection.	Pest infestation detected.
Ballester et al. [74]	Water stress detection on citrus and persimmon trees.	Water stress detected.
Bellett-Travers & Morris [61]	Relationship between surface temperature and radial wood thickness.	No apparent relationship in most of the trees; strong relationship when there was a gradual change in radial wood thickness caused by a cavity.
Burcham et al. [59]	Effect of mechanically induced internal voids on *Dracaena fragrans* L. stem temperature.	Only able to identify reductions temperature in internal defects with at least 76% of the stem cross-sectional area.
Burcham et al. [70]	Evaluate the relationship between the internal defects and trunk surface temperature in *Casuarina equisetifolia* L.	Does not provide accurate results about the internal condition of trees.
Catena & Catena [60]	Review in order to assess the accuracy, reliability, and costs.	Does not automatically distinguish between different kinds of alteration; does not accurately provide the extent of the damages found; provides enough information to decide regarding the need for remedial action or a more detailed kind of assessment; non-invasive, fast, reasonable prices, in real time.
Catena, G. [68,71]	Internal cavities in trees detection.	Enables the detection of cavities.
Catena, A. [79] and Catena et al. [69]	Damages in the roots.	Enables that damages in the roots can be deduced in real time.
Crisóstomo et al. [32]	Considerations over IRT as applied to the state of the tree healthiness.	Enables healthiness evaluation.
Crisóstomo et al. [52]	Quercus pygmentosis tree Analysis.	Evaluating its healthiness status.
García-Tejero et al [76]	Water stress detection in almond trees.	Water stress detected.
Giuliani & Flore [77]	Water stress detection in apple trees.	Water stress detected.
Goh et al. [29]	Review of the current sensing methods used for decay detection in trees.	Comparing methods concerning the fundamental of measurements, hardware implementation, damage caused to the tree and the ease of use.
Hoffmann et al. [73]	Detecting the larval stage of goat moth´s larvae in young tree species.	Was not able to detect.
Jones et al. [75]	Water stress detection on grapevine.	Water stress detected.
Leong et al. [31]	Evaluating the current tree decay detection tools.	Classifying the tree decay detection tools in terms of measurement speed, resolution and accuracy.

6. Conclusions

The IRT was presented as a tool to inspect tree health, and it has proved to be an efficient tool in the early detection of damages even though it is not possible to identify the type of damage. As well as other non-invasive and non-destructive techniques, the causative agent can be identified when resorting to invasive methods. However, the comparison of inspection methods has shown that IRT has great advantages in terms of the capacity to differentiate functional tissue from dysfunctional tissue, and thus to inspect the vitality and health status of trees. It allows monitoring the evolution of pathologies in a fast, economical, and non-destructive way. Result improvement can be achieved with the evolution and increased application of IRT.

7. Recommendations

Despite its merits, IRT is still a relatively new technique in assessing tree health and remains residually implemented in agriculture, but it needs more detailed studies to establish a solid application basis, which can guide practitioners. This would allow developing the IRT potential for applicability on a larger scale. Some challenges such as the complexity of the technique, the new and atypical aspects of the problem, and several knowledge gaps related to technical issues and applicability specificities the affect of the technique performance. The latest high-definition thermal cameras record thermal images of high resolution and sensitivity will also contribute to overcoming these challenges, which will help to turn the IRT technique as a decision-making tool to assess the health status of trees.

Author Contributions: D.V. and R.P. developed the methodology, carried out the review, made the analysis and wrote the manuscript.

Acknowledgments: The authors would thank João Crisóstomo for technical support and some study materials.

References

1. Pitarma, R.; Crisóstomo, J.; Ferreira, M.E. Learning About Trees in Primary Education: Potentiality of IRT Technology in Science Teaching. In Proceedings of the EDULEARN18 Conference, Palma, Spain, 2–4 July 2018; pp. 208–213.
2. Ferreira, M.; Crisóstomo, J.; Pitarma, R. Infrared thermography technology to support science teaching-meaningful learning about trees with university students. In Proceedings of the 13th International Technology, Education and Development Conference (INTED2019), Valencia, Spain, 11–13 March 2019.
3. Ferreira, M.E.; André, A.C.; Pitarma, R. Potentialities of Thermography in Ecocentric Education of Children: An Experience on Training of Future Primary Teachers. *Sustainability* **2019**, *11*, 2668. [CrossRef]
4. Lier, M.; Parviainen, J. *Integration of Nature Protection in Forest Policy in Finland*; INTEGRATE Country Report; EFICENT-OEF: Freiburg, Germany, 2013.
5. Ancient Tree Forum & The Woodland Trust. *Ancient Tree Guide No. 4: What Are Ancient, Veteran and Other Trees of Special Interest*; The Woodland Trust: Grantham, UK, 2008.
6. National Tree Safety Group. *Common Sense Risk Management of Trees: Guidance on Trees and Public Safety in the UK for Owners, Managers and Advisers*; Forestry Commission: Edinburgh, Scotland, 2011; ISBN 978-0-85538-840-9.
7. Health and Safety Executive. *Management of the Risk from Falling Trees*; Health & Safety Executive/Local Authorities Enforcement Liaison Committee (HELA): Bootle, UK, 2007.
8. *FLIR Tools+ User´s Guide*; Flir Systems, Inc.: Wilsonville, OR, USA, 2016.
9. Barr, E.S. Historical Survey of the Early Development of the Infrared Spectral Region. *Am. J. Phys.* **1960**, *28*, 42–54. [CrossRef]
10. Kylili, A.; Fokaides, P.A.; Christou, P.; Kalogirou, S.A. Infrared thermography (IRT) applications for building diagnostics: A review. *Appl. Energy* **2014**, *134*, 531–549. [CrossRef]
11. Catena, A. Thermography Reveals Hidden Tree Decay. *Arboric. J.* **2003**, *27*, 27–42. [CrossRef]

12. United Nations. *World Urbanization Prospects: The 2018 Revision*; United Nations, Department of Economic and Social Affairs, Population Division: New York, NY, USA, 2018.
13. Pokorny, J.; O'Brien, J.; Hauer, R.; Johnson, G.; Albers, J.; Bedker, P.; Mielke, M. *Urban Tree Risk Management: A Community Guide to Program Design and Implementation*; USDA Forest Service Northeastern Area State and Private Forestr: St. Paul, MN, USA, 2003.
14. Rotherham, I.D. Editorial: Trees In A Changing World. *Arboric J.* **2010**, *33*, 1–2. [CrossRef]
15. Kuo, F.E.; Sullivan, W.C. Environment and Crime in the Inner City: Does Vegetation Reduce Crime? *Environ. Behav.* **2001**, *33*, 343–367. [CrossRef]
16. Coder, K.D. *Identified Benefits of Community Trees and Forests*; University of Georgia School of Forest Resources: Athens, GA, USA, 1996.
17. *International Society of Arboriculture Benefits of Trees*; International Society of Arboriculture: Champaign, IL, USA, 2011.
18. Roy, S.; Byrne, J.; Pickering, C. A systematic quantitative review of urban tree benefits, costs, and assessment methods across cities in different climatic zones. *Urban For. Urban Green.* **2012**, *11*, 351–363. [CrossRef]
19. Song, X.P.; Tan, P.Y.; Edwards, P.; Richards, D. The economic benefits and costs of trees in urban forest stewardship: A systematic review. *Urban For. Urban Green.* **2018**, *29*, 162–170. [CrossRef]
20. Johnston, M.; Hirons, A. Urban Trees. In *Horticulture: Plants for People and Places*; Dixon, G.R., Aldous, D.E., Eds.; Springer: Dordrecht, The Netherlands, 2014; Volume 2, ISBN 978-94-017-8580-8.
21. Shigo, A.L.; Marx, H.G. *Compartmentalization of Decay in Trees*; U. S. Government Printing Office: Washington, DC, USA, 1977; pp. 4–15.
22. Shigo, A.L. Compartmentalization: A Conceptual Framework for Understanding How Trees Grow and Defend Themselves. *Annu. Rev. Phytopathol.* **1984**, *22*, 189–214. [CrossRef]
23. Shortle, W.C.; Dudzik, K.R. *Wood Decay in Living and Dead Trees: A Pictorial Overview*; U.S. Department of Agriculture, Forest Service, Northern Research Station: Newtown Square, PA, USA, 2012.
24. Sherwood Forest. Available online: https://www.visitsherwood.co.uk/things-to-do/the-major-oak/ (accessed on 18 June 2019).
25. Undiscovered Scotland. Available online: https://www.undiscoveredscotland.co.uk/blairgowrie/meikleourhedge/index.html (accessed on 18 June 2019).
26. Woodland Trust. Available online: https://www.woodlandtrust.org.uk/visiting-woods/trees-woods-and-wildlife/woodland-habitats/ancient-trees/ (accessed on 18 June 2019).
27. National Trust. Available online: https://www.nationaltrust.org.uk/ashridge-estate/features/looking-after-our-veteran-trees-at-ashridge-estate (accessed on 18 June 2019).
28. Mattheck, C.; Breloer, H. Field guide for visual tree assessment (Vta). *Arboric. J.* **1994**, *18*, 1–23. [CrossRef]
29. Goh, C.L.; Abdul Rahim, R.; Fazalul Rahiman, M.H.; Mohamad Talib, M.T.; Tee, Z.C. Sensing wood decay in standing trees: A review. *Sens. Actuators A Phys.* **2018**, *269*, 276–282. [CrossRef]
30. Hellier, C. Introduction to Nondestructive Testisng. In *Handbook of Nondestructive Evaluation*; Hellier, C., Ed.; McGraw-Hill: New York, NY, USA, 2003; ISBN 978-0-07-139947-0.
31. Leong, E.-C.; Burcham, D.C.; Fong, Y.-K. A purposeful classification of tree decay detection tools. *Arboric. J.* **2012**, *34*, 91–115. [CrossRef]
32. Crisóstomo, J.; Pereira, C.; Roque, E.; Jorge, L.; Pitarma, R. Considerações na Observação do Estado de Salubridade de Árvores Através da Termografia por Infravermelhos. In Proceedings of the 1st Iberic Conference on Theoretical and Experimental Mechanics and Materials/11th National Congress on Experimental Mechanics, Porto, Portugal, 4–7 November 2018; pp. 745–748.
33. Mattheck, C.; Bethge, K.; Albrecht, W. How To Read The Results Of Resistograth M. *Arboric. J.* **1997**, *21*, 331–346. [CrossRef]
34. Shigo, A.L.; Shortle, W.C. *Spruce Budworms Handbook: Shigometry–A Reference Guide*; Agric. Handb.; U.S. Department of Agriculture, Forest Service, Cooperative State Research Service: Washington, DC, USA, 1985.
35. Monk, B. Evaluation of Decay Detection Equipment in Standing Trees. Available online: https://www.fs.fed.us/t-d/programs/im/tree_decay/tree_decay_detect_equip.shtml (accessed on 17 June 2019).
36. Ross, R.J.; Pellerin, R.F. Inspection of Timber Structures Using Stress Wave Timing Nondestructive Evaluation Tools. In *Wood and Timber Condition Assessment Manual: Second Edition*; White, R.H., Ross, R.J., Eds.; U.S. Department of Agriculture, Forest Service, Forest Products Laboratory: Madison, WI, USA, 2014.

37. Oliva, J.; Romeralo, C.; Stenlid, J. Accuracy of the Rotfinder instrument in detecting decay on Norway spruce (Picea abies) trees. *For. Ecol. Manag.* **2011**, *262*, 1378–1386. [CrossRef]
38. Nicolotti, G.; Socco, L.V.; Martinis, R.; Godio, A.; Sambuelli, L. Application And Comparison Of Three Tomographic Techniques For Detection Of Decay In Trees. *J. Arboric.* **2003**, *29*, 66–78.
39. Bogosanovic, M.; Al Anbuky, A.; Emms, G.W. Overview and comparison of microwave noncontact wood measurement techniques. *J. Wood Sci.* **2010**, *56*, 357–365. [CrossRef]
40. Wang, P.C.; Chang, S.J. Nuclear Magnetic Resonance Imaging of Wood. *Wood Fiber Sci.* **1986**, *18*, 308–314.
41. Baietto, M.; Wilson, A.; Bassi, D.; Ferrini, F. Evaluation of Three Electronic Noses for Detecting Incipient Wood Decay. *Sensors* **2010**, *10*, 1062–1092. [CrossRef]
42. Habermehl, A.; Ridder, H.-W. Computerised Tomographic Investigationa Of Street And Park Trees. *Arboric. J.* **1995**, *19*, 419–437. [CrossRef]
43. Usamentiaga, R.; Venegas, P.; Guerediaga, J.; Vega, L.; Molleda, J.; Bulnes, F. Infrared Thermography for Temperature Measurement and Non-Destructive Testing. *Sensors* **2014**, *14*, 12305–12348. [CrossRef]
44. Maldague, X.P.V.; Streckert, H.H.; Trimm, M.W. Introduction to Infrared and Thermal Testing. In *Infrared and Thermal Testing*; Maldague, X.P.V., Moore, P.O., Eds.; Nondestructive Testing Handbook; American Society for Nondestructive Testing: Columbus, OH, USA, 2001; ISBN 978-1-57117-044-6.
45. Snell, J.R., Jr. Thermal Infrared Testing. In *Handbook of Nondestructive Evaluation*; Hellier, C., Ed.; McGraw-Hill: New York, NY, USA, 2003; ISBN 978-0-07-139947-0.
46. Meola, C. (Ed.) Carosena Origin and Theory of Infrared Thermography. In *Infrared Thermography Recent Advances and Future Trends*; Bentham Science Publishers: New York, NY, USA, 2012; pp. 3–28. ISBN 978-1-60805-143-4.
47. Ibarra-Castanedo, C.; Maldague, X.P.V. Infrared Thermography. In *Handbook of Technical Diagnostics*; Czichos, H., Ed.; Springer: Berlin/Heidelberg, Germany, 2013; pp. 175–220. ISBN 978-3-642-25849-7.
48. Holst, G.C. *Common Sense Approach to Thermal Imaging*; JCD Pub.; Co-Published by SPIE Optical Engineering Press: Winter Park, FL, USA; Bellingham, WA, USA, 2000; ISBN 978-0-9640000-7-0.
49. Pitarma, R.; Crisóstomo, J.; Jorge, L. Analysis of Materials Emissivity Based on Image Software. In *New Advances in Information Systems and Technologies*; Rocha, Á., Correia, A.M., Adeli, H., Reis, L.P., Mendonça Teixeira, M., Eds.; Springer International Publishing: Cham, Switzerland, 2016; Volume 444, pp. 749–757. ISBN 978-3-319-31231-6.
50. Crisóstomo, J.; Pitarma, R. The Importance of Emissivity on Monitoring and Conservation of Wooden Structures Using Infrared Thermography. In *Advances in Structural Health Monitoring*; Hassan, M., Ed.; IntechOpen: London, UK, 2019. [CrossRef]
51. Maldague, X.P.V. *Nondestructive Evaluation of Materials by Infrared Thermography*; Springer: London, UK, 1993; ISBN 978-1-4471-1997-5.
52. Crisóstomo, J.; Pereira, C.; Roque, E.; Jorge, L.; Pitarma, R. *Análise da Salubridade de Árvores Através da Termografia por Infravermelhos*; Gomes, J.F.S., Ed.; INEGI/FEUP: Porto, Portugal, 2018; pp. 749–758.
53. Meola, C.; Carlomagno, G.M. Recent advances in the use of infrared thermography. *Meas. Sci. Technol.* **2004**, *15*, R27–R58. [CrossRef]
54. Wyckhuyse, A.; Maldague, X. A Study of Wood Inspection by Infrared Thermography, Part I: Wood Pole Inspection by Infrared Thermography. *Res. Nondestruct. Eval.* **2001**, *13*, 1 12. [CrossRef]
55. Conde, M.J.M.; Liñán, C.R.; de Hita, P.R.; Gálvez, F.P. Infrared Thermography Applied to Wood. *Res. Nondestruct. Eval.* **2012**, *23*, 32–45. [CrossRef]
56. Rodríguez-Liñán, C.; Morales-Conde, M.J.; Rubio-de Hita, P.; Pérez-Gálvez, F. Análisis sobre la influencia de la densidad en la termografía de infrarrojos y el alcance de esta técnica en la detección de defectos internos en la madera. *Mater. de Construcción* **2012**, *62*, 99–113. [CrossRef]
57. Tanaka, T.; Divós, F. Wood Inspection by Thermography. In Proceedings of the 12th International Symposium on Nondestructive Testing, Sopron, Hungary, 13–15 September 2000.
58. Pereira, J.C.A. *Contribuição para a Análise de Manifestações Patológicas em Madeira na Construção com Recurso à Termografia*; Instituto Politécnico de Castelo Branco: Castelo Branco, Portugal, 2014.
59. Burcham, D.C.; Leong, E.-C.; Fong, Y.-K. Passive infrared camera measurements demonstrate modest effect of mechanically induced internal voids on Dracaena fragrans stem temperature. *Urban For. Urban Green.* **2012**, *11*, 169–178. [CrossRef]

60. Catena, A.; Catena, G. Overview of Thermal Imaging For Tree Assessment. *Arboric. J.* **2008**, *30*, 259–270. [CrossRef]
61. Bellett-Travers, M.; Morris, S. The Relationship Between Surface Temperature And Radial Wood Thickness Of Twelve Trees Harvested In Nottinghamshire. *Arboric. J.* **2010**, *33*, 15–26. [CrossRef]
62. López, G.; Basterra, L.-A.; Ramón-Cueto, G.; Diego, A. de Detection of Singularities and Subsurface Defects in Wood by Infrared Thermography. *Int. J. Archit. Herit.* **2014**, *8*, 517–536. [CrossRef]
63. Meinlschmidt, P. Thermographic Detection of Defects in Wood and Wood-based Materials. In Proceedings of the 14th International Symposium of Nondestructive Testing of Wood, Hannover, Germany, 2–4 May 2005.
64. Rodríguez Liñán, C.; Morales Conde, M.J.; Rubio de Hita, P.; Pérez Gálvez, F. Inspección mediante técnicas no destructivas de un edificio histórico: Oratorio San Felipe Neri (Cádiz). *Inf. de la Construcción* **2011**, *63*, 13–22. [CrossRef]
65. Kandemir-Yucel, A.; Tavukcuoglu, A.; Caner-Saltik, E.N. In situ assessment of structural timber elements of a historic building by infrared thermography and ultrasonic velocity. *Infrared Phys. Technol.* **2007**, *49*, 243–248. [CrossRef]
66. Grossman, J.L. Advanced techniques in IR thermography as a tool for the pest management professional. In Proceedings of the SPIE, Orlando, FL, USA, 18 April 2006; Volume 6205.
67. Grossman, J.L. Trestles anyone? A Thermographic Nightmare. In Proceedings of the SPIE, Orlando, FL, USA, 9 April 2007; Volume 6541.
68. Catena, G. A new application of thermography. *Atti Della Fond. Giorgio Ronchi* **1990**, *6*, 947–952.
69. Catena, A.; Catena, G.; Lugaresi, D.; Gasperoni, R. La Termografia rivela la presenza di danni anche nell'apparato radicale degli alberi. *Agric. Ric.* **2002**, 81–100.
70. Burcham, D.C.; Leong, E.-C.; Fong, Y.-K.; Tan, P.Y. An Evaluation of Internal Defects and Their Effect on Trunk Surface Temperature in *Casuarina equisetifolia* L. (Casuarinaceae). *Arboric. Urban For.* **2012**, *38*, 277–286.
71. Catena, G. Une Appication De La Thermographie En Phytopathologie. *Phytoma-La Défense Des Végétaux* **1992**, *439*, 46–48.
72. Al-doski, J.; Mansor, S.B.; Shafri, H.Z.B.M. Thermal Imaging For Pests Detecting-A Review. *Int. J. Agric. For. Plant.* **2016**, *2*, 10–30.
73. Hoffmann, N.; Schröder, T.; Schlüter, F.; Meinlschmidt, P. Potenzial von Infrarotthermographie zur Detektion von Insektenstadien und -schäden in Jungbäumen. *J. Für Kult.* **2013**, *65*, 2013.
74. Ballester, C.; Jiménez-Bello, M.A.; Castel, J.R.; Intrigliolo, D.S. Usefulness of thermography for plant water stress detection in citrus and persimmon trees. *Agric. For. Meteorol.* **2013**, *168*, 120–129. [CrossRef]
75. Jones, H.G.; Serraj, R.; Loveys, B.R.; Xiong, L.; Wheaton, A.; Price, A.H. Thermal infrared imaging of crop canopies for the remote diagnosis and quantification of plant responses to water stress in the field. *Funct. Plant Biol.* **2009**, *36*, 978. [CrossRef]
76. García-Tejero, I.; Durán-Zuazo, V.H.; Arriaga, J.; Hernández, A.; Vélez, L.M.; Muriel-Fernández, J.L. Approach to assess infrared thermal imaging of almond trees under water-stress conditions. *Fruits* **2012**, *67*, 463–474. [CrossRef]
77. Giuliani, R.; Flore, J.A. Potential Use Of Infra-Red Thermometry For The Detection Of Water Stress In Apple Trees. *Acta Hortic.* **2000**, *537*, 383–392. [CrossRef]
78. Ibarra-Castanedo, C.; Tarpani, J.R.; Maldague, X.P.V. Nondestructive testing with thermography. *Eur. J. Phys.* **2013**, *34*, S91–S109. [CrossRef]
79. Catena, A. Thermography Shows Damaged Tissue and Cavities Present in Trees. In *Nondestructive Characterization of Materials XI.*; Green, R.E., Djordjevic, B.B., Hentschel, M.P., Eds.; Springer: Berlin/Heidelberg, Germany, 2002; pp. 515–522. ISBN 978-3-540-40154-4.
80. Sharma, S.; Kaushik, A. Views of Irish Farmers on Smart Farming Technologies: An Observational Study. *AgriEngineering* **2019**, *1*, 164–187.
81. Bellett-Travers, M. A Risk Assessment Methodology For Trees In Parkland Based On Comparative Population Analysis. *Arboric. J.* **2010**, *33*, 3–14. [CrossRef]

4

Multi-Pig Part Detection and Association with a Fully-Convolutional Network

Eric T. Psota [1,*], Mateusz Mittek [1], Lance C. Pérez [1], Ty Schmidt [2] and Benny Mote [2]

1 Department of Electrical and Computer Engineering, University of Nebraska–Lincoln, Lincoln, NE 68505, USA; mmittek@gmail.com (M.M.); lperez@unl.edu (L.C.P.)
2 Department of Animal Science, University of Nebraska–Lincoln, Lincoln, NE 68588, USA; tschmidt4@unl.edu (T.S.); benny.mote@unl.edu (B.M.)
* Correspondence: epsota@unl.edu

Abstract: Computer vision systems have the potential to provide automated, non-invasive monitoring of livestock animals, however, the lack of public datasets with well-defined targets and evaluation metrics presents a significant challenge for researchers. Consequently, existing solutions often focus on achieving task-specific objectives using relatively small, private datasets. This work introduces a new dataset and method for instance-level detection of multiple pigs in group-housed environments. The method uses a single fully-convolutional neural network to detect the location and orientation of each animal, where both body part locations and pairwise associations are represented in the image space. Accompanying this method is a new dataset containing 2000 annotated images with 24,842 individually annotated pigs from 17 different locations. The proposed method achieves over 99% precision and over 96% recall when detecting pigs in environments previously seen by the network during training. To evaluate the robustness of the trained network, it is also tested on environments and lighting conditions unseen in the training set, where it achieves 91% precision and 67% recall. The dataset is publicly available for download.

Keywords: computer vision; deep learning; image processing; pose estimation; animal detection; precision livestock

---

1. Introduction

Researchers have established that changes in animal behavior correlate with changes in health [1–3], however, the labor-intensive methods used to monitor behaviors do not lend themselves well to modern commercial swine facilities where just two seconds of daily observation per pig is recommended [4]. Because industry caretakers are typically responsible for thousands of pigs, it is nearly impossible for them to thoroughly assess the health and well-being of individuals using manual observation. This limitation is further compounded by the fact that the effectiveness of human visual assessments are inherently limited by both the attention span and subjectivity of observers [5].

By adapting modern technology to commercial operations, precision livestock farming aims to ease the burden on individual caregivers and provide a solution for continuous automated monitoring of individual animals [6–8]. Over the past two decades, researchers have approached this problem from a multitude of different angles [9,10]. These include 3D tracking via wearable ultra-wide band (UWB) devices [11,12], GPS motes [13,14], inertial measurement unit (IMU) activity trackers [15–18], RFID ear tags [19–21], and detection-free tracking using depth cameras [22,23].

This work aims to provide a generalizable vision-based solution for animal detection in group-housing environments. While animal tracking methods using radio identification devices directly provide data on individual animals, there are several disadvantages to using wearable methods when compared to video-based approaches [24,25]. Wearables must withstand harsh environments, they require costly installation and maintenance on a per animal basis, and the localization accuracy of both UWB and GPS systems is typically too low to detect animal orientation, activities, and social behaviors. In contrast, video already provides information-rich data that allows humans to identify precisely what each animal is doing at all times. However, converting digital video to meaningful data without human intervention is an ongoing pursuit in the research community. Existing solutions require sophisticated computer vision algorithms and they have begun to rely increasingly on advancements in machine learning and artificial intelligence [26,27].

Computer vision researchers experienced a dramatic shift in 2012 when Krizhevsky et al. demonstrated that a deep convolutional neural network, trained from scratch without any hand-crafted heuristics, could easily outperform all existing methods for image classification [28]. This shift was made possible by efficient implementations of convolutional neural network (CNN) training [29,30], improvements in computational resources via graphics processing units (GPUs) [31], and large (100K+ images) human-annotated datasets like PASCAL [32], MS-COCO [33], and CityScapes [34]. Modern computer vision applications that locate and classify objects in images often leverage established methods like YOLO [35,36], SSD [37], and Faster R-CNN [38,39] in their pipeline. While common tasks like pedestrian detection lend themselves well to pre-trained networks and existing datasets, there still exists unique challenges that require custom datasets and accompanying solutions.

This work introduces a fully convolutional neural network used to identify the location and orientation of multiple group-housed pigs. The target output of the network is an image-space representation of each pig's body part locations along with a method for associating them to form complete instances. To train the network, a new dataset is used that contains 2000 images with 24,842 uniquely labeled pig instances. The dataset is divided into a training set and a testing set, and the testing set is subdivided into two sets: one with images depicting the same environments as the training set, and another with images of new environments not represented in the training set. This dataset design allows the robustness of detection algorithms to be tested against novel animal presentations and environments. The three key contributions of this work are (1) a fully convolutional instance detection method, (2) a public dataset for training and evaluation, and (3) metrics that can be used to measure the performance of methods that detect both location and orientation.

The remainder of this paper is organized as follows. Section 2 discusses related work in vision-based animal detection and modern deep learning detection methods. Section 3 introduces the proposed method and the design of the fully-convolutional neural network. The dataset, evaluation methodology, and results are presented in Sections 4 and 4.6 discusses some key insights and additional practical considerations. Finally, concluding remarks are provided in Section 5.

2. Background

Visual detection of multiple moving targets with a static camera often begins with segmentation of foreground objects using background subtraction. If sufficient separation between targets exists, traditional computer vision methods like connected components can be used to easily identify unique instances [40]. However, this is hardly the case for group-housed animals that constantly engage each other socially and often prefer to lie in groups to preserve warmth.

Using Otsu's method [41] to adaptively threshold background from foreground and Kashiha's ellipse-fitting method [42], Nasirahmadi et al. [43] used computer vision to identify the lying behavior of group-housed pigs as a function of temperature. While their method achieved approximately 95% accuracy when locating pigs in the images, it was only tested on a single environment where the animals had consistent age and appearance, so it is difficult to predict how well the method generalizes. Ahrendt et al. [44] used a continuously updated model and a 5D Mahalanobis distance between foreground and background pixels to isolate and track the targets, however, the authors point out that the method struggles to identify separate instances when animals are moving quickly and/or overlapping one another. Nilsson et al. [45] used a region-based method to segment group-housed pigs using supervised learning, demonstrating accurate results when the pigs are fully visible (i.e., non-occluded) from a top-down perspective.

A popular approach to dealing with the difficulties of background subtraction using color images is to incorporate depth-imaging cameras [22,46–54]. However, while foreground extraction is made simple, depth image processing is still faced with the challenge of separating abutting pigs. To address this challenge, Ju et al. [27] introduced a top-down method that first detects bounding boxes around pigs using the YOLO object detector [36]. When YOLO inevitably detects multiple touching pigs with a single bounding box, their method proceeds by separating segmented images (obtained using a Gaussian mixture model) to establish boundaries between instances. While their overall method achieves 92% accuracy, it is limited to three touching pigs and results were obtained within a single environment over a ten minute window, so generalizability is difficult to assess.

Some approaches using the Kinect depth-sensing camera explicitly define the region of interest (pen) as a three-dimensional bounding box to separate background from foreground. Mittek et al. [23] use this approach to track pigs as ellipsoidal collections of points in the foreground. While their method achieves an average of 20 min of continuous tracking without errors, it requires manual initialization and does not introduce a detection method. The method introduced by Matthew et al. [55] uses regional grouping of surface normals to detect individual instances. They do not perform a formal assessment of the detection accuracy, but the resulting tracking algorithm maintains accurate tracking for an average of 22 s without requiring manual initialization.

Beginning in 2014 with the introduction of R-CNN by Girshick et al. [56], visual detection methods have predominantly used deep convolutional neural networks. Furthermore, they generally fall into one of two categories: (1) top-down approaches that define regions of interest before performing subsequent segmentation or keypoint annotation, or (2) bottom-up approaches that directly segment pixels or detect keypoints without explicitly detecting regions of interest. Mask R-CNN [57], for example, is a state-of-the-art top-down approach for performing instance-level object segmentation and keypoint detection. However, because it relies on a priori region proposal, it is inherently unable to separate objects with significant bounding box overlap—a common occurrence among group-housed animals. In contrast, bottom-up detection and classification is directly done per-pixel in the image space [58]. Two notable methods, OpenPose [59] and PersonLab [60], directly identify keypoints and affinity metrics that allow them to be clustered into whole instances.

Bottom-up keypoint detection was adapted to cow tracking by Ardö et al. [26]. They identified a set of landmarks visible from a top-down view to represent each cow's location and orientation, and trained a fully-convolutional neural network to detect them in the image space. A post-processing network was then used to convert the annotations to per-pixel orientation classification outputs, resulting in 95% accuracy in correctly labeling all cows in a given image. In a follow-up experiment, they applied the previously trained network to a new environment and observed that it only succeeded on 55% of images, indicating that the network was overfitting to a particular environment.

The proposed method introduces a new bottom-up strategy that identifies multiple pig instances in images as a collection of keypoints. Unlike the approach by Ardö et al. [26], this method preserves the association between keypoints and instances, making it possible to evaluate the performance of the method directly as a keypoint detection method. Furthermore, keypoints provide a precise representation of the pose of each animal, making it possible to identify activities and interactions between animals.

3. Proposed Method

The goal of the proposed method is to detect the location and orientation of all visible pigs in the pen environment. The first stage of the process aims to find the location of all pertinent body parts, while the second stage aims to associate them with one another to form whole instances. The following two sections (Sections 3.1 and 3.2) describe the method used to represent parts and associations within the image space. Section 3.3 then presents a method that interprets these image space representations and produces a set of unique animal instances. Finally, Section 3.4 introduces the fully-convolutional network that takes, as input, an image of the pen environment and attempts to produce the image space outputs described in Sections 3.1 and 3.2.

3.1. Representation of Body Part Location

The proposed method assumed that images of the pen environment were captured from a downward-facing camera mounted above the pen. When trying to detect and differentiate multiple animals in a group-housed setting, a top-down view has three distinct advantages over alternative visual perspectives. Firstly, animals are generally nonoccluded from the top-down perspective unless they are crawling over (or lying on top of) one another. Secondly, the size and appearance of animals is consistent from a top-down perspective, making it easier for a system to identify the animals. Thirdly, one can reliably approximate the 3D position of each animal from their projection onto the 2D image plane by assuming a constant height above the pen floor plane. Approximation is often necessary if 3D coordinates are desired and the single-camera system being used lacks the ability to measure depth.

From a top-down perspective, the part of the animal most likely to be visible was the surface of the back. Thus, in order to represent both the position and orientation of each pig, the proposed method used the image-space location of the tail and shoulder belonging to each animal. Assuming there are N animals in the pen, the tail and shoulder position of animal $n \in \{1, \ldots, N\}$ is denoted $\mathbf{t}_n = (x_{\mathbf{t}_n}, y_{\mathbf{t}_n})$ and $\mathbf{s}_n = (x_{\mathbf{s}_n}, y_{\mathbf{s}_n})$, respectively. More specifically, "tail" refers to a surface point along the center ridge of the back that is between the left and right ham. The term "shoulder" refers to a surface point along the center ridge of the back between the shoulder blades. The chosen representation also includes the 2D position of the left and right ears, denoted $\mathbf{l}_n = (x_{\mathbf{l}_n}, y_{\mathbf{l}_n})$ and $\mathbf{r}_n = (x_{\mathbf{r}_n}, y_{\mathbf{r}_n})$, respectively. While their visibility is not guaranteed, such as when the animal lies on its side or positions its head in a feeder, their locations can be used to approximate the pose of the head and/or assign animals with a unique visual marker in the form of an ear tag.

Figure 1a illustrates an example of an input image depicting a single pig. The location of the left ear, right ear, shoulder, and tail are represented by the red, green, blue, and yellow spots in the target mapping shown in Figure 1b, respectively. Finally, the superimposed visualization given in Figure 1c illustrates the locations of the four parts in reference to the original image. Note that the target mapping in Figure 1b has the same spatial dimensions (rows and columns) as the input image, but it contains four channels with each corresponding to a single body part type.

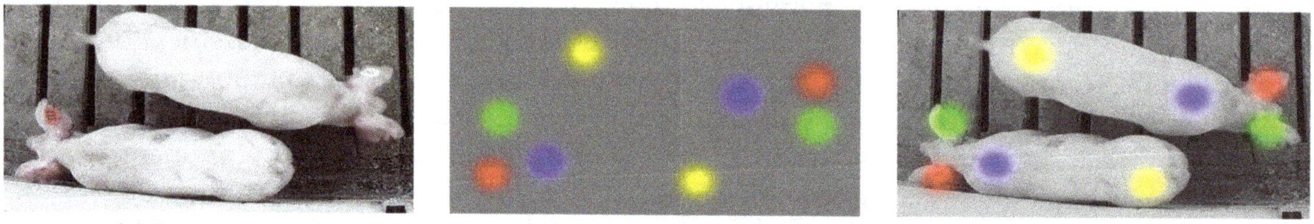

(a) Input image (b) Target mapping (c) Visualization

Figure 1. The original image (**a**) is mapped to a four-channel output (**b**), where the location of the left ear, right ear, shoulder, and tail are represented by Gaussian kernels in channels 1–4 of the output, respectively. Note that the four colors used in (**b**) are purely for visualization of the four separate channels of the target image. The overlay in (**c**) is provided to illustrate the locations relative to the original image.

To approximate the level of uncertainty inherent in the user annotations of each body part location, parts within the target mapping were each represented by 2D Gaussian kernels. While the distribution of the 2D Gaussian kernels is defined by a 2×2 covariance matrix, here we have only considered symmetric 2D Gaussian kernels that can be characterized by a single standard deviation multiplied by a 2×2 identity matrix. This standard deviation will be denoted σ_n for each animal $n \in \{1, \ldots, N\}$.

This mapping of uncertainty was meant to approximate the probability distribution of part locations annotated by a human given the original image $\mathbf{I}$. The kernels were also scaled so that the magnitude at the center is 1.0. This allowed for a straightforward translation between kernels and 2D image coordinates via a simple thresholding operation. The first four channels of the target output, as defined in Table 1, were thus proportional to the probability that parts $\{\mathbf{l}, \mathbf{r}, \mathbf{s}, \mathbf{t}\}$ exist at each spatial location in the image.

Table 1. Channels 1–4 of the image-space representation used to represent pig locations and orientations. Each channel corresponds to a different body part. The locations of each part are marked with Gaussian kernels meant to represent the distribution of part locations provided by a human annotator.

Channel	1	2	3	4
Encoding	$\propto P(\mathbf{l}\|\mathbf{I})$	$\propto P(\mathbf{r}\|\mathbf{I})$	$\propto P(\mathbf{s}\|\mathbf{I})$	$\propto P(\mathbf{t}\|\mathbf{I})$

3.2. *Representation of Body Part Association*

Even if every body part location is detected correctly, parts must be associated with each other in order to identify individual whole-animal instances. A naive approach would be to associate each body part with its nearest neighbor in terms of Euclidian distance using an optimal bipartite assignment method, such as the Hungarian algorithm. However, due to the elongated shape of pigs, this approach was prone to failure in cluttered environments, as illustrated in Figure 2.

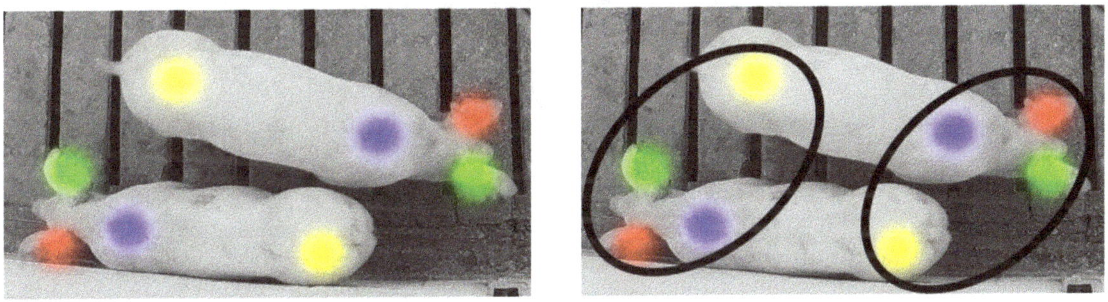

(**a**) Image with part detections (**b**) Optimal grouping using Euclidean distance

Figure 2. Two nearby pigs with body parts properly annotated are depicted in (**a**). Whole instances must be formed by joining the parts together through part association. An optimal Euclidean nearest-neighbor part association is prone to failure when the animals are in close proximity, as illustrated in (**b**).

The proposed method used additional channels in the target output to encode body part locations with 2D vector offsets to other body parts belonging to the same animal. These offsets represented the direction and distance in pixels from one body part to another. While there were a total of $\binom{4}{2} = 6$ part pairs that exist between the four parts, the target output only represented three in order to reduce unnecessary redundancy (e.g., vectors joining tail-to-left-ear can be closely approximated by combining a vector joining the tail-to-shoulder and then the shoulder-to-left-ear). Specifically, 12 channels were used to represent three part pair associations, as listed in Table 2. The three part pairs and their associated channels are given below:

- Channels 1–4: Left Ear $\leftrightarrow$ Shoulder
- Channels 5–8: Right Ear $\leftrightarrow$ Shoulder
- Channels 9–12: Shoulder $\leftrightarrow$ Tail

Table 2. Channels 5–16 of the image-space representation used to represent pig locations and orientations. Pairs of neighboring channels corresponds to the x and y offset between neighboring parts. Overall, these 12 channels represent bidirectional vectors linking three pairs of body parts.

Channel	5	6	7	8	9	10
Encoding	$\propto (x_{\mathbf{l}_n} - x_{\mathbf{s}_n})$	$\propto (y_{\mathbf{l}_n} - y_{\mathbf{s}_n})$	$\propto (x_{\mathbf{s}_n} - x_{\mathbf{l}_n})$	$\propto (y_{\mathbf{s}_n} - y_{\mathbf{l}_n})$	$\propto (x_{\mathbf{r}_n} - x_{\mathbf{s}_n})$	$\propto (y_{\mathbf{r}_n} - y_{\mathbf{s}_n})$
Channel	11	12	13	14	15	16
Encoding	$\propto (x_{\mathbf{s}_n} - x_{\mathbf{r}_n})$	$\propto (y_{\mathbf{s}_n} - y_{\mathbf{r}_n})$	$\propto (x_{\mathbf{s}_n} - x_{\mathbf{t}_n})$	$\propto (y_{\mathbf{s}_n} - y_{\mathbf{t}_n})$	$\propto (x_{\mathbf{t}_n} - x_{\mathbf{s}_n})$	$\propto (y_{\mathbf{t}_n} - y_{\mathbf{s}_n})$

Each of the 12 channels encoded a real-valued offset from one point to another. Much like the part detection mappings, these vectors were encoded regionally into the spatial dimensions of the image. Figure 3 illustrates this encoding for a pair of side-by-side pigs. The diameter of the circular regions is denoted d_n for each pig n in the image, and it is proportional to the standard deviation used for the Gaussian kernel used in the previous section. For visualization purposes, each of the six images in Figure 3b,c represent the direction and distance between part pairs as a color, where the hue represents the direction and the saturation represents the magnitude (encoding provided in Figure 3d). Figure 3c further illustrates the lines joining the parts to one another.

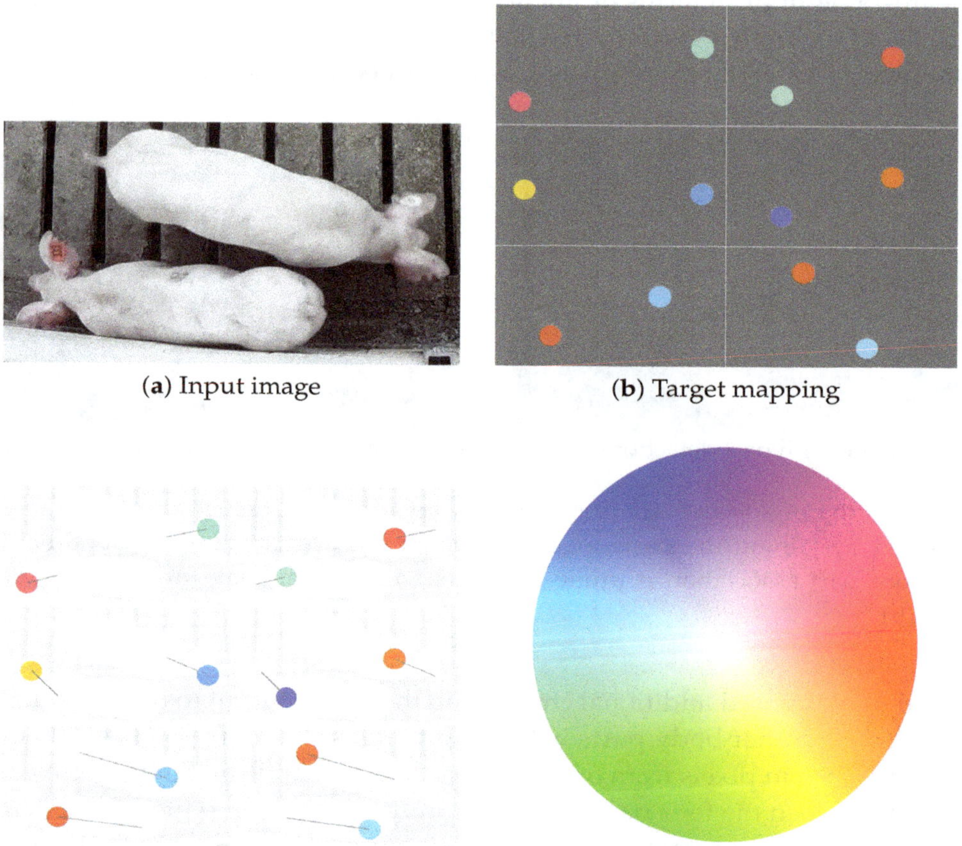

(**a**) Input image (**b**) Target mapping

(**c**) Visualization (**d**) 2D vector → color

Figure 3. The original image (**a**) is mapped to a 12-channel output (**b**), where vectors joining three pairs of body parts are encoded into circular regions in channels 5–16 of the output. Note that the four colors used in (**b**) are purely for visualization of the direction and magnitude of the vectors, where (**d**) provides a mapping between vectors and colors. The overlay in (**c**) is provided to illustrate the locations of the vector encodings and their magnitude and direction (illustrated by the gray line) in relation to the original image.

3.3. Instances from Part Detection and Association

The goal of the proposed method was to detect all visible parts and group them together in order to form whole-animal instances. Sections 3.1 and 3.2 provided a means to represent body part locations and vectors associated them to one another in the form of a 16-dimensional image-space mapping. To translate this image-space mapping to a set of discrete instance locations, one must follow a sequence of steps (illustrated in Figure 4).

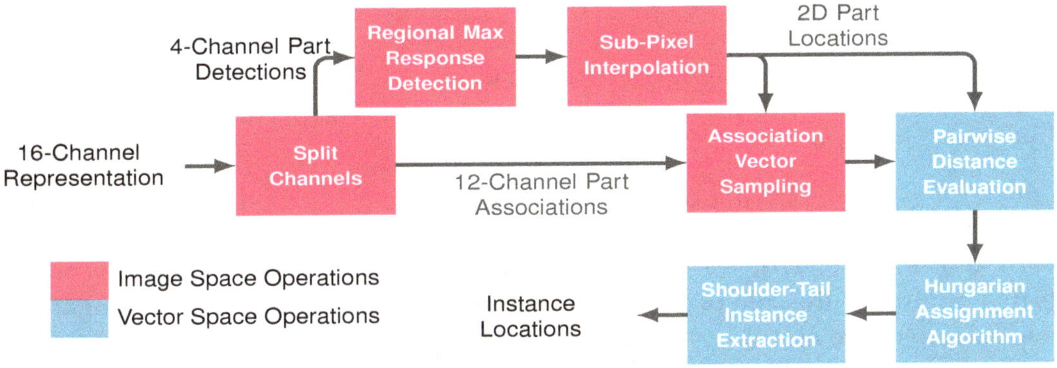

Figure 4. Flow diagram of the proposed method for converting the 16-channel image space representation to a set of 2D coordinates of each visible instance.

First, each 2D body part locations must have been extracted from the Gaussian kernels contained in channels 1–4 of the 16-channel representation image-space representation. Thus, the first step was to split channels into 4-channel part detections and 12-channel part associations. The precise 2D part locations were represented by the peaks of the Gaussian kernels in the image space mapping. Let $\mathbf{M}_\mathbf{p}$ be the $R \times C$ image space map for body part $\mathbf{p} \in \{\mathbf{l}, \mathbf{r}, \mathbf{s}, \mathbf{t}\}$, which corresponds with the left ear, right ear, shoulder, and tail, respectively. Note that it was assumed that the number of rows and columns in the input image and output mappings are R and C. The part locations can be extracted from the image space using a form of regional max response detection defined by

$$\{\mathbf{p}\} = \left\{(x,y) | \mathbf{M}_\mathbf{p}(x,y) \geq \mathbf{M}_\mathbf{p}(x',y') \text{ for all } (x',y') \in \mathcal{R}_{(x,y)}\right\} \quad \text{for } \mathbf{p} \in \{\mathbf{l}, \mathbf{r}, \mathbf{s}, \mathbf{t}\}, \tag{1}$$

where $\mathcal{R}_{(x,y)}$ was a region surrounding image space location (x,y). Part locations are therefore only detected if their value in the image space mapping was greater than that of its neighbors. This worked well for detecting the peak pixel coordinates of Gaussian kernels, but it can be further refined by using quadratic sub-pixel interpolation. Here, interpolation was performed by replacing the original integer coordinates (x,y) with real number coordinates using

$$(x,y) \leftarrow \left(x + \frac{\mathbf{M}_\mathbf{p}(x-1,y) - \mathbf{M}_\mathbf{p}(x+1,y)}{2(\mathbf{M}_\mathbf{p}(x+1,y) + \mathbf{M}_\mathbf{p}(x-1,y) - 2\mathbf{M}_\mathbf{p}(x,y))}, y + \frac{\mathbf{M}_\mathbf{p}(x,y-1) - \mathbf{M}_\mathbf{p}(x,y+1)}{2(\mathbf{M}_\mathbf{p}(x,y+1) + \mathbf{M}_\mathbf{p}(x,y-1) - 2\mathbf{M}_\mathbf{p}(x,y))}\right). \tag{2}$$

Given the complete set of detected body part locations

$$\{\mathbf{p}_1, \ldots, \mathbf{p}_{N_\mathbf{p}}\} = \{(x_{\mathbf{p}_1}, y_{\mathbf{p}_1}), \ldots, (x_{\mathbf{p}_{N_\mathbf{p}}}, y_{\mathbf{p}_{N_\mathbf{p}}})\} \quad \text{for } \mathbf{p} \in \{\mathbf{l}, \mathbf{r}, \mathbf{s}, \mathbf{t}\}, \tag{3}$$

the next step was to estimate the locations of associated parts using an association vector sampling of the 12-channel part associations mapping. The 12 dimensions of the association mapping will be denoted $[\mathbf{M}^x_{\mathbf{l}\to\mathbf{s}} \mathbf{M}^y_{\mathbf{l}\to\mathbf{s}} \mathbf{M}^x_{\mathbf{s}\to\mathbf{l}} \mathbf{M}^y_{\mathbf{s}\to\mathbf{l}} \mathbf{M}^x_{\mathbf{r}\to\mathbf{s}} \mathbf{M}^y_{\mathbf{r}\to\mathbf{s}} \mathbf{M}^x_{\mathbf{s}\to\mathbf{r}} \mathbf{M}^y_{\mathbf{s}\to\mathbf{r}} \mathbf{M}^x_{\mathbf{s}\to\mathbf{t}} \mathbf{M}^y_{\mathbf{s}\to\mathbf{t}} \mathbf{M}^x_{\mathbf{t}\to\mathbf{s}} \mathbf{M}^y_{\mathbf{t}\to\mathbf{s}}]$ and the estimated location of an associated part $\mathbf{q}$ from location $\mathbf{p}_n$ can be obtained using

$$(\mathbf{p} \to \mathbf{q})_n = (x_{\mathbf{p}_n} - \mathbf{M}^x_{\mathbf{p}\to\mathbf{q}}(x_{\mathbf{p}_n}, y_{\mathbf{p}_n}), y_{\mathbf{p}_n} - \mathbf{M}^y_{\mathbf{p}\to\mathbf{q}}(x_{\mathbf{p}_n}, y_{\mathbf{p}_n})) \quad \text{for all} \quad n = 1, \ldots, N_\mathbf{p}. \tag{4}$$

To join parts together, the distance between the estimated part locations and the actual locations were first computed using a pairwise distance evaluation. Specifically, the association distance between two parts $\mathbf{p}_n$ and $\mathbf{q}_m$ is given by

$$d(\mathbf{p}_n, \mathbf{q}_m) = \frac{|(\mathbf{p} \rightarrow \mathbf{q})_n - \mathbf{q}_m| + |(\mathbf{q} \rightarrow \mathbf{p})_m - \mathbf{p}_n|}{2}, \quad (5)$$

where $|\mathbf{a}|$ denotes the L2-norm of vector $\mathbf{a}$. Overall, this collection of part-to-part distances forms a set of three unique distance matrices

$$\mathbf{D}_{\mathbf{p},\mathbf{q}} = \begin{bmatrix} d(\mathbf{p}_1, \mathbf{q}_1) & d(\mathbf{p}_1, \mathbf{q}_2) & \dots & d(\mathbf{p}_1, \mathbf{q}_{N_\mathbf{q}}) \\ d(\mathbf{p}_2, \mathbf{q}_1) & d(\mathbf{p}_2, \mathbf{q}_2) & \dots & d(\mathbf{p}_2, \mathbf{q}_{N_\mathbf{q}}) \\ \vdots & \vdots & \ddots & \vdots \\ d(\mathbf{p}_{N_\mathbf{p}}, \mathbf{q}_1) & d(\mathbf{p}_{N_\mathbf{p}}, \mathbf{q}_2) & \dots & d(\mathbf{p}_{N_\mathbf{p}}, \mathbf{q}_{N_\mathbf{q}}) \end{bmatrix}, \quad (6)$$

where $(\mathbf{p} = \mathbf{l}, \mathbf{q} = \mathbf{s})$, $(\mathbf{p} = \mathbf{r}, \mathbf{q} = \mathbf{s})$, and $(\mathbf{p} = \mathbf{s}, \mathbf{q} = \mathbf{t})$. An optimal assignment between pairs of body parts that minimizes the sum of distances can be obtained by applying the Hungarian assignment algorithm to each distance matrix.

Finally, individual animals are identified as those that contain a joined shoulder and tail. The shoulder-tail instance extraction method began by identifying matches from the Hungarian assignment algorithm's output for $\mathbf{D}_{\mathbf{s},\mathbf{t}}$. Once all instances have been identified, the left and right ear detections can be joined to the shoulder locations of all instances via the Hungarian assignment algorithm's output for $\mathbf{D}_{\mathbf{l},\mathbf{s}}$ and $\mathbf{D}_{\mathbf{r},\mathbf{s}}$.

3.4. *Fully-Convolution Network for Part Detection and Association Mapping*

A fully-convolutional neural network was used to approximate the 16-channel target output, given a red-blue-green (RGB) image as input. Specifically, the hourglass network illustrated in Figure 5 was proposed. Hourglass networks with symmetry in the downsampling and upsampling stages have previously been used for pose estimation [61] and image segmentation [62]. Specifically, the proposed network was largely based on the popular SegNet architecture [62], which introduced a way to improve upsampling by sharing the indices of each max pooling layer with a corresponding max unpooling layer. This approach has been shown to achieve impressive performance in segmentation tasks by removing the burden of "learning to upsample" from the network.

The network architecture used by the proposed method also incorporated skip-connections in the form of depth concatenation immediately after max unpooling layers. Skip-connections have been demonstrated to encourage feature-reuse, thus improving performance with a fixed number of network coefficients [63]. They were also shown to drastically decrease the amount of training time required by the network. The U-net architecture, introduced by Ronneberger et al. [64], further demonstrated the power of skip-connections for hourglass-shaped networks.

During training, the objective function attempted to minimize the mean-squared error between the network output and the target ground truth. For the first four channels that correspond to part detections, gradients were back-propagated for all pixel locations regardless of their value. For the last 12 channels, gradients were back-propagated exclusively for pixel locations where the target output is assigned (non-zero). Therefore, a specific output was only encouraged in the regions surrounding the point locations. This type of selective training helped to ensure that the vector outputs did not tend toward zero in areas where part detections were uncertain. In essence, this approach was meant to separate the tasks of detection and association mapping.

Receptive Field

When designing neural networks for visual tasks, it is important that the network was able to "see" the entirety of the objects it is considering. This viewable area is often referred to as the receptive field. To derive the receptive field, the effective stride length between adjacent coordinates in the feature map is calculated using

$$s_{l_{\text{effective}}} = s_{l-1_{\text{effective}}} \times s_l, \tag{7}$$

where s_l is equal to the stride length at layer block l in the network and $s_0 = 1$. Note that, in the proposed network, all max pooling layers have $s_l = 2$ and all max unpooling layers have $s_l = 0.5$, while all other layers have $s_l = 1$. The overall stride length essentially relates the resolution of a downsampled feature map to the original input size. Given s_l for all l in the network, the receptive field size can be calculated using

$$r_l = r_{l-1} + (w_l - 1) \times s_{l-1}, \tag{8}$$

where w_l is the convolutional kernel width at layer l and $r_0 = 1$. In the proposed network, each convolutional kernel had width $w_l = 3$. Because of the stochastic nature of max pooling and max unpooling operations, it was difficult to define their effective kernel size. Therefore, in this analysis, we have used the lower bound of $w_l = 1$ for all pooling operations.

Table 3 provides the receptive field of the network as a function of layer block for a subset of the 41 layer blocks featured in Figure 5. The receptive field represented the width of a square region in the original image space that affected a single pixel location at the output. While the receptive field of the proposed network was 363, the distance between any two image locations that can affect each others' outputs was 181 (the radius of the receptive field). In practice, it was recommended that the receptive field size should be considerably larger than the maximum object size due to a decaying effect that has been observed on the outer extremes of the square region [65]. It will be shown in Section 4 that the chosen image scale results in pigs that are typically much smaller than the receptive field radius.

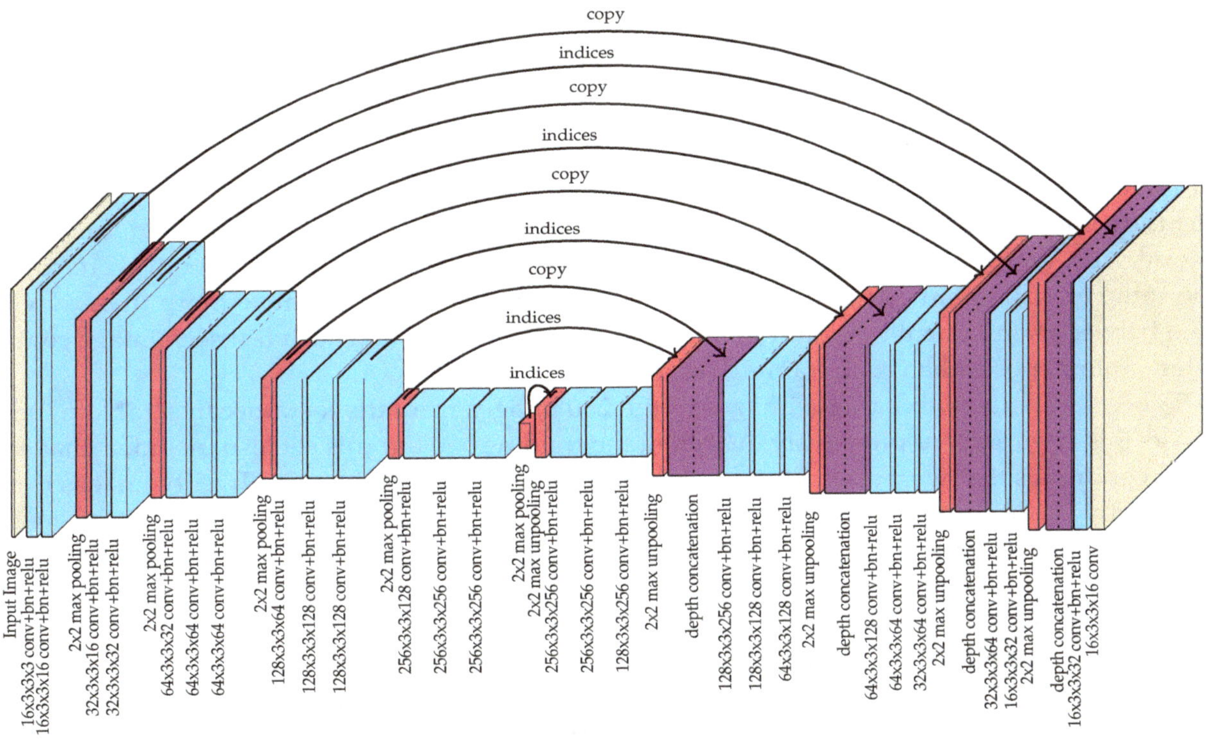

Figure 5. The hourglass-shaped network used by the proposed method to convert images to 16-channel image-space instance detection maps.

Table 3. Sampling of the receptive field calculations at the output of every layer of the proposed network. The different types of layers are abbreviated with the following notation: I: input image, C: convolution block, M: max pooling, U: max unpooling, D: depth concatenation, O: output image

Layer Type	I	C	C	M	C	...	C	M	U	C	...	C	U	D	C	O
l	1	2	3	4	5	...	18	19	20	21	...	37	38	39	40	41
s_l	1	1	1	2	1	...	1	2	0.5	1	...	1	0.5	1	1	1
$s_{l_{\text{effective}}}$	1	1	1	2	2	...	16	32	16	16	...	2	1	1	1	1
w_l	1	3	3	1	3	...	3	1	1	3	...	3	1	1	3	3
r_l	1	3	5	5	9	...	181	181	181	213	...	359	359	359	361	363

4. Experimental Results

4.1. Dataset

To the best of our knowledge, no open-source dataset exists for pig detection in group-housing environments. Therefore, to enable quantitative evaluation, a new dataset with 2000 annotated images of pigs was introduced. The dataset (http://psrg.unl.edu/Projects/Details/12-Animal-Tracking) depicts 17 different pen locations and includes pigs ranging in age from 1.5 to 5.5 months. Each unique image was randomly extracted from video recordings spanning multiple weeks in each location. More than two hours, on average, existed between samples at each location. Thus, a wide range of unique animal poses were represented in the dataset.

The dataset was divided into two subsets: 1600 images for training and 400 images for testing. Furthermore, the 400 testing images were subdivided into two additional subsets: 200 captured in the same environments seen in the training set (test:seen), and 200 images from environments previously unseen in the training set (test:unseen). The cameras used to capture the images included both a Microsoft Kinect v2 color camera with resolution 1080 × 1920 and Lorex LNE3162B and LNE4422 color/IR cameras with resolution 1520 × 2688. All of the environments were captured with the camera mounted above the pen looking down. The distance between the pen floor and the camera varied between 2.5 and 6.0 m, and the specific poses of the cameras ensured that the animal pen of interest was centered and entirely contained within the field of view. Variations in environment and image capture technology were used to ensure that the analysis emphasizes robustness.

Figure 6 shows sample images from the training set, depicting 13 different pen locations with color-coded annotations for each hand-labeled body part. Note that the last two images in Figure 6 depict the same environment, but one was captured with full color in the daytime and the other was captured with active IR at night. The first 200 images of the testing set (test:seen) were captured in the same environment as the training set, but at different times. Because more than two hours existed between subsequent randomly sampled images, it is likely that each test:seen image contained different animal poses than each training set image.

Figure 7 illustrates six sample images of the 200 images from the test:unseen set. Note that, not only was this environment previously unseen in the training set, but this set also included challenging lighting conditions that were also not represented among the training images. Twenty images from the training set were captured where the camera's IR night vision was activated, but all of the remaining 1580 training set images (and all of the test:seen images) were captured with overhead lights on. To achieve the challenging lighting conditions present in the test:unseen set, the lights were turned on at 6 am and off at 6 pm every day. For a short duration between approximately 6 pm and 8 pm, ambient lighting dimly illuminated the pens. After 8 pm, the cameras activated night-vision mode and captured IR images while actively illuminating the scene with built-in IR lights. Two of the four pens presented in the test:unseen set were also illuminated with IR flood lights. This had the effect of creating well lit scenes with harsh shadows and side-lighting.

In each of the images, a user manually annotated the location of the left ear (red), right ear (green), shoulder (blue), and tail (yellow) for each visible animal in that order. Annotations belonging to the same instance are connected with a continuous black line. If ears were not visible, they were not annotated, however, emphasis was placed on annotating both shoulders and tail for each instance even when these locations are occluded, i.e., both shoulder and tail were annotated as long as they are located in the pen of interest and their estimated positions were within the field of view of the camera.

It should be noted that, in some cases, pigs from adjacent pens were partially visible through bars that separate the pens. These partially visible pigs were not annotated. It was assumed that a camera placed above a pen is responsible for detecting only animals in that pen and, while some areas of the image belonging to the adjacent pen were masked out, it was difficult to remove partially visible pigs from the image without unintentionally masking out pigs within the pen of interest. In practice, areas of interests were defined by polygons for each image in the dataset and masking out was done by setting all pixels outside the area of interest to pure black. Examples of masked out regions can been seen in Figures 6 and 7, where the blacked out regions correspond to areas with pigs in adjacent pens.

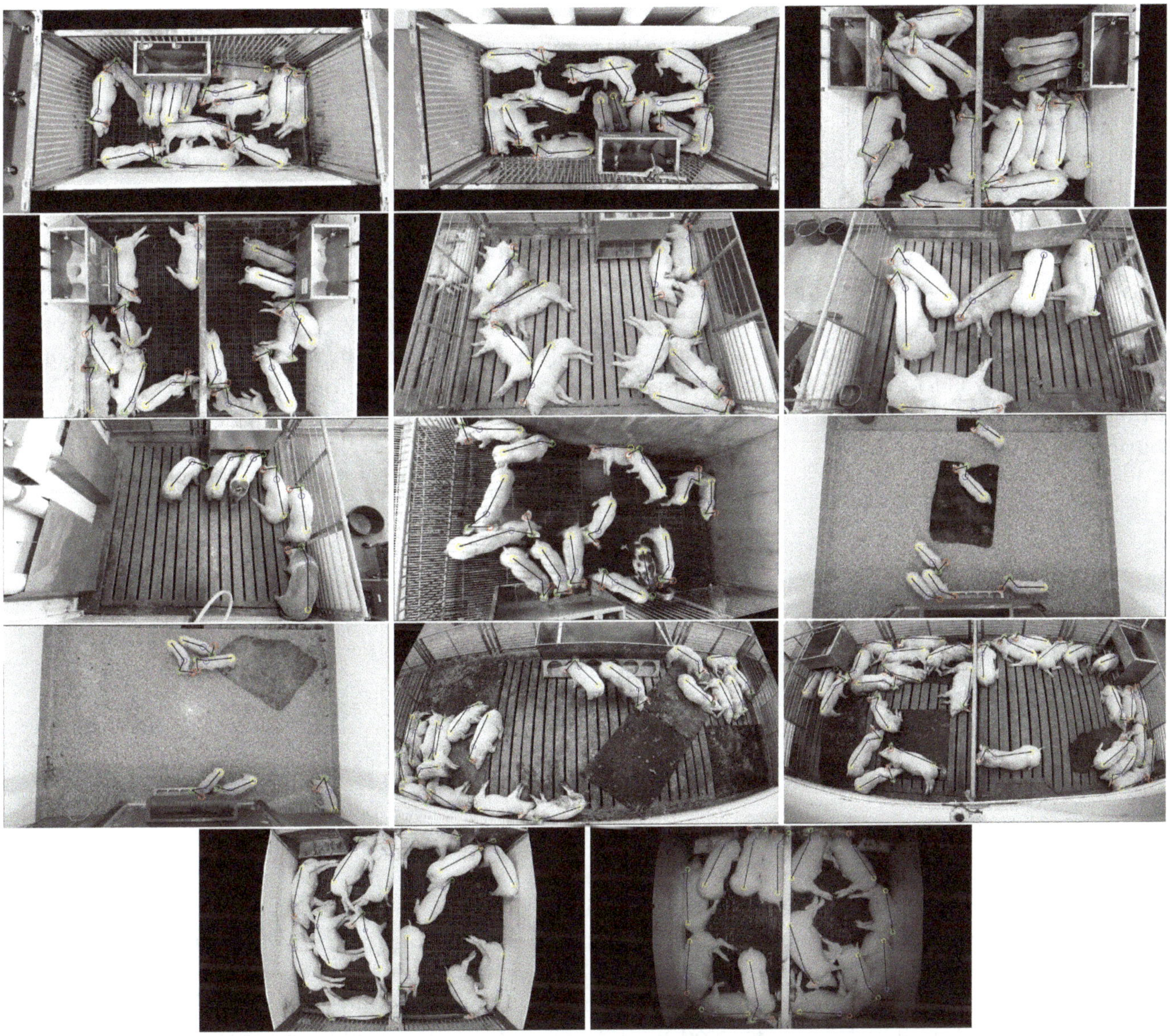

Figure 6. Sample images depicting different environments represented in the training set. The first 13 images (left-to-right, top-to-bottom) were captured during daylight hours with lights on. The last image (from the same environment as the 13th image) was captured using the infrared night vision mode used by the Lorex LNE3162B with active IR illumination.

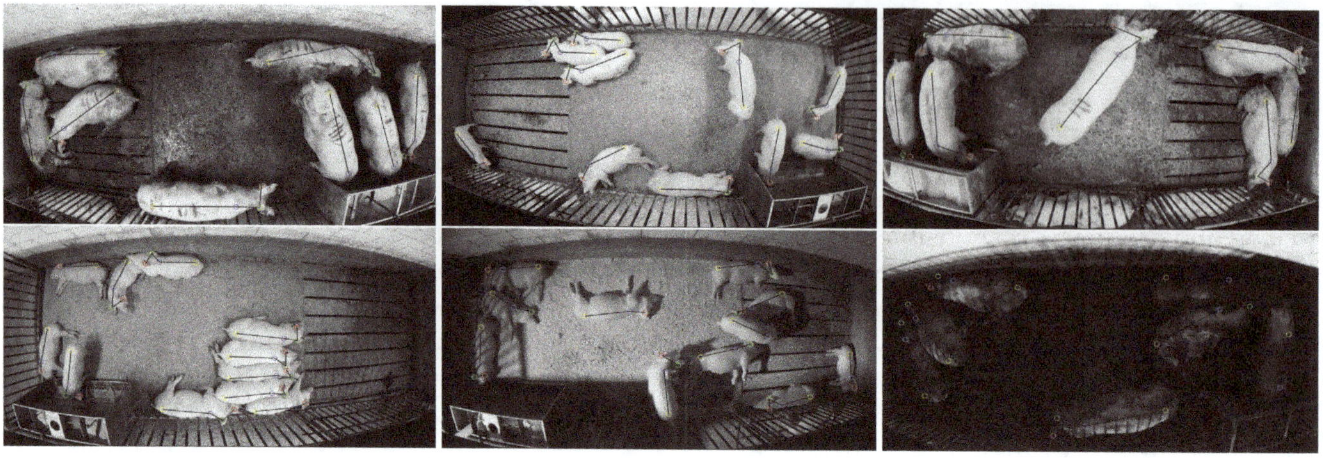

Figure 7. Sample images from the "unseen" portion of the testing set (test:unseen). These images depict environments and lighting conditions not represented in the training set.

4.2. Training Details

Prior to training the network, images were downsampled so that the number of columns was 480. This was empirically deemed to be a sufficient resolution for discerning the parts of interest while remaining small enough for the computing hardware to process multiple images per second. The average length of pigs in each image after downsampling is presented in the histogram of Figure 8. While the majority of the pigs had a body length of less than 100 pixels, there were some that exceed 140 pixels in length. For these pigs, it was important that the network was able to "see" the entirety of the pig as it estimated the body part locations and vector associations. In Section 3.4, the radius of the receptive field was found to be 181 using the proposed network. Therefore, the network was capable of observing the entire animal even in the most extreme images where the shoulder-to-tail distance approached 140 pixels.

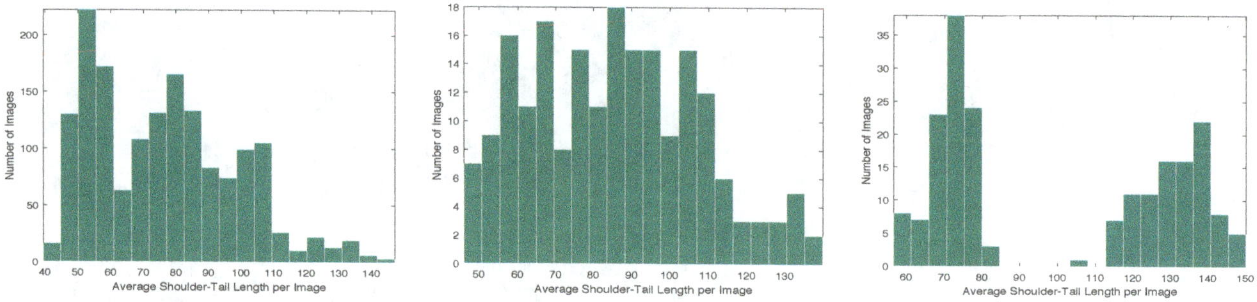

(**a**) Training (**b**) Test:Seen (**c**) Test:Unseen

Figure 8. Distribution of the average length from shoulder to tail in each partition of the dataset.

Target images for training the fully-convolutional network were created by adapting the size of the Gaussian kernels used to mark each part in channels 1–4 (Figure 1b) to the size of the animals. This adaptation encouraged continuity of image-space annotations between different environments and ages/sizes of pigs. Specifically, this was done by first computing the average distance between the shoulder and tail for all instances, denoted $\mu_{\mathbf{s}\rightarrow\mathbf{t}}$, to provide a numerical representation of the average size of pigs in the image space. Then, the shoulder-to-tail distance for instance n, given by $\delta_{\mathbf{s}\rightarrow\mathbf{t}}$, was combined with the average distance in order to compute the Gaussian kernel standard deviation, defined as $\sigma_n = 0.16 \times (\mu_{\mathbf{s}\rightarrow\mathbf{t}} + \delta_{\mathbf{s}\rightarrow\mathbf{t}})$. This combination was used to prevent unusual animal poses from shrinking the size of the kernels too much, while still allowing some adaptation to individual size variations. It should be noted that the scale factor of 0.16 was determined empirically to provide a suitable approximation to the variability of human annotations. If σ_n was too large, kernels belonging to nearby pigs interfered with each other and often resulted in a single part location being extracted by regional maximum detection. When σ_n was too small, the network training unfairly applied a penalty to kernels that were not exactly equal to the location provided by the human annotator, even if the kernel's location was within the natural variance of human annotations. Finally, the Gaussian kernels were then multiplied by a scalar value in order to set their maximum value to 1.0 and, in cases where two nearby Gaussian kernels for the same body part intersected, the output was simply assigned to the maximum of the two Gaussian kernel values. Scaling the kernels so that the peak is 1.0 helped to ensure that a fixed threshold can be used in peak detection regardless of σ_n.

The circular regions used to assign association vectors between parts in channels 5–16 (Figure 3b) should ideally have covered all possible pixel locations where the part might be detected. In practice, this area can be sufficiently covered by considering regions where the Gaussian kernel for each part had a magnitude greater than 0.2. In situations where one region intersected with another, the target output vector was composed of a weighted combination of the two intersecting vectors. The weights in these circumstances came from the corresponding Gaussian kernel magnitude at each pixel location.

The network was trained using heavy augmentation of both input and target images. Augmentations included random left-right flipping, random rotations sampled from a uniform distribution ranging from 0 to 360°, random scaling sampled uniformly between 0.5 and 1.5, and XY shifts uniformly sampled from the range of ± 20 pixels along both dimensions. Careful consideration was needed for augmenting the 16-channel target image. Rotations and scaling were applied spatially to both the association vector regions and also the output values along pairs of channels that correspond to XY offsets between body parts. Left-right flips were handled by switching the labels for left and right ears.

4.3. Processing Details

After obtaining the 16-channel mapping from the trained network, each of the part location maps (channels 1–4) and the association maps (channels 5–16) were smoothened using a 5×5 averaging box filter. This step would not be necessary to extract information from ground truth mappings, but it was beneficial for reducing the effects of noise on regional maximum response detection. In practice, box filtering was done by adding an average pooling layer to the end of the neural network. The size of regions $\mathcal{R}_{(x,y)}$ used in (1) consisted of a 15×15 window surrounding each pixel (x, y).

The method was implemented in Matlab 2018b using the deep learning toolbox [66]. A desktop computer, equipped with an Intel i7-6700K CPU, 32 GB of RAM, and an NVIDIA GTX1070 GPU was used for training and inference. The approximate time required by the fully-convolutional neural network to perform forward inference is 0.24 s and it took an additional 0.01 s to find instance locations. Thus, the system was capable of fully processing four frames per second.

4.4. Instance Detection Performance Metric

The goal of the proposed method was to identify the location and orientation of each pig in a given image. Although the method generated detections and associations for four body parts, only the shoulder and tail location were used to identify a complete instance. This decision was based on two factors. Firstly, they are sufficient for approximating the center-of-mass location and orientation of each animal and, second, special emphasis was placed on ensuring their labeling by human annotators. Given a complete set of N ground truth shoulder-tail pairs $\{(\mathbf{s}_1, \mathbf{t}_1), \ldots, (\mathbf{s}_N, \mathbf{t}_N)\}$ and a set of M estimated shoulder-tail pairs $\{(\tilde{\mathbf{s}}_1, \tilde{\mathbf{t}}_1), \ldots, (\tilde{\mathbf{s}}_N, \tilde{\mathbf{t}}_M)\}$, an association method was needed to determine if an estimate corresponded to the ground truth, since both sets of pixel coordinates were unlikely to contain exactly the same values.

Bipartite matching problems are commonly solved with a Hungarian assignment, however, this can sometimes lead to matches between far-away pairs in order to minimize a global cost. For this particular matching problem where shoulder-tail pairs are associated with each other, there was likely to be very little distance between the ground truth and detected positions. Setting the maximum distance allowed between matching pairs can fix this issue, but it comes at the cost of introducing additional parameters that depended on image resolution and the relative sizes of animals. To avoid parameterization, the strict cross-check matching criteria was used here to assign estimates to the ground truth only when they were each others' minimum cost matches. More formally, two instances n and m matched if and only if

$$m = \underset{m \in \{1,\ldots,M\}}{\operatorname{argmax}} \ |\mathbf{s}_n - \tilde{\mathbf{s}}_m| + |\mathbf{t}_n - \tilde{\mathbf{t}}_m| \tag{9}$$

and

$$n = \underset{n \in \{1,\ldots,N\}}{\operatorname{argmax}} \ |\mathbf{s}_n - \tilde{\mathbf{s}}_m| + |\mathbf{t}_n - \tilde{\mathbf{t}}_m|, \tag{10}$$

where $||$ denotes the L2 norm. Figure 9 illustrates the advantage of using the cross-check method instead of the unparameterized Hungarian algorithm.

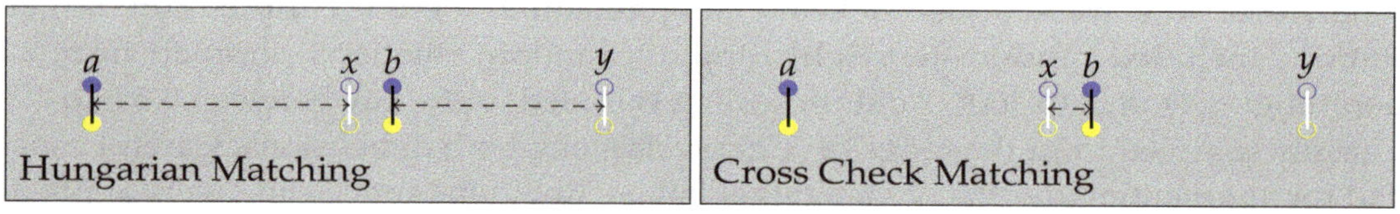

Figure 9. Example with two ground truth instance locations (solid circles connected with black lines) and two detected instances (empty circles connected with white lines) using matching results achieved with both the Hungarian algorithm and the proposed cross-check matching of Equations (9) and (10). While the detection x and ground truth location b in the middle are clearly nearest neighbors of one another, they are not matched by the Hungarian algorithm. Instead, in an effort to minimize the global matching cost, the Hungarian algorithm will assign a to x and b to y. In contrast, the cross-check matching method leaves the outer detection y and ground truth location a unmatched while assigning the two in the middle, x and b, together.

4.5. Instance Matching Results

In order to evaluate the effectiveness of the proposed vector part association method, it was compared to an alternative Euclidean part association method. The Euclidean part association method joins parts together by simply minimizing their Euclidean distance. This method, previously illustrated in Figure 2, removes the effects of part association vectors on detection performance and allows for a partial ablation study. Figure 10 presents the precision and recall for all three partitions of the dataset and Table 4 presents full numerical results over the training set. Each sample along the curve corresponds to a different threshold for part detection, where parts are detected only if the neural network output in channels 1–4 exceeds the threshold value.

The results show that the proposed vector part association method provided a significant boost to matching precision when compared to Euclidean part association. Less than 0.1% of detections were false positives compared to more than 5% when using Euclidean matching, regardless of threshold. Figure 10a,b illustrate nearly identical results across training and test:seen sets. This provides a strong indication that the method was not overfitting to the specific animal poses presented in the training set. Both Figure 10a,b demonstrate a minimum precision of $\approx$0.91 at a recall of $\approx$0.42 for the Euclidean matching method. This was because, at this threshold less than half of the animal parts were being detected but the ones that were detected are matched to their nearest neighbor. As a result, there was a relatively high likelihood that only a shoulder is detected, but not a tail, or vice versa. In an effort to form whole instances, the Euclidean method simply joined together nearest neighbors and many of these instances were not aligned with the ground truth. When the threshold was adjusted higher or lower, there was a higher likelihood that either both shoulder and tail were detected or neither was detected. In either case, this lead to improved precision, because the instances that were identified were more likely to be true positives.

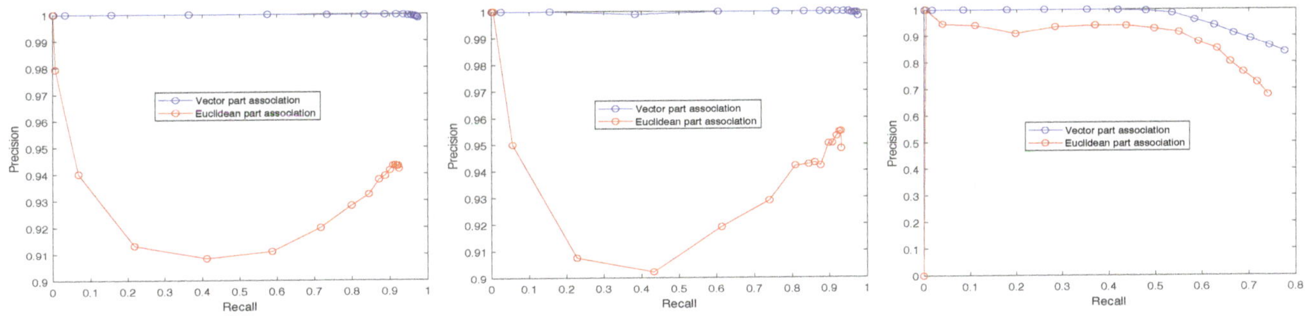

(**a**) Training (**b**) Test: Seen (**c**) Test: Unseen

Figure 10. Precision-recall curves for both the proposed method and an alternative association strategy that assigns parts to one another by minimizing Euclidean distance. The results are nearly identical across the training set (**a**) and test:seen set (**b**), and both illustrate a dramatic improvement achieved by joining parts together using the proposed vector mapping. The results on the test:unseen set (**c**) illustrate the limitations of the method when operating on different environments than those used for training.

Table 4. Detailed results obtained with the proposed method and an alternative association strategy that assigns parts to one another by minimizing Euclidean distance. The table includes true positives (TP), false positives (FP), false negatives (FN), precision, recall, and F-measure for different part detection thresholds.

Part Detection Threshold	Vector Matching						Euclidean Matching					
	TP	FP	FN	Recall	Precision	F-Measure	TP	FP	FN	Recall	Precision	F-Measure
0.10	20,217	27	525	0.975	0.999	0.987	19,170	1181	1572	0.924	0.942	0.933
0.15	20,160	20	582	0.972	0.999	0.985	19,127	1156	1615	0.922	0.943	0.932
0.20	20,092	17	650	0.969	0.999	0.984	19,058	1154	1684	0.919	0.943	0.931
0.25	19,999	13	743	0.964	0.999	0.981	18,971	1141	1771	0.915	0.943	0.929
0.30	19,865	10	877	0.958	0.999	0.978	18,851	1135	1891	0.909	0.943	0.926
0.35	19,675	7	1067	0.949	1.000	0.973	18,675	1162	2067	0.900	0.941	0.920
0.40	19,413	3	1329	0.936	1.000	0.967	18,418	1190	2324	0.888	0.939	0.913
0.45	19,029	2	1713	0.917	1.000	0.957	18,077	1195	2665	0.872	0.938	0.904
0.50	18,408	2	2334	0.887	1.000	0.940	17,526	1269	3216	0.845	0.932	0.887
0.55	17,287	2	3455	0.833	1.000	0.909	16,568	1281	4174	0.799	0.928	0.859
0.60	15,227	2	5515	0.734	1.000	0.847	14,871	1294	5871	0.717	0.920	0.806
0.65	11,929	0	8813	0.575	1.000	0.730	12,184	1189	8558	0.587	0.911	0.714
0.70	7565	0	13,177	0.365	1.000	0.534	8543	860	12,199	0.412	0.909	0.567
0.75	3261	0	17,481	0.157	1.000	0.272	4523	430	16,219	0.218	0.913	0.352
0.80	692	0	20,050	0.033	1.000	0.065	1414	90	19,328	0.068	0.940	0.127
0.85	53	0	20,689	0.003	1.000	0.005	143	3	20,599	0.007	0.979	0.014
0.90	1	0	20,741	0.000	1.000	0.000	5	0	20,737	0.000	1.000	0.000

In Table 5, the results are compared across all three partitions of the dataset with a fixed threshold of 0.25. While the F-measure was 0.981 at threshold 0.25, which was lower than the peak F-measure of 0.987 achieved at a threshold of 0.1, the decreased threshold produced more than twice the number of false positives. When F-measure values were nearly identical, the choice of threshold depended on how sensitive an application was to false positives and false negatives. The comparison at threshold 0.25 highlighted both the performance similarities across the training and test:seen sets and the discrepancies between both of those sets and the test:unseen set. One interpretation is that the discrepancy illuminates the importance of environment and lighting variations when training the neural network. The test:seen results were nearly identical to the training results, even though the specific poses of each animal were previously unseen. However, due to the use of heavy image augmentation, similar poses were likely represented during training. In contrast, the test:unseen results were much worse, likely due to the novel environments and challenging lighting conditions not present in the training set images.

Table 5. Instance detection results for the training set (1600 images), the test set with images of the same environments in the training set (200 images), and the test set with images of new environments and challenging lighting conditions (200 images). The part detection threshold was fixed at 0.25.

Evaluation Set	Vector Matching					
	TP	FP	FN	Recall	Precision	F-Measure
Training	19,999	13	743	0.964	0.999	0.981
Test: Seen	2273	1	94	0.960	1.000	0.980
Test: Unseen	1150	112	573	0.667	0.911	0.771

By digging deeper into the results and looking at specific examples, it is possible to learn more about the performance discrepancies. An example of 100% successful detections from both test:seen and test:unseen sets are shown in Figure 11. Here, the neural network output is illustrated for each of the 16 channels, and the final detections are shown below, where a green line connecting the shoulder to the tail defines a true positive detection. Note that, unlike the target part association maps illustrated in Figure 3, the outputs of the neural network do not clearly conform to the part locations. This is because the network was only trained to produce the correct vector (illustrated by color) at part locations and, at all other locations, the network essentially tried to minimize the cost with hypothetical part association vectors in case a part was present in that location. This attempt to minimize cost "in case" was most visible when comparing the part association maps of shoulder-to-tail and tail-to-shoulder (the bottom two part association maps in Figure 11). Even when the network was highly confident that a location belonged to a tail, it produced an association vector pointing from shoulder-to-tail at that location, just in case a shoulder at that location was mistaken for a tail.

Due to the similar lighting and overall appearance of test:seen image in Figure 11a, the method was able to identify every instance within the pen environment with high confidence (as indicated by the first four channel outputs of the neural network). However, in the test:unseen image (Figure 11b), the pig behind bars in the adjacent pen caused some confusion in the network. This was likely due to the fact that the network had never been exposed to this particular pen environment, and thus it had not been trained to ignore partial animals on the other side.

Alternatively, Figure 12 illustrates failure cases for both test:seen and test:unseen images. Each of the failures in the test:seen image occurred because of occlusions that made it difficult to discern the location of the shoulders and/or tail. In this case, it was even difficult for a human observer to confidently assign the ground truth locations. On the other hand, failures on the test:unseen image were not due to occlusions. They can instead by attributed to the unusual lighting conditions and the relatively large presentation of the animals in the image. Both of these properties were not represented in the training set, making it difficult for the neural network to interpret the image.

Figures 13 and 14 illustrate 24 failures from the test:seen and test:unseen set, respectively. In the test:seen sample set of Figure 13, 17 of the 23 false negatives can be attributed to occlusions or lack of visibility when the pig approached the edge of the image. Some other causes of error include unusual poses where the head was hidden, and situations where the pig had atypical markings. In contrast, only four false negatives out of 21 from the test:unseen sample set (Figure 14) can be attributed to occlusion. At least 10 can likely be attributed to lighting conditions. All three false positives occurred when a pig in an adjacent pen was laying next to the dividing bars. The outline of the bars on the pig's body appeared to confuse the network into interpreting this as a smaller body pointed in the orthogonal direction.

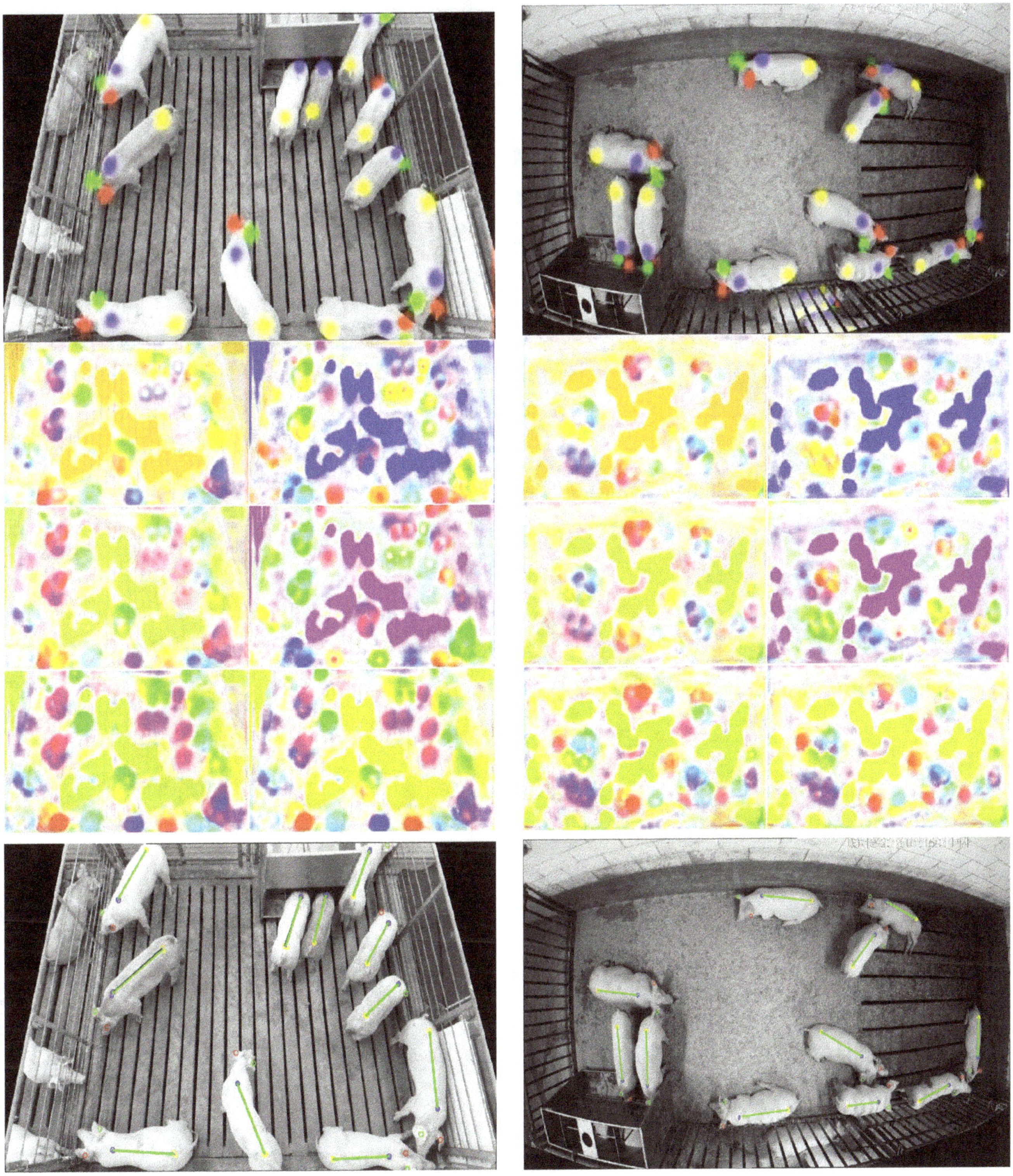

(**a**) Test:Seen (**b**) Test:Unseen

Figure 11. Examples of successful instance detection from both the test:seen set (**a**) and the test:unseen set (**b**). The top images depict the first four channels of the neural network output. The middle image composed of six sub-images depicts the color-coded vector associations from the last 12 channels of the neural network output. The bottom images depict both ground truth locations and estimates using the following color coding: false negative (blue), and true positive (green). Note that these images depict 100% successful detections, so only true positives are present.

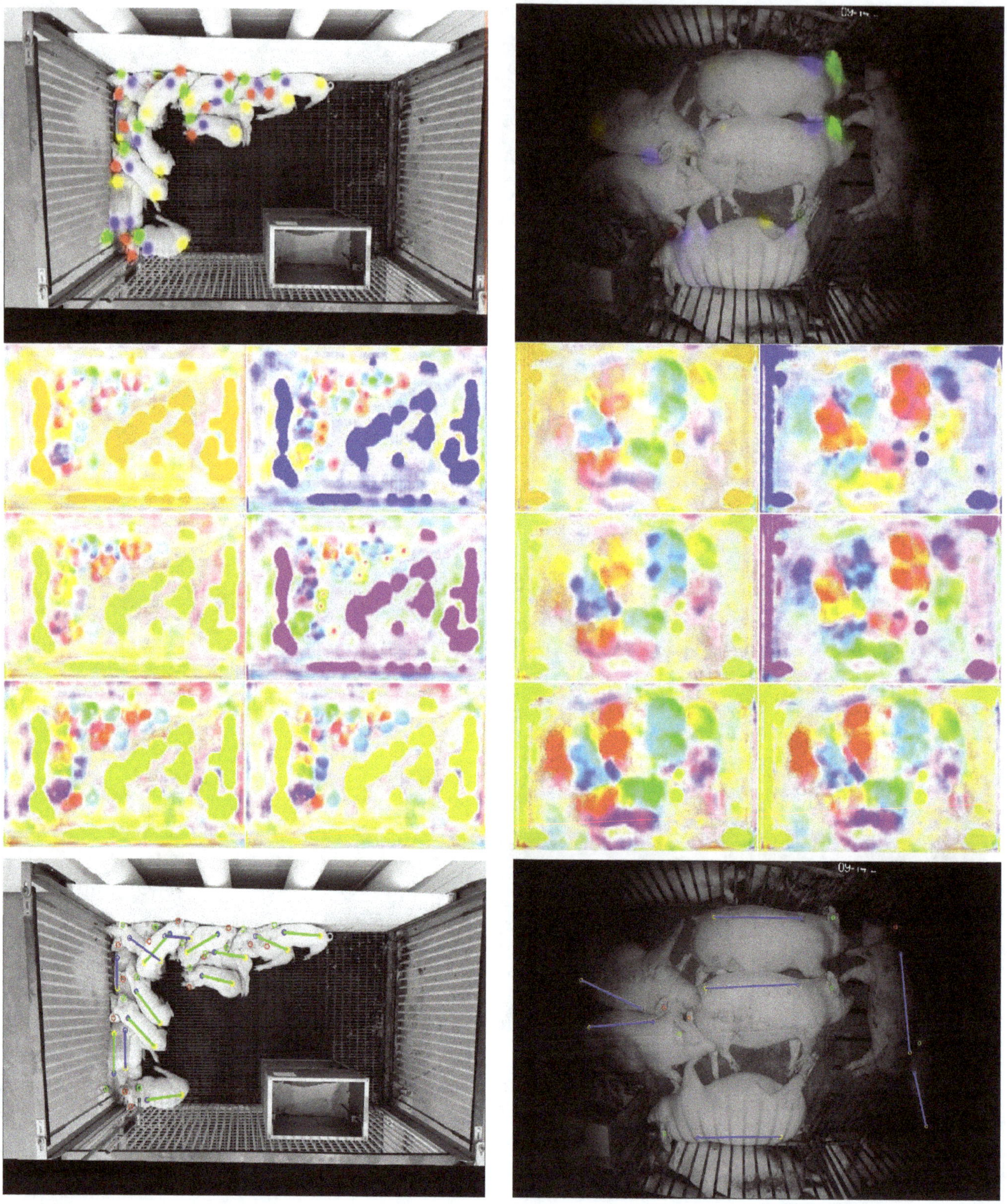

(**a**) Test:Seen (**b**) Test:Unseen

Figure 12. Examples of unsuccessful instance detection from both the test:seen set (**a**) and the test:unseen set (**b**). The top images depict the first four channels of the neural network output. The middle image composed of six sub-images depicts the color-coded vector associations from the last 12 channels of the neural network output. The bottom images depict both ground truth locations and estimates using the following color coding: false negative (blue), and true positive (green).

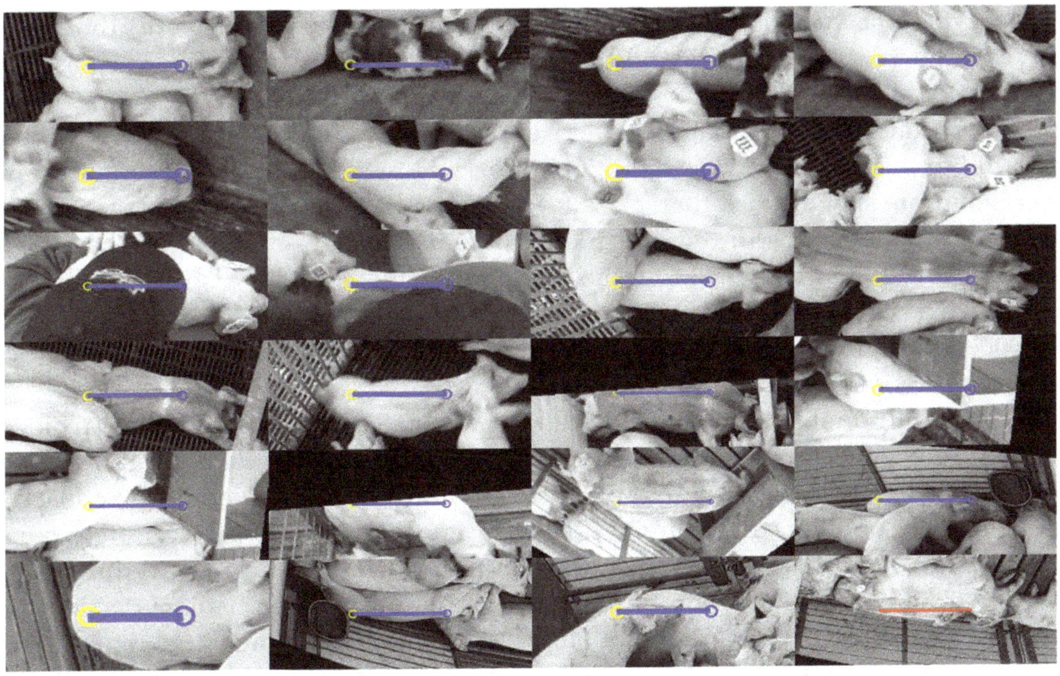

Figure 13. Twenty-four random samples of unsuccessful instance detections from the test:seen set using the following color coding: false negative (blue), and false positive (red).

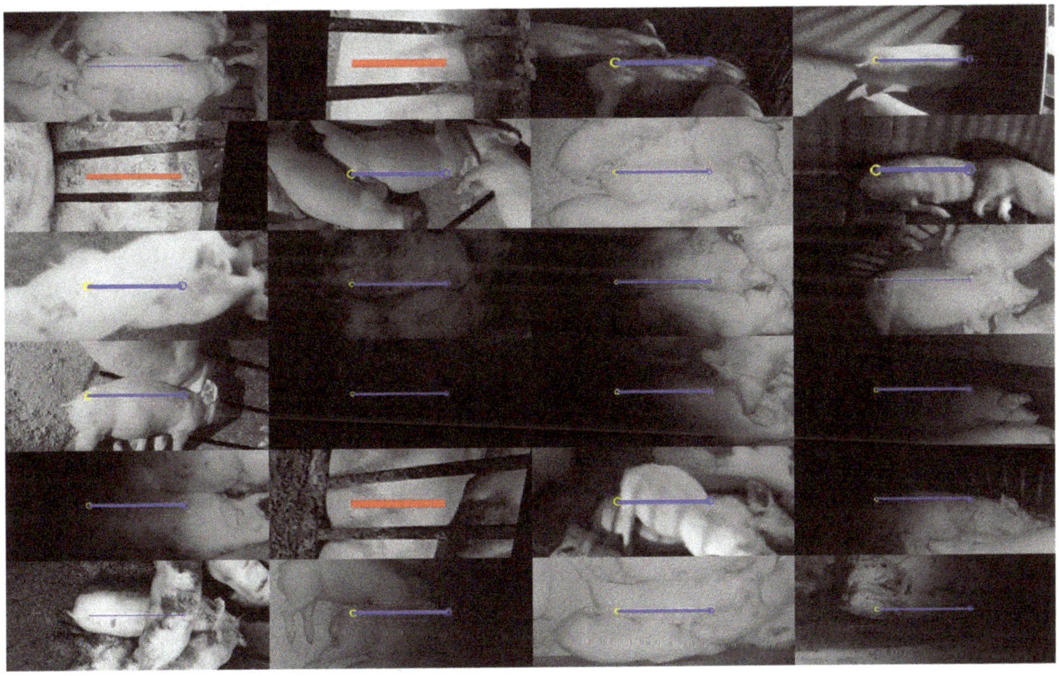

Figure 14. Twenty-four random samples of unsuccessful instance detections from the test:unseen set using the following color coding: false negative (blue), and false positive (red).

4.6. Discussion

The proposed method focuses on detecting the location and orientation of individual pigs in group-housing environments. Due to the lack of an available public dataset, it was necessary to create a new collection of annotated images. This presented a challenge in terms of capturing the appropriate level of variability and, while we believe that the chosen images sufficiently represent a variety of environments and ages of pigs, it would likely have been beneficial to include more camera angles and more than three types of cameras. Furthermore, the four body part locations were arbitrarily chosen as representatives of the location and orientation of each animal instance. A different set of body parts might have been equally effective.

Compared to datasets like ImageNet [67] and COCO [33], 2000 images may seem like an insufficient number of images to train a deep network. However, pig detection from an overhead camera is a much more specific task than classifying images into one of 1000 categories. With nearly 25,000 different animal poses captured, it is likely that any new pose presented to the network will strongly resemble one that already exists in the dataset. Augmentation is also critical to the success of network training. The chosen network contains nearly 4,000,000 coefficients, so it might be possible to overfit to 25,000 static animal poses, but it is much more difficult to overfit when the angle, size, and left-right orientation is randomized.

The fully-convolutional network introduced in Section 3.4 to estimate body part locations and association vectors was designed with sufficient complexity and a wide enough receptive field to achieve high performance levels in terms of precision and recall. However, the chosen hourglass architecture using max pooling/unpooling with skip connections and depth concatenations is almost certainly one of many network architectures capable of producing the desired 16-dimensional output representation. Atrous convolution, for example has been shown to be particular effective for creating a large depth of field [68] and spatial pyramid pooling has further been demonstrated to achieve excellent performance on multi-scale tasks [69]. A thorough investigation into the suitability of different network architectures might lead to a more accurate or efficient network for instance detection.

By inspecting the specific outputs of the network and the instance formation process, it becomes evident that errors are most commonly caused when the shoulder or tail of one pig occludes the same body part on another pig. Due to the network's inability to represent multiple part instances in the same image space location, it is only possible for one part instance to be detected in these situations. The vector associations inherently estimate the location of adjacent body parts, therefore, the occlusion can be inferred from the existing output of the network. Alternatively, it might also be possible to augment the dataset to explicitly label occlusions and build the network to detect such events.

In addition to shoulders and tails, the left and right ear were annotated in the dataset and explicitly detected by the network. While the results for instance-level detection do not evaluate the quality of these detections, we foresee them being integrated into future systems as a way to uniquely identify each instance. Ear tags are a common way for livestock to be identified in commercial facilities, and this may provide a convenient way to differentiate between individuals in an otherwise homogeneous population.

Perhaps the most likely application of this detection method would be within a larger tracking system, where detection serves as the first stage for video processing. To this end, the proposed per-frame detection method naturally lends itself to multi-object tracking (MOT). Specifically, a sub-category known as tracking-by-detection MOT methods directly process the outputs of per-frame detection methods, and their performance is often strongly tied to the quality of the detector [70]. For this reason, it is expected that high quality detection methods will eventually contribute to more reliable methods for multi-object tracking.

5. Conclusions

The proposed method and accompanying dataset introduced in this paper attempt to provide a robust solution to instance-level detection of multiple pigs in group-housing environments. A major contribution of this work is the introduction of an image space representation of each pig as a collection of body parts along with a method to join parts together to form full instances. The method for estimating the desired image space representation leverages the power of deep learning using a fully-convolutional neural network. Through gradual downsampling and upsampling, the network is able to consider large regions in the image space with a receptive field that covers even the largest pigs in the dataset.

Results demonstrate that the method is capable of achieving over 99% precision and over 95% recall at the task of instance detection when the network is tested and trained under the same environmental conditions. When testing on environments and lighting conditions that the network had not been trained to handle, the results drop significantly to 91% precision and 67% recall. These results can be interpreted in one of three ways: (1) networks should be fine-tuned to handle new environments, (2) a larger number and variety of images should be included in the dataset, or (3) the design and/or training methodology should be revised to improve the robustness to environmental variability. As the dataset and the number of environments grows, eventually there might be enough variety such that new environments add little to the network's ability to handle novel presentations. Regarding the third interpretation, while significant augmentations were applied to the input and output images during training, it is impossible for spatial transformations to mimic variations in lighting conditions. Therefore, a new set of non-uniform color-space transformations may provide a solution that improves the robustness of the trained network.

By introducing a new dataset and providing an accompanying method for instance-level detection, we hope that this work inspires other researchers to introduce competing methods and perform objective evaluations on the dataset.

Author Contributions: E.P., M.M., L.C.P., T.S. and B.M. conceived and designed the experiments and assisted in data acquisition. E.P. designed and implemented the neural network algorithm and performed the formal analysis. E.P., M.M., L.C.P., T.S. and B.M. wrote the paper.

Acknowledgments: We would like to thank Lucas Fricke and Union Farms for their cooperation and allowing us to capture data in their swine facility.

References

1. Matthews, S.G.; Miller, A.L.; Clapp, J.; Plötz, T.; Kyriazakis, I. Early detection of health and welfare compromises through automated detection of behavioural changes in pigs. *Vet. J.* **2016**, *217*, 43–51. [CrossRef]
2. Wedin, M.; Baxter, E.M.; Jack, M.; Futro, A.; D'Eath, R.B. Early indicators of tail biting outbreaks in pigs. *Appl. Anim. Behav. Sci.* **2018**, *208*, 7–13. [CrossRef]
3. Burgunder, J.; Petrželková, K.J.; Modrý, D.; Kato, A.; MacIntosh, A.J. Fractal measures in activity patterns: Do gastrointestinal parasites affect the complexity of sheep behaviour? *Appl. Anim. Behav. Sci.* **2018**, *205*, 44–53. [CrossRef]
4. PIC North America. Standard Animal Care: Daily Routines. In *Wean to Finish Manual*; PIC: Hendersonville, TN, USA, 2014; pp. 23–24.
5. Tuyttens, F.; de Graaf, S.; Heerkens, J.L.; Jacobs, L.; Nalon, E.; Ott, S.; Stadig, L.; Van Laer, E.; Ampe, B. Observer bias in animal behaviour research: can we believe what we score, if we score what we believe? *Anim. Behav.* **2014**, *90*, 273–280. [CrossRef]
6. Wathes, C.M.; Kristensen, H.H.; Aerts, J.M.; Berckmans, D. Is precision livestock farming an engineer's daydream or nightmare, an animal's friend or foe, and a farmer's panacea or pitfall? *Comput. Electron. Agric.* **2008**, *64*, 2–10. [CrossRef]
7. Banhazi, T.M.; Lehr, H.; Black, J.; Crabtree, H.; Schofield, P.; Tscharke, M.; Berckmans, D. Precision livestock farming: An international review of scientific and commercial aspects. *Int. J. Agric. Biol. Eng.* **2012**, *5*, 1–9.
8. Tullo, E.; Fontana, I.; Guarino, M. Precision livestock farming: An overview of image and sound labelling. In *European Conference on Precision Livestock Farming 2013:(PLF) EC-PLF*; KU Leuven: Leuven, Belgium, 2013; pp. 30–38.
9. Kim, S.H.; Kim, D.H.; Park, H.D. Animal situation tracking service using RFID, GPS, and sensors. In Proceedings of the 2010 IEEE Second International Conference on Computer and Network Technology (ICCNT), Bangkok, Thailand, 23–25 April 2010; pp. 153–156.
10. Stukenborg, A.; Traulsen, I.; Puppe, B.; Presuhn, U.; Krieter, J. Agonistic behaviour after mixing in pigs under commercial farm conditions. *Appl. Anim. Behav. Sci.* **2011**, *129*, 28–35. [CrossRef]

11. Porto, S.; Arcidiacono, C.; Giummarra, A.; Anguzza, U.; Cascone, G. Localisation and identification performances of a real-time location system based on ultra wide band technology for monitoring and tracking dairy cow behaviour in a semi-open free-stall barn. *Comput. Electron. Agric.* **2014**, *108*, 221–229. [CrossRef]
12. Giancola, G.; Blazevic, L.; Bucaille, I.; De Nardis, L.; Di Benedetto, M.G.; Durand, Y.; Froc, G.; Cuezva, B.M.; Pierrot, J.B.; Pirinen, P.; et al. UWB MAC and network solutions for low data rate with location and tracking applications. In Proceedings of the 2005 IEEE International Conference on Ultra-Wideband, Zurich, Switzerland, 5–8 September 2005; pp. 758–763.
13. Clark, P.E.; Johnson, D.E.; Kniep, M.A.; Jermann, P.; Huttash, B.; Wood, A.; Johnson, M.; McGillivan, C.; Titus, K. An advanced, low-cost, GPS-based animal tracking system. *Rangel. Ecol. Manag.* **2006**, *59*, 334–340. [CrossRef]
14. Schwager, M.; Anderson, D.M.; Butler, Z.; Rus, D. Robust classification of animal tracking data. *Comput. Electron. Agric.* **2007**, *56*, 46–59. [CrossRef]
15. Taylor, K. Cattle health monitoring using wireless sensor networks. In Proceedings of the Communication and Computer Networks Conference, Cambridge, MA, USA, 8–10 November 2004.
16. Ruiz-Garcia, L.; Lunadei, L.; Barreiro, P.; Robla, I. A Review of Wireless Sensor Technologies and Applications in Agriculture and Food Industry: State of the Art and Current Trends. *Sensors* **2009**, *9*, 4728–4750. [CrossRef]
17. Escalante, H.J.; Rodriguez, S.V.; Cordero, J.; Kristensen, A.R.; Cornou, C. Sow-activity classification from acceleration patterns: A machine learning approach. *Comput. Electron. Agric.* **2013**, *93*, 17–26. [CrossRef]
18. Alvarenga, F.A.P.; Borges, I.; Palkovič, L.; Rodina, J.; Oddy, V.H.; Dobos, R.C. Using a three-axis accelerometer to identify and classify sheep behaviour at pasture. *Appl. Anim. Behav. Sci.* **2016**, *181*, 91–99. [CrossRef]
19. Voulodimos, A.S.; Patrikakis, C.Z.; Sideridis, A.B.; Ntafis, V.A.; Xylouri, E.M. A complete farm management system based on animal identification using RFID technology. *Comput. Electron. Agric.* **2010**, *70*, 380–388. [CrossRef]
20. Feng, J.; Fu, Z.; Wang, Z.; Xu, M.; Zhang, X. Development and evaluation on a RFID-based traceability system for cattle/beef quality safety in China. *Food Control* **2013**, *31*, 314–325. [CrossRef]
21. Floyd, R.E. RFID in animal-tracking applications. *IEEE Potentials* **2015**, *34*, 32–33. [CrossRef]
22. Mittek, M.; Psota, E.T.; Pérez, L.C.; Schmidt, T.; Mote, B. Health Monitoring of Group-Housed Pigs using Depth-Enabled Multi-Object Tracking. In Proceedings of the International Conference on Pattern Recognition, Workshop on Visual observation and analysis of Vertebrate and Insect Behavior, Cancun, Mexico, 4 December 2016; pp. 9–12.
23. Mittek, M.; Psota, E.T.; Carlson, J.D.; Pérez, L.C.; Schmidt, T.; Mote, B. Tracking of group-housed pigs using multi-ellipsoid expectation maximisation. *IET Comput. Vis.* **2017**, *12*, 121–128. [CrossRef]
24. Neethirajan, S. Recent advances in wearable sensors for animal health management. *Sens. Bio-Sens. Res.* **2017**, *12*, 15–29. [CrossRef]
25. Schleppe, J.B.; Lachapelle, G.; Booker, C.W.; Pittman, T. Challenges in the design of a GNSS ear tag for feedlot cattle. *Comput. Electron. Agric.* **2010**, *70*, 84–95. [CrossRef]
26. Ardö, H.; Guzhva, O.; Nilsson, M.; Herlin, A.H. Convolutional neural network-based cow interaction watchdog. *IET Comput. Vis.* **2017**, *12*, 171–177. [CrossRef]
27. Ju, M.; Choi, Y.; Seo, J.; Sa, J.; Lee, S.; Chung, Y.; Park, D. A Kinect-Based Segmentation of Touching-Pigs for Real-Time Monitoring. *Sensors* **2018**, *18*, 1746. [CrossRef]
28. Krizhevsky, A.; Sutskever, I.; Hinton, G.E. Imagenet classification with deep convolutional neural networks. In Proceedings of the Advances in Neural Information Processing Systems, Lake Tahoe, NV, USA, 3–6 December 2012; pp. 1097–1105.
29. LeCun, Y.; Bottou, L.; Bengio, Y.; Haffner, P. Gradient-based learning applied to document recognition. *Proc. IEEE* **1998**, *86*, 2278–2324. [CrossRef]
30. Jia, Y.; Shelhamer, E.; Donahue, J.; Karayev, S.; Long, J.; Girshick, R.; Guadarrama, S.; Darrell, T. Caffe: Convolutional architecture for fast feature embedding. In Proceedings of the 22nd ACM International Conference on Multimedia, Orlando, FL, USA, 3–7 November 2014; pp. 675–678.
31. Kirk, D. NVIDIA CUDA software and GPU parallel computing architecture. In Proceedings of the 6th international symposium on Memory management, Montreal, QC, Canada, 21–22 October, 2007; Volume 7, pp. 103–104.

32. Everingham, M.; Eslami, S.A.; Van Gool, L.; Williams, C.K.; Winn, J.; Zisserman, A. The pascal visual object classes challenge: A retrospective. *Int. J. Comput. Vis.* **2015**, *111*, 98–136. [CrossRef]
33. Lin, T.Y.; Maire, M.; Belongie, S.; Hays, J.; Perona, P.; Ramanan, D.; Dollár, P.; Zitnick, C.L. Microsoft coco: Common objects in context. In Proceedings of the European Conference on Computer Vision, Zurich, Switzerland, 6–12 September 2014; pp. 740–755.
34. Cordts, M.; Omran, M.; Ramos, S.; Rehfeld, T.; Enzweiler, M.; Benenson, R.; Franke, U.; Roth, S.; Schiele, B. The cityscapes dataset for semantic urban scene understanding. In Proceedings of the IEEE Conference on Computer Vision and Pattern Recognition, Las Vegas, NV, USA, 27–30 June 2016; pp. 3213–3223.
35. Redmon, J.; Divvala, S.; Girshick, R.; Farhadi, A. You only look once: Unified, real-time object detection. In Proceedings of the IEEE Conference on Computer Vision and Pattern Recognition, Las Vegas, NV, USA, 27–30 June 2016; pp. 779–788.
36. Redmon, J.; Farhadi, A. YOLO9000: Better, Faster, Stronger. In Proceedings of the IEEE Conference on Computer Vision and Pattern Recognition, Honolulu, HI, USA, 21–26 July 2017; pp. 7263–7271.
37. Liu, W.; Anguelov, D.; Erhan, D.; Szegedy, C.; Reed, S.; Fu, C.Y.; Berg, A.C. Ssd: Single shot multibox detector. In Proceedings of the European Conference on Computer Vision, Amsterdam, The Netherlands, 8–16 October 2016; Springer: Berlin/Heidelberg, Germany, 2016; pp. 21–37.
38. Girshick, R. Fast r-cnn. In Proceedings of the IEEE International Conference on Computer Vision, Santiago, Chile, 7–13 December 2015; pp. 1440–1448.
39. Ren, S.; He, K.; Girshick, R.; Sun, J. Faster r-cnn: Towards real-time object detection with region proposal networks. In Proceedings of the 28th International Conference on Neural Information Processing Systems, Montreal, QC, Canada, 7–12 December 2015; pp. 91–99.
40. Di Stefano, L.; Bulgarelli, A. A simple and efficient connected components labeling algorithm. In Proceedings of the 10th International Conference on Image Analysis and Processing, Venice, Italy, 27–29 September 1999; p. 322.
41. Otsu, N. A threshold selection method from gray-level histograms. *IEEE Trans. Syst. Man Cybern.* **1979**, *9*, 62–66. [CrossRef]
42. Kashiha, M.A.; Bahr, C.; Ott, S.; Moons, C.P.; Niewold, T.A.; Tuyttens, F.; Berckmans, D. Automatic monitoring of pig locomotion using image analysis. *Livest. Sci.* **2014**, *159*, 141–148. [CrossRef]
43. Nasirahmadi, A.; Richter, U.; Hensel, O.; Edwards, S.; Sturm, B. Using machine vision for investigation of changes in pig group lying patterns. *Comput. Electron. Agric.* **2015**, *119*, 184–190. [CrossRef]
44. Ahrendt, P.; Gregersen, T.; Karstoft, H. Development of a real-time computer vision system for tracking loose-housed pigs. *Comput. Electron. Agric.* **2011**, *76*, 169–174. [CrossRef]
45. Nilsson, M.; Ardö, H.; Åström, K.; Herlin, A.; Bergsten, C.; Guzhva, O. Learning based image segmentation of pigs in a pen. In Proceedings of the Visual Observation and Analysis of Vertebrate and Insect Behavior, Stockholm, Sweden, 24–28 August 2014; pp. 1–4.
46. Kongsro, J. Estimation of pig weight using a Microsoft Kinect prototype imaging system. *Comput. Electron. Agric.* **2014**, *109*, 32–35. [CrossRef]
47. Zhu, Q.; Ren, J.; Barclay, D.; McCormack, S.; Thomson, W. Automatic Animal Detection from Kinect Sensed Images for Livestock Monitoring and Assessment. In Proceedings of the 2015 IEEE International Conference on Computer and Information Technology, Ubiquitous Computing and Communications, Dependable, Autonomic and Secure Computing, Pervasive Intelligence and Computing, Liverpool, UK, 26–28 October 2015; pp. 1154–1157.
48. Stavrakakis, S.; Li, W.; Guy, J.H.; Morgan, G.; Ushaw, G.; Johnson, G.R.; Edwards, S.A. Validity of the Microsoft Kinect sensor for assessment of normal walking patterns in pigs. *Comput. Electron. Agric.* **2015**, *117*, 1–7. [CrossRef]
49. Lee, J.; Jin, L.; Park, D.; Chung, Y. Automatic Recognition of Aggressive Behavior in Pigs Using a Kinect Depth Sensor. *Sensors* **2016**, *16*, 631. [CrossRef]
50. Lao, F.; Brown-Brandl, T.; Stinn, J.; Liu, K.; Teng, G.; Xin, H. Automatic recognition of lactating sow behaviors through depth image processing. *Comput. Electron. Agric.* **2016**, *125*, 56–62. [CrossRef]
51. Choi, J.; Lee, L.; Chung, Y.; Park, D. Individual Pig Detection Using Kinect Depth Information. *Kips Trans. Comput. Commun. Syst.* **2016**, *5*, 319–326. [CrossRef]
52. Kim, J.; Chung, Y.; Choi, Y.; Sa, J.; Kim, H.; Chung, Y.; Park, D.; Kim, H. Depth-Based Detection of Standing-Pigs in Moving Noise Environments. *Sensors* **2017**, *17*, 2757. [CrossRef]

53. Pezzuolo, A.; Guarino, M.; Sartori, L.; González, L.A.; Marinello, F. On-barn pig weight estimation based on body measurements by a Kinect v1 depth camera. *Comput. Electron. Agric.* **2018**, *148*, 29–36. [CrossRef]
54. Fernandes, A.; Dórea, J.; Fitzgerald, R.; Herring, W.; Rosa, G. A novel automated system to acquire biometric and morphological measurements, and predict body weight of pigs via 3D computer vision. *J. Anim. Sci.* **2018**, *97*, 496–508. [CrossRef]
55. Matthews, S.G.; Miller, A.L.; PlÖtz, T.; Kyriazakis, I. Automated tracking to measure behavioural changes in pigs for health and welfare monitoring. *Sci. Rep.* **2017**, *7*, 17582. [CrossRef]
56. Girshick, R.; Donahue, J.; Darrell, T.; Malik, J. Rich feature hierarchies for accurate object detection and semantic segmentation. In Proceedings of the IEEE Conference on Computer Vision and Pattern Recognition, Portland, OR, USA, 23–28 June 2014; pp. 580–587.
57. He, K.; Gkioxari, G.; Dollár, P.; Girshick, R. Mask r-cnn. In Proceedings of the 2017 IEEE International Conference on Computer Vision (ICCV), Venice, Italy, 22–29 Ocotober 2017; pp. 2980–2988.
58. Long, J.; Shelhamer, E.; Darrell, T. Fully convolutional networks for semantic segmentation. In Proceedings of the IEEE Conference on Computer Vision and Pattern Recognition, Boston, MA, USA, 7–12 June 2015; pp. 3431–3440.
59. Cao, Z.; Simon, T.; Wei, S.E.; Sheikh, Y. Realtime Multi-Person 2D Pose Estimation using Part Affinity Fields. In Proceedings of the IEEE Conference on Computer Vision and Pattern Recognition, Honolulu, HI, USA, 21–26 July 2017.
60. Papandreou, G.; Zhu, T.; Chen, L.C.; Gidaris, S.; Tompson, J.; Murphy, K. PersonLab: Person Pose Estimation and Instance Segmentation with a Bottom-Up, Part-Based, Geometric Embedding Model. In Proceedings of the European Conference on Computer Vision, Munich, Germany, 8–14 September 2018.
61. Newell, A.; Yang, K.; Deng, J. Stacked hourglass networks for human pose estimation. In Proceedings of the European Conference on Computer Vision, Zurich, Switzerland, 6–12 September 2016; Springer: Berlin/Heidelberg, Germany, 2016; pp. 483–499.
62. Badrinarayanan, V.; Kendall, A.; Cipolla, R. SegNet: A Deep Convolutional Encoder-Decoder Architecture for Image Segmentation. *IEEE Trans. Pattern Anal. Mach. Intell.* **2017**, *39*, 2481–2495. [CrossRef]
63. Huang, G.; Liu, Z.; Van Der Maaten, L.; Weinberger, K.Q. Densely connected convolutional networks. In Proceedings of the CVPR, Honolulu, HI, USA, 21–26 July 2017; Volume 1, p. 3.
64. Ronneberger, O.; Fischer, P.; Brox, T. U-net: Convolutional networks for biomedical image segmentation. In Proceedings of the International Conference on Medical Image Computing and Computer-Assisted Intervention, Munich, Germany, 5–9 Ocotber 2015; Springer: Berlin/Heidelberg, Germany, 2015; pp. 234–241.
65. Luo, W.; Li, Y.; Urtasun, R.; Zemel, R. Understanding the effective receptive field in deep convolutional neural networks. In Proceedings of the 30th International Conference on Neural Information Processing Systems, Barcelona, Spain, 5–10 December 2016; pp. 4905–4913.
66. *MATLAB*, version 9.5.0 (R2018b); The MathWorks Inc.: Natick, MA, USA, 2018.
67. Deng, J.; Dong, W.; Socher, R.; Li, L.J.; Li, K.; Fei-Fei, L. Imagenet: A large-scale hierarchical image database. In Proceedings of the 2009 IEEE Conference on Computer Vision and Pattern Recognition (CVPR 2009), Miami, FL, USA, 20–25 June 2009; pp. 248–255.
68. Chen, L.C.; Papandreou, G.; Kokkinos, I.; Murphy, K.; Yuille, A.L. Deeplab: Semantic image segmentation with deep convolutional nets, atrous convolution, and fully connected crfs. *IEEE Trans. Pattern Anal. Mach. Intell.* **2018**, *40*, 834–848. [CrossRef]
69. Chen, L.C.; Zhu, Y.; Papandreou, G.; Schroff, F.; Adam, H. Encoder-Decoder with Atrous Separable Convolution for Semantic Image Segmentation. In Proceedings of the ECCV, Munich, Germany, 8–14 September 2018.
70. Luo, W.; Xing, J.; Milan, A.; Zhang, X.; Liu, W.; Zhao, X.; Kim, T.K. Multiple object tracking: A literature review. *arXiv* **2014**, arXiv:1409.7618.

5

Early Detection of *Zymoseptoria tritici* in Winter Wheat by Infrared Thermography

Yuxuan Wang [1], Shamaila Zia-Khan [1,*], Sebastian Owusu-Adu [1], Thomas Miedaner [2] and Joachim Müller [1]

1 Tropics and Subtropics Group (440e), Institute of Agricultural Engineering, Universität Hohenheim, Garbenstraße 9, 70593 Stuttgart, Germany

2 State Plant Breeding Institute (720), Universität Hohenheim, Fruwirthstr. 21, 70593 Stuttgart, Germany

* Correspondence: shamaila.ziakhan@uni-hohenheim.de

Abstract: The use of thermography as a means of crop water status estimation is based on the assumption that a sufficient amount of soil moisture enables plants to transpire at potential rates resulting in cooler canopy than the surrounding air temperature. The same principle is applied in this study where the crop transpiration changes occur because of the fungal infection. The field experiment was conducted where 25 wheat genotypes were infected with *Zymoseptoria tritici*. The focus of this study was to predict the onset of the disease before the visual symptoms appeared on the plants. The results showed an early significant increase in the maximum temperature difference within the canopy from 1 to 7 days after inoculation (DAI). Biotic stress associated with increasing level of disease can be seen in the increasing average canopy temperature (ACT) and maximum temperature difference (MTD) and decreasing canopy temperature depression (CTD). However, only MTD ($p \leq 0.01$) and CTD ($p \leq 0.05$) parameters were significantly related to the disease level and can be used to predict the onset of fungal infection on wheat. The potential of thermography as a non-invasive high throughput phenotyping technique for early fungal disease detection in wheat was evident in this study.

Keywords: IR imaging; canopy temperature; maximum temperature difference; fungal infection; wheat genotypes

1. Introduction

Wheat is a staple food for about two billion people in the world [1]. Yield losses in agricultural production are caused by several biotic and abiotic stresses. Abiotic stresses such as frost, salinity, heat, and drought contribute to about more than 50% yield loss and are identified as the primary cause of yield loss worldwide [2–4]. Biotic factors such as pests and diseases also cause substantial damage to the crops. For example, in wheat, yield loss caused by diseases varies between 14 and 27%, depending on the different diseases and region [5]. High yield losses are mostly observed in susceptible genotypes, which usually display high disease severity [6]. The most common fungal diseases in temperate wheat growing regions are leaf rust (*Puccinia triticina*), stripe or yellow rust (*P. striiformis*), Fusarium head blight (*Fusarium* spp.), powdery mildew (*Blumeria graminis*), and *Septoria tritici* blotch (*Zymoseptoria tritici*) [7].

The symptoms of fungal infection usually appear after a certain period of time depending on weather conditions. Early detection and diagnosis of plant pathogens can provide adequate information to predict the onset of disease and adequate measures can be taken to protect the crop before the disease is widespread [8]. There is a need for an advanced technique for rapid, accurate, and reliable detection of plant diseases especially at the time when symptoms are not yet visible on the crop [9]. Though studies have been conducted to apply imaging technologies such as fluorescence imaging [10], multispectral or hyperspectral imaging [11,12], and nuclear magnetic resonance (NMR) spectroscopy [13] to detect fungal diseases. However, these methods work only on a certain wavelength, which are plant and disease specific [9]. For example, Bauriegel et al. [14] detected head blight in wheat using hyper-spectral imaging in the wavelength range of 400–1000 nm. In contrast, infrared thermography works in a wide range of wavelength and can be applied on large number of plant types and varieties. Application of infrared thermography to determine crop water status for irrigation scheduling has been established [15,16]. Crop water stress index (CWSI) is widely used to quantify the plant water stress based on the principle that, under water stress, plants close their stomata and this leads to the increase in leaf temperature [17,18]. The same principle is applied in this study, i.e., the plant transpiration and photosynthetic activity is influenced by *Z. tritici* infection and this results in the change in canopy temperature which can be detected by thermography. The primary host penetration of *Z. tritici* occurs by penetrating the stomata 24 to 48 hours after contact with the leaf surface. About 12 days later a rapid change from the biotrophic growth stage to necrotrophic growth takes place with appearance of lesions on the leaves [19]. In cooler climates, like Germany, the occurrence of the first symptoms can be as late as 21 days after infection. This long latency period makes it difficult for the farmer to decide the correct fungicide application date. The chlorotic and necrotic symptoms of *Z. tritici* can lead to a decrease in leaf photosynthesis [20].

Recently, a number of studies have been conducted to determine the suitability of thermal imaging to detect biotic stress, both in the field and greenhouse: for example, the effect of downy mildew on transpiration of cucumber [21] and the effects of *Z. tritici* on wheat leaf gas exchange [20]. However, further studies need to be carried out in order to assess the potential of using thermography for other crops and for different locations with varying environmental conditions. The objective of this study was to detect fungal colonization at canopy level in field conditions at an early stage. The sensitivity of the methodology and the technology will be tested on different wheat genotypes. Finally, the earliest possible time of *Z. tritici* attack will be determined.

2. Material and Methods

2.1. Experimental Setup and Wheat Varieties

Twenty-five wheat varieties were planted in the field at Stuttgart-Hohenheim in a split-plot design with three replications. The main plots consisted of the two treatments (non-inoculated, artificially inoculated by *Z. tritici*) that were arranged in a complete randomised block design, the subplots of the 25 wheat varieties were randomized according to a 5 × 5 lattice design. Each plot was sown by 60 seeds in two rows 1.2 m long with a distance of 0.21 m between rows resulting in an area of 0.5 m^2 per variety. Each main plot was separated by a strip of tall-growing winter triticale ("Cando") to prevent the spreading of the pathogen by spray drift during inoculation or secondary spore production. Likewise, a strip of winter rye around the experimental field as a border was planted.

The seeds were sown on 6 October 2011 by using a small tractor with a planter. No irrigation was applied throughout the growing period, as precipitation was sufficient to meet the crop water requirement. The selection of wheat varieties was done by the State Plant Breeding Institute, University of Hohenheim. In Table 1 the varieties were marked according their resistance against *Z. tritici* as classified by the German Federal Plant Variety Office [22].

Table 1. Tested wheat varieties and their susceptibility to *Z. tritici* as classified by the German Federal Plant Variety Office (Bundessortenamt, 2012); 1 = low susceptibility, 9 = high susceptibility, n.i. = no information.

Variety	Country of Origin	Susceptibility
Akratos	G	5
Apache	F	n.i.
Arina	CH	n.i.
Batis	G	4
Biscay	G	7
Bussard	G	7
Cubus	G	6
Dream	G	n.i.
Egoist	G	4
F201-R	ROM	n.i.
Florett	G	n.i.
History	G	n.i.
Impression	G	4
Julius	G	3
MES130	C	n.i.
Meteor	G	4
Naturastar	G	6
Nelson	G	3
Pamier	G	3
Rubens	G	n.i.
Sailor	G	5
Skalmeje	G	4
Solitär	G	n.i.
Toras	G	4
Tuareg	G	5

C = China, CH = Switzerland, F = France, G = Germany, ROM = Romania.

2.2. *Inoculation with Z. tritici*

Fungal spores were collected from infected wheat leaves, which were shaken before inoculation to remove the old spores. To prepare a medium for inoculation, *Z. tritici* was cultivated on culture medium made of 4 g yeast extract, 4 g malt extract (powdered), 4 g of glucose, 15 g of agar-agar, and 1000 mL distilled water. After autoclaving at 120 °C for 20 min the agar was poured into petri dishes and cooled down. For propagation, *Z. tritici* was placed on agar and exposed to white- and UV-light (16 h daytime at 18 °C and 8 h night time at 12 °C) for 3–5 days of incubation. For the mass reproduction of *Z. tritici* 150 mL fluid culture, made of 4 g yeast extract, 4 g malt extract, 4 g glucose, and 1000 mL distilled water, was added into a 300 mL Erlenmeyer flask and closed with an aluminium foil at 120 °C for 30 min. When the medium had cooled after autoclaving, *Z. tritici* was taken by an inoculation loop from the plate on which the spores were well developed and brought into the Erlenmeyer flask. The flask was placed on a shaker for 5 days under white- and UV-light in the same condition as mentioned above. The spores were counted and adjusted to a concentration of 5×10^6 spores/mL and later used for inoculation. Finally, inoculation of *Z. tritici* was done for all varieties on 21 May 2012 in the field. Spores were than sprayed on the fully developed flag leaves of the wheat plants by using a knapsack sprayer with a constant air pressure.

2.3. *Acquiring Thermal Images*

An infrared camera (VarioCAM®, InfraTec GmbH, Dresden, Germany) was used for thermal imagining. Images were acquired from 17 May until 28 June 2012 between 13:00 and 14:00 to minimize the influence of changing solar angle. Images were taken only on the sunny days when transpiration rate of the crop was high. Each thermal image comprised two wheat varieties as shown in Figure 1A,B. The thermal images were analysed using thermography software IRBIS (InfraTec GmbH, Dresden, Germany) and the chosen value for emissivity was 0.95. To differentiate between canopy and soil surface, an upper temperature threshold was defined and pixels with higher temperature were excluded by image processing, Figure 1C. A polygon was manually drawn along the canopy area for each variety representing the crop temperature, Figure 1D. The average crop temperature (ACT) was calculated from the pixels within the canopy polygon and calculated as

$$ACT = \frac{\sum T_{Pixel}}{n_{Pixel}}$$

The maximum temperature difference (MTD) within the canopy polygon for each variety was calculated as

$$MTD = T_{Pixel,\ max} - T_{Pixel,\ min}$$

The canopy temperature difference (CTD) was defined as

$$CTD = T_{air} - ACT$$

where, air temperature T_{air} was measured by the weather station installed at the station.

Figure 1. Canopy temperature measurement, (**A**) visual image, (**B**) thermal image, (**C**) histogram with upper temperature threshold, (**D**) polygons for calculating crop temperature.

2.4. Visual Scoring

The visual scoring was done by an experienced staff on plot basis. *Septoria tritici* blotch was scored according to the severity of symptoms observed on the flag leaf and ranged between 0% and 100%. Visual scoring was done on 13 June (23 DAI), 18 June (28 DAI), 21 June (31 DAI), 25 June (35 DAI), 28 June (38 DAI), and a last time at 3 July (43 DAI).

2.5. Data Analysis

SPSS version 16.0 for Windows (SPSS Inc., Chicago, IL, USA) was used for analysis of variance (ANOVA) to evaluate the effect of *Z. tritici* infection on ACT, MTD, and CTD. A linear regression analysis was performed between the disease level and ACT, MTD, and CTD for the last day of the experiment (DAI 38) across all varieties. Furthermore, a *t*-test for mean comparison was conducted individually for each variety to determine the date when the first significant difference of ACT, MTD, and CTD to control appeared at a least significant level of 0.05.

3. Results

3.1. Disease Level and Temperature Effects across 25 Wheat Varieties

Figure 2 shows summarized results across all varieties at the last day of the experiment (38 DAI). The percentage of *Z. tritici* disease levels on inoculated plants was higher than the control treatment. At the end of the experiment, all the wheat varieties were *Z. tritici* infested irrespective of their susceptibility level to the disease. Variety "Solitär" proved to be the most *Z. tritici* resistant showing only 18% of disease level whereas, variety "MES130" showed the highest disease level of 93%. ACT of the *Z. tritici* inoculated group showed a 2.3 °C higher value than the control group, which was highly significant ($p \leq 0.001$). MTD of *Z. tritici* treatment was 1.4 °C higher than that of the control group ($p \leq 0.001$). CTD from *Z. tritici* treatment was below zero for all the varieties except of "Solitär" and "Toras", showing values of 0.27 °C and 0.22 °C. In contrast, CTD for the control treatment was always above zero. The difference of means between treatment and control was significant ($p \leq 0.001$).

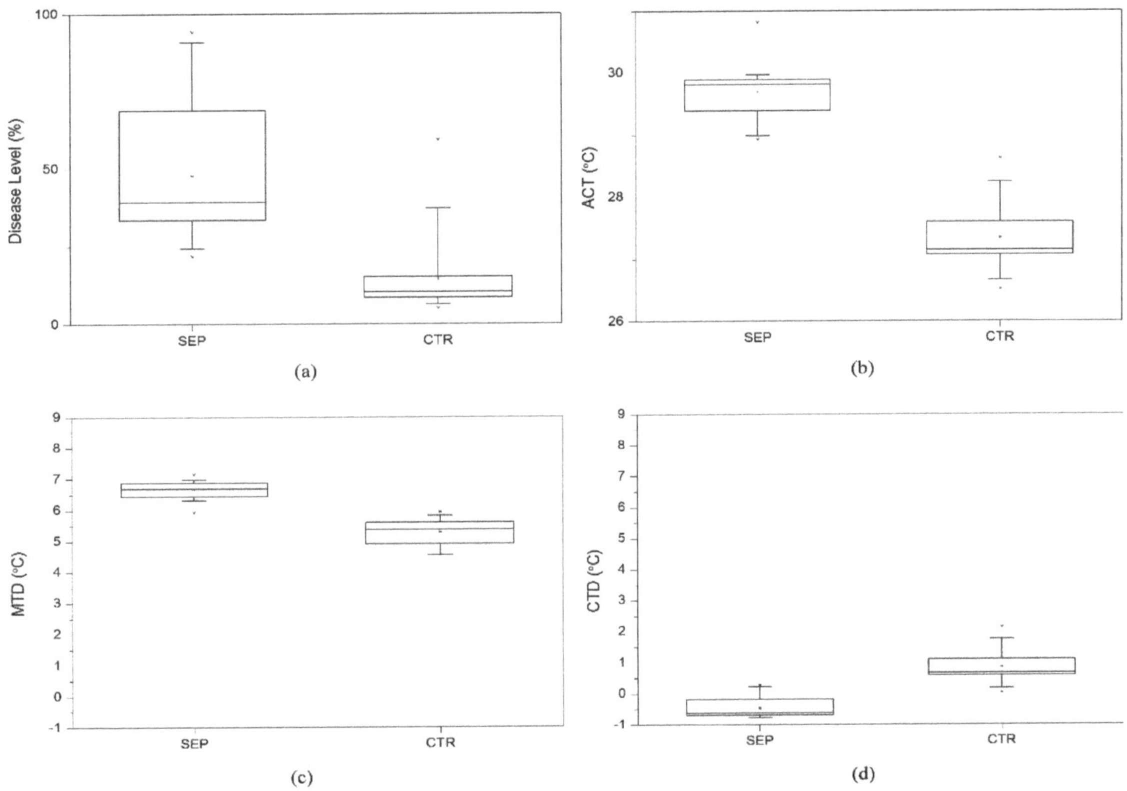

Figure 2. Crop parameters at the end of the experiment (38 DAI), summarized across 25 wheat varieties: (**a**) disease level, (**b**) average canopy temperature (ACT), (**c**) maximum temperature difference (MTD), (**d**) canopy temperature difference (CTD); SEP = *Z. tritici* treatment and CTR = control.

3.2. Temperature Effects Related to Disease Level for 25 Wheat Varieties

ACT, CTD, and MTD on the last day of the experiment were plotted against disease level of the varieties as shown in Figure 3. It can be observed that ACT and MTD increased as the disease level increased, whereas CTD decreased with the disease severity. A stepwise regression analysis showed that CTD and MTD are the only parameters which can be used to explain the diseases level in all 25 wheat varieties. The significance level of MTD was however higher ($p \leq 0.01$) compared to that of CTD ($p \leq 0.05$). The estimation of the disease level of *Z. tritici* based on MTD and CTD is expressed by following equation (Adj.R^2 = 0.7):

$$P_{\text{infested area}}\,(\%) = -183.426 + 35.485 \times \text{MTD} - 17.192 \times \text{CTD}$$

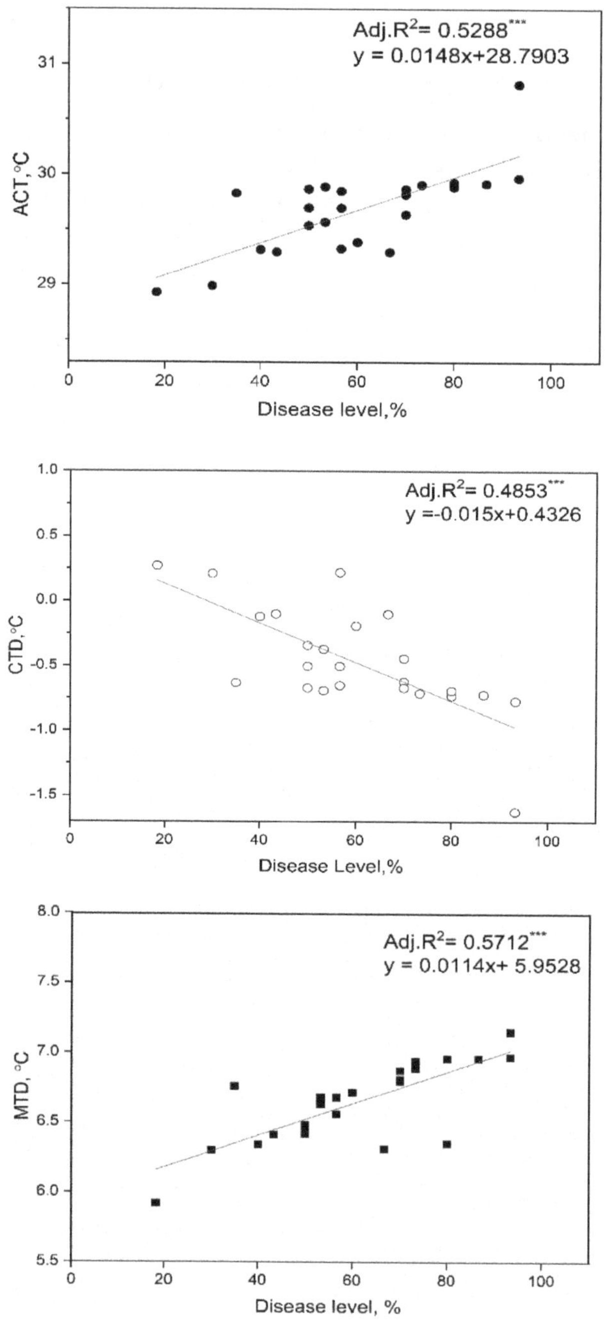

Figure 3. Regression analyses between Average Canopy Temperature (ACT), Canopy Temperature Depression (CTD), and Maximum Temperature Depression (MTD) and disease level at the end of the experiment (DAI 38) for 25 wheat varieties.

3.3. *Temperature Effect on Wheat Varieties*

As shown in Table 2, ANOVA test was conducted to compare the difference between the control and *Z. tritici* treatments in terms of CTD and MTD. It was observed that throughout the experiment the *Z. tritici* treatment had lower CTD than the control group and 12 varieties out of 25 showed significant difference at 0.05 levels. Before inoculation MTD of the control treatment was lower than the *Z. tritici* treatment however, these differences were not statistically significant. Later, after inoculation from 2 DAI till 28 DAI significant differences were observed in most of the varieties. An early significant difference in CTD and MTD were observed, this shows that these parameters can describe the health of the plant long before the naked eye observed.

Table 2. ANOVA test between non-inoculated and *Z. tritici* treatment on the last day of the field experiment (38 DAI), difference in canopy temperature difference (CTD), maximum temperature difference (MTD) and the maximum difference from control treatment (DAI = days after inoculation).

Variety	First Visual Symptoms	ANOVA Test		Maximum Difference from Control				First Significant Difference from Control			
	DAI	CTD	MTD	ΔCTD_{max}	DAI	ΔMTD_{max}	DAI	ΔCTD	DAI	ΔMTD	DAI
Akratos	23	-	**	2.74	7	5.42	38	-	-	1.06	3
Apache	23	**	-	3.23	9	1.06	5	2.06	4	-	-
Arina	28	**	***	3.28	5	2.33	28	2.87	7	1.92	7
Batis	23	**	***	4.6	7	2.12	28	4.60	7	0.91	5
Biscay	28	**	**	4.09	9	1.64	38	3.82	8	0.51	4
Bussard	23	**	***	4.31	28	2.35	38	2.61	7	0.90	5
Cubus	23	**	**	3.74	12	1.35	38	1.91	5	0.78	4
Dream	23	-	***	4.88	3	2.05	29	-	-	0.81	4
Egoist	23	***	***	3.63	5	2.21	28	2.98	4	0.45	3
F201-R	23	***	*	4.25	7	1.51	38	2.34	4	0.29	3
Florett	23	**	-	3.20	9	1.06	5	1.70	4	-	-
History	23	**	*	3.45	9	1.69	38	1.75	5	1.04	5
Impression	23	**	-	3.50	8	0.98	26	1.52	3	-	-
Julius	28	-	**	1.69	9	1.43	8	-	-	1.09	4
MES130	23	***	**	3.45	38	2.12	38	1.83	4	0.86	7
Meteor	23	**	**	3.41	3	1.91	38	2.24	1	1.04	5
Naturastar	23	**	***	4.03	7	2.72	3	2.22	4	1.65	4
Nelson	23	***	***	4.54	9	2.57	12	3.51	4	1.17	4
Pamier	23	**	**	3.35	8	1.86	5	2.27	4	1.86	5
Rubens	23	*	*	3.06	29	1.7	29	1.34	1	0.84	3
Sailor	23	***	**	3.78	9	1.15	9	3.77	7	0.95	5
Skalmeje	28	-	**	2.86	26	1.49	1	-	-	1.23	3
Solitär	28	*	-	2.67	26	1.28	38	2.03	4	-	-
Toras	23	*	**	3.24	3	1.91	12	2.50	8	0.90	4
Tuareg	23	-	**	2.76	38	2.01	7	-	-	1.87	5

- Not significant, * Significant at 0.1 levels, ** Significant at 0.05 levels, *** Significant at 0.01 levels.

3.4. *Temporal Development of Disease Level and Temperature Effects*

The development of CTD, MTD, and disease level is displayed in Figure 4. Two varieties were selected, viz. the one with the highest and the one with the lowest difference from the control during the experimental study. The variety "Dream" had the highest ΔCTD value with a twice-higher disease level (20%) at the end of the experiment than that of Julius (10%), which had the lowest value in ΔCTD. The CTD of the control treatment was clearly higher than that of the *Z. tritici* treatment throughout the experimental study of both varieties, i.e., "Dream" and "Julius".

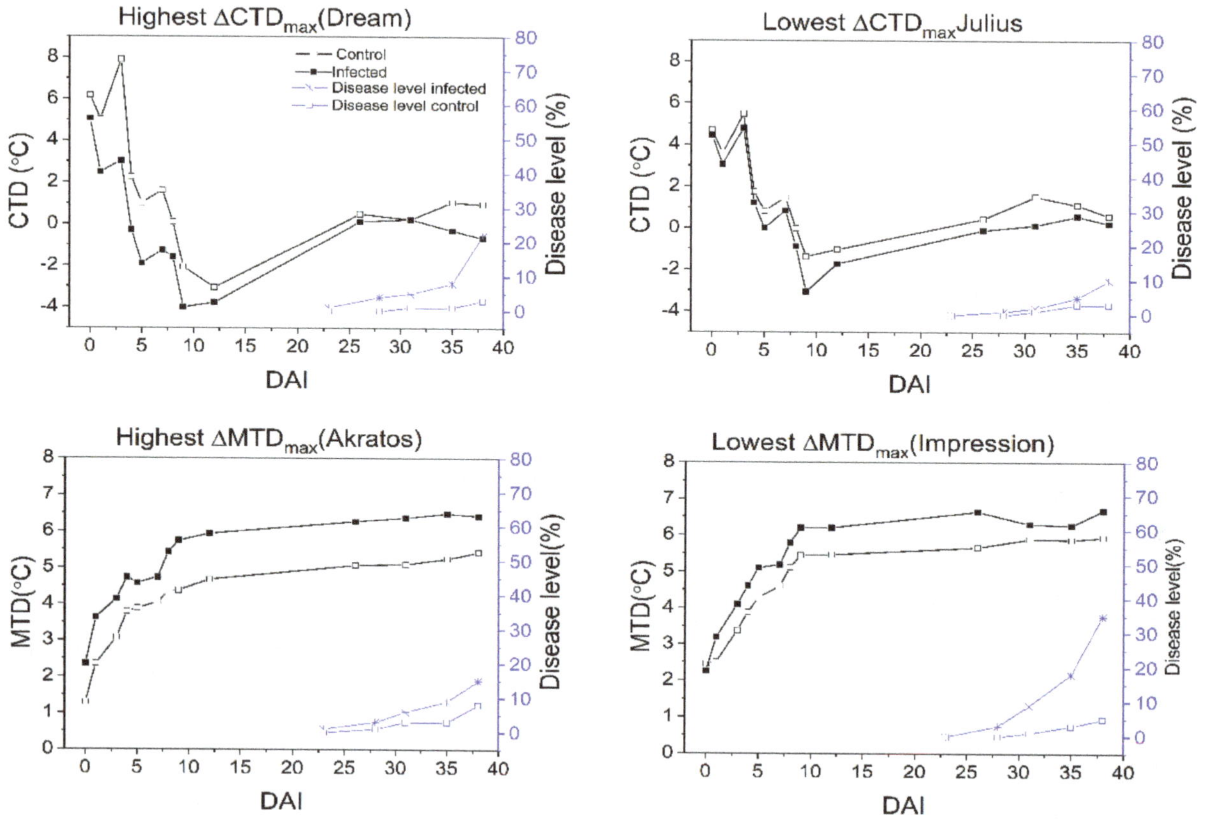

Figure 4. Effects of *Z. tritici* inoculation on Canopy Temperature Depression (CTD), Maximum Temperature Difference (MTD), Crop Water Stress Index (CWSI) and disease level.

The variety "Impression" had the lowest value of ΔMTD_{max} and has nearly more than doubled the disease level (35%) of variety "Akratos" (15%) which was the highest in ΔMTD_{max}. Even though the variety "Akratos" has a similar resistance level as the variety "Impression" according to Bundessortenamt (2012) in Table 1, in *Z. tritici* treatment, the MTD parameter of varieties "Akratos" and "Impression" had a similar trend up to 4 DAI. However, from 5 DAI onwards, the MTD of variety "Impression" was always higher than that of variety Akratos until the end of the experiment. Though it can be seen that the difference of MTD between the varieties was not very high, it did range between 0.27 °C and 0.53 °C. In addition, the MTD parameter of the *Z. tritici* was always higher than that of the control treatment in both varieties throughout the experiment.

4. Discussion

In this paper, Infrared thermography as a useful tool for early fungal disease detection in winter wheat was studied. Significant differences in ACT, CTD, and MTD between *Z. tritici* and control treatment were observed. It was interesting to find out that a significant difference in CTD, MTD was observed within a week from inoculation (Table 1) which is far earlier than the disease symptoms appeared on the plant. This is due to the fact that *Z. tritici* can germinate within 12 h and penetrate through stomata within 24–48 h after coming in contact depending on the climatic conditions [19]. Robert et al. [20] observed the chlorotic symptoms of *Z. tritici* on the flag leaves from 10 DAI and on the second leaves on 15 DAI. Plants resist the invasion of these pathogens by adopting various mechanisms such as reinforcing the cell wall, and producing compounds and inhibitors as a resistance reaction [23]. These processes could affect some of the physiological processes and could cause stress on the plant [24]. The chlorotic and necrotic symptoms of *Z. tritici* leads to a reduction in leaf photosynthesis [20]. However, how early the significant difference in CTD and MTD appeared was not related to the susceptibility level of wheat varieties. It agrees with the finding of Kema et al. [25] who found out that the spore germination is not influenced by the susceptibility of wheat varieties.

A decrease in CTD of *Z. tritici* treatment occurred as the disease level increased (Figure 3) which is caused by the reduction in the photosynthetic activity. This is in line with the results of Rosyara et al. [26] who mentioned that the stress from spot blotch caused by *Cochliobolus sativus* lead to a decrease in CTD because of the reduction in photosynthetic activity in infected plants whose green area is less than the healthy leaves. In addition, the invasion of these pathogens could negatively affect transpiration [27,28]. Reduction in the transpiration rate will result in higher temperature, and this could account for the reason why most of the wheat varieties showed significant differences in CTD and MTD towards the end of the experiment. Similarly, Shtienberg [28] mentioned that the colonisation and the level of damage caused by the fungi results in the reduction of photosynthesis and transpiration, and reduced transpiration would lead to a rise in the canopy temperature. The highest correlation of $R^2 = 0.6$ (Figure 3) was found between disease severity and MTD. This implies that the parameter MTD can be used as an indicator of early detection of *Z. tritici* in winter wheat. Also Oerke et al. [29] found out a strong correlation ($R^2 = 0.8$) between MTD and diseased severity of cucumber with inoculation of *Pseudoperonospora cubensi*. In our study the result of stepwise regression showed that only MTD and CTD parameters are related with the disease level with $R^2 = 0.7$ (Equation (7)). This is better than only considering one parameter for the prediction of disease level of *Z. tritici*. However, the reliability of CTD and MTD for evaluating the disease severity depends on the environmental conditions. CTD is related to ambient VPD and net radiation according to the energy balance model [30]. MTD of the leaf is also influenced by the environmental conditions such as ambient temperature and relative humidity which have an effect on the transpiration and hence on the leaf surface temperature distribution. In addition, as the disease severity increased, the CTD and MTD parameters of the healthy (control) and *Z. tritici* treatment became more prominent (Figure 4). This can be explained by the relationship between CTD and transpiration rate [31]. As the transpiration rate of the infected treatment decreases, the canopy temperature increases.

5. Conclusions

In this study, all the wheat varieties with *Z. tritici* showed significant differences in ACT, CTD, and MTD compared to the healthy plants (control) throughout the study period. However, only CTD and MTD parameters were significantly related to the disease level and therefore, can be used to predict the onset of fungal disease. The results showed that in some varieties, the earliest disease symptoms according to the MTD parameter could be determined as early as 3 DAI ($p \leq 0.01$). Similarly, according to the CTD parameter, the earliest disease symptom was found as early as 4 DAI ($p \leq 0.01$). However, in the same varieties, the first visual symptoms appeared 23 DAI. Hence, it can be concluded that thermography can be used as high throughput to accelerate monitoring of fungal disease in the field. When mounted on a drone, breeders could highly benefit from thermography for selecting disease-resistant varieties from thousands of progenies in a high-throughput manner.

Author Contributions: Data curation, Y.W. and S.O.-A.; Formal analysis, Y.W.; Funding acquisition, J.M.; Investigation, Y.W., S.Z.-K. and S.O.-A.; Methodology, S.Z.-K., T.M. and J.M.; Project administration, S.Z.-K. and J.M.; Resources, T.M. and J.M.; Software, Y.W., S.Z.-K. and S.O.-A.; Validation, S.Z.-K. and J.M.; Visualization, S.Z.-K. and J.M.; Writing–original draft, Y.W.; Writing–review & editing, S.Z.-K., T.M. and J.M.

Acknowledgments: We are very grateful to B. Lieberherr of the State Plant Breeding Institute for her cooperation and effort in supporting the experiment.

References

1. Siddiqui, K. Green biotechnology at the crossroads of nanobiotechnology, globalization, poverty alleviation and food sovereignty. *Indian J. Crop. Sci.* **2007**, *2*, 1–5.
2. Dhlamini, Z.; Spillane, C.; Moss, J.P.; Ruane, J.; Urquia, N.; Sonnino, A. *Status of Research and Application of Crop Biotechnologies in Developing Countries: Preliminary Assessment*; Food and Agriculture Organization of the United Nations: Rome, Italy, 2005.
3. Jarvis, D.; Mar, I.; Sears, L. (Eds.) *Enhancing the Use of Crop Genetic Diversity to Manage Abiotic Stress in Agricultural Production Systems*; IPGRI: Rome, Italy, 2006.
4. Wang, W.; Vinocur, B.; Altman, A. Plant responses to drought, salinity and extreme temperatures: Towards genetic engineering for stress tolerance. *Planta* **2003**, *218*, 1–14. [CrossRef] [PubMed]
5. Kosina, P.; Reynolds, M.; Dixon, J.; Joshi, A. Stakeholder perception of wheat production constraints, capacity building needs, and research partnerships in developing countries. *Euphytica* **2007**, *157*, 475–483. [CrossRef]
6. Ahmad, S.; Afzal, M.; Noorka, I. Prediction of yield losses in wheat (*Triticum aestivum* L.) in relation to epidemiological factors in Faisalabad. *Pak. J. Bot.* **2010**, *42*, 401–407.
7. Duveiller, E.; Singh, P.R.; Nicol, J.M. The challenges of maintaining wheat productivity: Pests, diseases, and potential epidemics. *Euphytica* **2007**, *157*, 417–430. [CrossRef]
8. Lucas, R.; Ubeda, A.; Paya, M.; Alves, M.; del Olmo, E.; Lopez, J.L.; San Feliciano, A. Synthesis and enzyme inhibitory activities of a series of lipidic diamine and amino alcohol derivatives on cytosolic and secretory phospholipases A2. *Bioorg. Med. Chem. Lett.* **2000**, *10*, 285–288. [CrossRef]
9. Sankaran, S.; Mishra, A.; Ehsani, R.; Davis, C. A review of advanced techniques for detecting plant diseases. *Comput. Electron. Agric.* **2010**, *72*, 1–13. [CrossRef]
10. Moshou, D.; Bravo, C.; Oberti, R.; West, J.; Ramon, H.; Vougioukas, S.; Bochtis, D. Intelligent multi–sensor system for the detection and treatment of fungal diseases in arable crops. *Biosyst. Eng.* **2011**, *184*, 311–321. [CrossRef]
11. Huang, W.; Lamb, D.W.; Niu, Z.; Zhang, Y.; Liu, L.; Wang, J. Identification of yellow rust in wheat using in-situ spectral reflectance measurements and airborne hyperspectral imaging. *Precis. Agric.* **2007**, *8*, 187–197. [CrossRef]
12. Mehl, P. Development of hyperspectral imaging technique for the detection of apple surface defects and contaminations. *J. Food Eng.* **2004**, *61*, 67–81. [CrossRef]
13. Choi, Y.; Angenot, L.; Harnischfeger, G. Analysis of *Strychnos nux-vomica, Strychnos icaja* and *Strychnos ignatii* extracts by 1H nuclear magnetic resonance spectrometry and multivariate analysis techniques. *Phytochemistry* **2004**, *65*, 1993–2001.
14. Bauriegel, E.; Giebel, A.; Geyer, M.; Schmidt, U.; Herppich, W.B. Early detection of fusarium infection in wheat using hyper–spectral imaging. *Comput. Electron. Agric.* **2011**, *75*, 304–312. [CrossRef]
15. Padhi, J.; Misra, R.K.; Payero, J. Prospects of using infrared thermography for irrigation scheduling of wheat crop. In Proceedings of the Australian Irrigation Conference and Exhibition 2010: One Water Many Futures, Sydney, Australia, 8–10 June 2010.
16. Zia, S.; Spohrer, K.; Merkt, N.; Wenyong, D.; He, X.; Müller, J. Non-invasive water status detection in grapevine (*Vitis vinifera* L.) by thermography. *Int. J. Agric. Biol. Eng.* **2009**, *2*, 46–54.
17. Jones, H.G. Use of infrared thermometry for estimation of stomatal conductance as a possible aid to irrigation scheduling. *Agric. For. Meterorol.* **1999**, *95*, 139–149. [CrossRef]
18. Jones, H.G.; Stoll, M.; Santos, T.; de Sousa, C.; Chaves, M.M.; Grant, O.M. Use of infrared thermography for monitoring stomatal closure in the field: Application to grapevine. *J. Exp. Bot.* **2002**, *53*, 2249–2260. [CrossRef] [PubMed]
19. Eyal, Z.; Sharen, A.L.; Prescott, J.M. *The Septoria Diseases of Wheat, Concepts and Methods of Disease Management*; CIMMYT: Texcoco, Mexico, 1987.
20. Robert, C.; Bancal, M.-O.; Lannou, C.; Ney, B. Quantification of the effects of *Septoria tritici* blotch on wheat leaf gas exchange with respect to lesion age, leaf number, and leaf nitrogen status. *J. Exp. Bot.* **2006**, *57*, 225–234. [CrossRef]

21. Lindenthal, M.; Steiner, U.; Dehne, H.W.; Oerke, E.C. Effect of downy mildew development on transpiration of cucumber leaves visualized by digital infrared thermography. *Phytopathology* **2005**, *95*, 233–240. [CrossRef]
22. *Beschreibende Sortenliste 2012–Getreide, Mais, Öl- und Faserpflanzen, Leguminosen, Rüben, Zwischenfrüchte*; Bundessortenamt: Hannover, Deutschland, 2012.
23. Ding, L.; Xu, H.; Yi, H.; Yang, L.; Kong, Z.; Zhang, L.; Xue, S.; Jia, H.; Ma, Z. Resistance to hemi-biotrophic *F. graminearum* infection is associated with coordinated and ordered expression of diverse defense signaling pathways. *PLoS ONE* **2011**, *6*, 1–17. [CrossRef]
24. Kuc, J. Phytoalexins, stress metabolism, and disease resistance in plants. *Ann. Rev. Phytopathol.* **1995**, *33*, 275–297. [CrossRef]
25. Kema, G.H.J.; Sayoud, R.; Annone, J.G.; Van Silfhout, C.H. Genetic variation for virulence and resistance in the wheat–*Mycosphaerella graminicola* pathosystem. II: Analysis of interactions between pathogen isolates and host cultivars. *Phytopathology* **1996**, *8*, 213–220. [CrossRef]
26. Rosyara, R.U.; Subedi, S.; Duveiller, E.; Sharma, R.C. The effect of spot blotch and heat stress on variation of canopy temperature depression, chlorophyll fluorescence and chlorophyll content of hexaploid wheat genotypes. *Euphytica* **2010**, *174*, 377–390. [CrossRef]
27. Rabbinge, R.; Jorritsma, I.T.M.; Schans, J. Damage components of powdery mildew in winter wheat. *Neth. J. Plant. Pathol.* **1985**, *91*, 235–247. [CrossRef]
28. Shtienberg, D. Effect of Folia Diseases on Gas Exchange Processes: A comparative Study. *Phytopathology* **1992**, *82*, 760–765. [CrossRef]
29. Oerke, E.C.; Steiner, U.; Dehne, H.W.; Lindenthal, M. Thermal imaging of cucumber leaves affected by downy mildew and environmental conditions. *J. Exp. Bot.* **2006**, *57*, 2121–2132. [CrossRef] [PubMed]
30. Jackson, R.G.; Idso, S.B.; Reginato, R.J.; Pinter, P.J. Canopy temperature as a crop water stress indicator. *Water Resour. Res.* **1981**, *17*, 1133–1138. [CrossRef]
31. Balota, M.; Payne, A.W.; Evett, R.S.; Lazar, D.M. Canopy Temperature Depression Sampling to Assess Grain Yield and Genotypic Differentiation in Winter Wheat. *Crop Sci.* **2007**, *47*, 1518–1529. [CrossRef]

6

Spatial Variability of Aroma Profiles of Cocoa Trees Obtained through Computer Vision and Machine Learning Modelling: A Cover Photography and High Spatial Remote Sensing Application

Sigfredo Fuentes [1,*], Gabriela Chacon [1], Damir D. Torrico [1,2], Andrea Zarate [1] and Claudia Gonzalez Viejo [1]

1 School of Agriculture and Food, Faculty of Veterinary and Agricultural Sciences, University of Melbourne, Melbourne, VIC 3010, Australia
2 Department of Wine, Food and Molecular Biosciences, Faculty of Agriculture and Life Sciences, Lincoln University, Lincoln 7647, New Zealand
* Correspondence: sfuentes@unimelb.edu.au.

Abstract: Cocoa is an important commodity crop, not only to produce chocolate, one of the most complex products from the sensory perspective, but one that commonly grows in developing countries close to the tropics. This paper presents novel techniques applied using cover photography and a novel computer application (VitiCanopy) to assess the canopy architecture of cocoa trees in a commercial plantation in Queensland, Australia. From the cocoa trees monitored, pod samples were collected, fermented, dried, and ground to obtain the aroma profile per tree using gas chromatography. The canopy architecture data were used as inputs in an artificial neural network (ANN) algorithm, with the aroma profile, considering six main aromas, as targets. The ANN model rendered high accuracy (correlation coefficient (R) = 0.82; mean squared error (MSE) = 0.09) with no overfitting. The model was then applied to an aerial image of the whole cocoa field studied to produce canopy vigor, and aroma profile maps up to the tree-by-tree scale. The tool developed could significantly aid the canopy management practices in cocoa trees, which have a direct effect on cocoa quality.

Keywords: leaf area index; cocoa beans; volatile compounds; artificial neural networks; VitiCanopy app

---

1. Introduction

Cocoa (*Theobroma cacao* L) is considered a major world commodity crop and is the seventh most exported food product [1]. Its primary use is for chocolate manufacture, with an estimate of 3981 thousand tons of cocoa bean production in 2016 globally [2]. However, as with any other crop, the quality of cocoa beans is determinant for consumers' acceptability and end-product cost. The cocoa beans intended for manufacturing must comply with specific quality parameters such as shape, size, color [3], flavor, aroma, cocoa butter, and protein content [4]. These are mainly related to the genotype of the clone, climatic and agricultural conditions, and agricultural practices (e.g., water management, fertilization, canopy management), as well as fermentation, drying, and industrialization processes [5–9]. The cocoa crop and its production are also very important for developing countries. For example, cocoa produce from Ecuador is highly valued internationally due to their recognized aroma intensity development [10].

Cocoa trees are grown in regions with high rainfall (1250–3000 mm/year), average temperatures between 18 and 32 °C, and calm wind conditions to avoid defoliation. Furthermore, shade is an important factor, especially in young trees, as it provides protection from excessive solar radiation and helps enhance aroma profiles that are desirable in the final chocolate product. They are usually planted at a density of 400–2500 trees per hectare and produce an average yield of between 200 and 800 kg of fermented dried cocoa per hectare [4,11,12].

There are three main cocoa tree varieties: (i) Criollo, (ii) Forastero, and (iii) Trinitario, the latter being a hybrid group. Varieties i and iii are considered as fine-quality or flavor-quality cocoa beans, while ii is used to produce bulk-quality cocoa. Specifically, the cocoa quality is mainly influenced by the growing conditions where the tree is grown, such as soil depth, pH (4.5–7), weather changes, and pest and disease control [13]. Some techniques to manage the cocoa trees to assure high-quality beans are: (i) Pest and disease control, (ii) soil nutrients management, (iii) weed control, and (iv) management of shade and canopy pruning [14]. The latter technique, (iv), is important due to the fact that the relationship between light and shade is critical, as the shade provides protection against sunlight and high winds, and lowers the temperature to manage the growth of the cocoa tree [4,11].

In cocoa plantations, some in-field factors that have been reported to affect cocoa quality are the exposure to solar radiation, which has effects on the final fruits' dry weight and the growth temperature, which has, in turn, an influence on the lipid content and melting point [15]. However, no relationship between the aroma quality of the cocoa beans and canopy architecture parameters, such as leaf area index (LAI), has been reported. A complex method proposed by researchers to assess cocoa beans quality from the field is through the genetic identification of the variety by using deoxyribonucleic acid (DNA) fingerprinting and utilizing microsatellite markers containing the information from leaf samples [13,16]. However, this method is time-consuming, non-practical, and cost ineffective. Furthermore, the analysis of the trees at the leaf level has shown to be challenging due to the difficulty of sampling leaves with the same growth conditions (e.g., shade, age) [11] and the complexity of the environment when plants are grown between other tree species (i.e., within a forest).

This paper presents the assessment of canopy architecture parameters of cocoa trees using the VitiCanopy computer application [17] as a ground-truth to validate information extracted from aerial images, to automatically obtain these parameters within a plantation and up to a tree-by-tree resolution. From the same ground-truth cocoa trees, the volatile aroma compounds were assessed from fermented and dried cocoa beans by using gas chromatography–mass spectroscopy (GC–MS) with a solid phase microextraction in the headspace (SPME-HS). The canopy architecture data were used as inputs to develop a machine learning model to predict six of the main volatile compounds in cocoa beans and their associated aromas. Since all the information obtained either through computer vision algorithms (canopy architecture) and machine learning modeling (aroma profiles) was at the tree-by-tree scale, spatial distribution aroma maps were constructed from the cocoa plantation used in this research.

2. Materials and Methods

2.1. Site Description

The cocoa trees used in this research are located in a commercial farm with Papua New Guinea Hybrids (PNH) at Mt Edna Maria Creeks, (17°49′14″ S 146°02′21″ E), North Queensland, Australia. A map showing the location can be found as supplementary material in Figure S1. The soil type is clay loam with a pH of 5, under a tree sprinkler irrigation system with sprinkler discharge of 40 L per hour. Sprinklers were located between tree rows at 4 m intervals, and minimal chemical usage for pests and disease management was applied when required.

Cocoa trees were planted in three blocks using a Tatura trellis pruning system, which requires a continuous shaping of the tree canopy. Measurements were only done in Blocks 1 and 2. Among Blocks 1 and 2, the planting density was 1000 trees ha^{-1} in a double-row layout, with distances of 5 m between rows and 2 m between trees. The blocks and rows were planted in different seasons for pruning test purposes. Block 1 was planted in March 2013, and Block 2 in October 2013. In both blocks, row A was planted first. The pattern area (Figure 1) was excluded from data collection because the trees were established as free-standing trees and were at the flowering stage. On the right side of Block 1 and at the lower lateral of the two blocks, the plantation is surrounded by *Erythrina variegata*, or coral tree, used as a windbreak, while Block 2 is surrounded by common vegetation of the site. The cocoa plantation was grown under full-sunlight conditions without intercropping species or shade species.

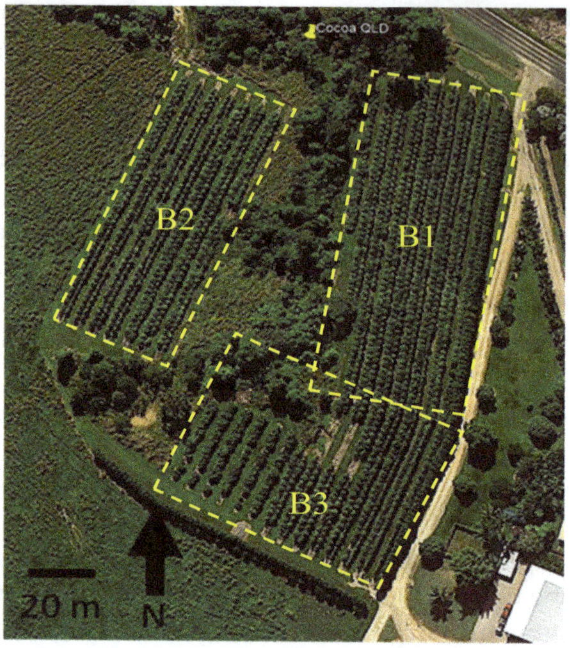

Figure 1. Aerial image of a cocoa plantation close to harvest from Mt Edna Maria Creeks, Queensland, Australia (Google Earth Pro™). Yellow dotted lines show the blocks used, numbered from 1 to 3.

The weather data of the site were obtained at an approximate distance of 30 km from the plantation from the Bureau of Meteorology (BOM) weather station located at South Johnstone (Australian BOM Station No. 032037, 17.61° S 146.00° E). Average rainfall of 3289.3 mm per year, and mean values for maximum and minimum temperatures of 28.1 and 19.2 °C, respectively, were recorded in the season studied (2015–2016), which is consistent with the climate classification for the site as a tropical monsoonal region.

2.2. *Cocoa Trees and Pods/Beans Sampling*

A total of 24 cocoa trees were used for this study. Each tree was measured nine times to use as replicates. The cocoa pods were randomly sampled from the mentioned trees according to the general guidelines on sampling CAC/GL 50-2004, as recommended by the Codex Alimentarius Commission. Two to three cocoa pods were harvested from each tree. Each pod was assigned a label as B1 or B2 according to the blocks from which they were collected (one or two), followed by the number of the tree sampled (e.g., A15) and the letter of the row (A or B). Every tree was geo-located using the VitiCanopy App, which will be described following. The crops for the two blocks from this trial were harvested on the 2 February 2016.

2.3. *Image Data Acquisition*

For each of the 24 sampled trees, a total of nine upward-looking canopy images were obtained using the VitiCanopy computer application [17] within three consecutive days (three per day). The digital images from the tree canopies were obtained with the front camera of an iPhone 5S (Apple Inc. Corp. Cupertino, CA, USA), which has 1.2 megapixels of resolution. The device was mounted on a Soniq Bluetooth Selfie Stick (SONIQ Australia Pty Ltd., Braeside, Victoria, Australia) facing upwards at 0.5 m height using a tripod. The device was aligned using a bubble level attached to the selfie stick to ensure it was in a horizontal position for each image, obtained at a 0° Zenith angle. It was positioned with its maximum length perpendicular to the direction of the trunk. The images were acquired by locating the device at 0.5 m from the trunk and 0.5 m to the left and right sides (Figure 2). Imagery were obtained four hours before noon to avoid direct sunlight into the lens of the device, as recommended by Poblete-Echeverría et al. [18]. The dates of image acquisition were 29 January to 2 February 2016. All the equipment used for image acquisition were ubiquitous (smartphones), low-cost, and suitable for fieldwork.

Figure 2. (**A**) Distance of image acquisition data in cocoa trees under a Tatura trellis system using the VitiCanopy application (app), and (**B**) image of canopy recorded using the VitiCanopy app.

2.4. Estimation of Canopy Architecture Parameters by the VitiCanopy App

Pictures with direct sunlight, weeds, posts, operators, and canopy from the other trees were considered unsuitable for analysis and discarded. Therefore, a total of 173 images were suitable for the analysis, with six to nine images analyzed from each tree. Canopy architecture parameters were obtained using the following algorithms to obtain the fractions of foliage projective cover (f_f), crown cover (f_c), and crown porosity (Φ), which were calculated from McFarlane et al. [19] as:

$$f_f = 1 - \frac{t_g}{t_p} \quad (1)$$

$$f_c = 1 - \frac{l_g}{t_p} \quad (2)$$

$$\Phi = 1 - \frac{f_f}{f_c} \quad (3)$$

where l_g = large gap pixels; t_g = total pixels in all gaps, and t_p = total gap pixels.

LAI is calculated from Beer's Law:

$$LAI = f_c + \frac{\ln \Phi}{k} \quad (4)$$

where k is the coefficient of light extinction ($k = 0.61$) [20], and the clumping index at the zenith, Ω (0), was calculated as follows:

$$\Omega(0) = \frac{(1-\Omega)\ln(1-f_f)}{\frac{\ln(\Phi)}{f_f}} \tag{5}$$

The clumping index is a correction factor to obtain the effective LAI (LAI_e), which is the product of:

$$LAI_e = LAI_M \times \Omega(0) \tag{6}$$

Equation (5) describes the non-random distribution of canopy elements. If $\Omega(0) = 1$, that means that the canopy displays random dispersion. For $\Omega(0) >$ or < 1, the canopy is defined as clumped.

This innovative app estimates canopy architecture parameters using upward-looking digital images based on a method developed by Fuentes and De Bei et al. [21]. The image can be automatically sub-divided into $m \times n$ (m = number of rows, n = number of columns) sub-images for gap analysis. For this research, a sub-division of 5 × 5 was used for the VitiCanopy App and to analyze each tree sub-image from the aerial image analysis. The algorithms used in the VitiCanopy app were calculated based on the formulas from Macfarlane et al. [19]. In the settings menu of the application, the gap fraction threshold was set as 75%, which means that a big gap was considered when 75% of the pixels from the images corresponded to the sky [21]. A value of the light extinction coefficient (k) of 0.7 was set based on the range of the standard mean values identified from the samples and according to the literature review data of cocoa tree studies made by Yapp [22] and Daymond et al. [23]. Since in these images the sky serves as a background to isolate non-leaf material, the app filters sky and clouds automatically by segmentation of images using the blue channel of the RGB images. In order to record the Global Positioning System (GPS) data of each cocoa tree, the location service of the app was enabled. The images were analyzed with the application individually as they were obtained. The results from the app, which included the canopy architecture parameters, date, GPS location, and settings used were exported via email by tapping the "Export" option, which generates a comma separated values (.csv) file with all the data.

2.5. Aerial Imagery and Processing

The aerial image used for this research was obtained from Google Earth Pro (ver. 7.3.2.5776; Google, Mountain View, CA, USA), which uses a composite between satellite and aerial images. The only available image closer to the harvest was for 27 May 2016. However, it is expected from the biological point of view that the LAI will be similar within this month and that the trends between plants should also be maintained. The image was saved as a maximum high-resolution (4800 × 2679 pixels) Joint Photographic Expert Group (JPEG) file. The aerial image was processed to recognize every single plant and extract information from each one using the methodology proposed and described in previous studies [24,25] through customized codes written in Matlab® (Mathworks Inc., Natick, MA, USA). In general, the automated procedure consists of the following steps:

1. Identification of the area designated by the plant (distance between plants and between rows). This was done by cropping the area for blocks and automatically sub-cropping each tree area for further analysis according to the distance between plants and the distance between rows [24,25] (Figure 3A).
2. Automatic Individual segmentation of tree biomass material using the CIELab (specifically the "a" band) color code to generate a binary image of sub-images analyzed per tree (Figure 3B).
3. Analysis of binary images per area assigned per plant using algorithms described previously (Equations (1)–(6)). Each sub-image is treated as an individual upward-looking image obtained by VitiCanopy, using the sky as a background. A subsequent subdivision of 5 × 5 was used for each tree image for gap analysis, as described by [17,21] (Figure 3C). From the aerial image,

it was assumed to be a downward-looking image at Nadir (0° angle) using the inter-row as a background [17,21].

4. The obtained canopy architecture data were compiled in a matrix form to preserve plant positions and for easier handling for mapping purposes.
5. If canopies were of low cover ($f_c > 0.1$), they were considered as missing trees [24].
6. Tree GPS positions were extracted from the GPS coordinates of the corners of blocks for data mapping purposes.

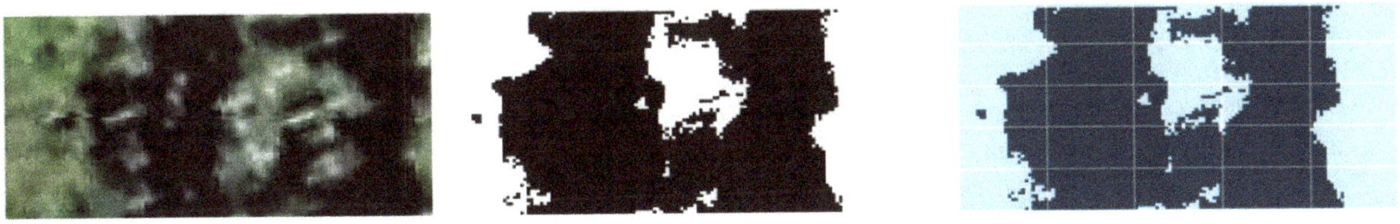

(A) (B) (C)

Figure 3. Analysis process from automatically cropped images per tree (**A**) and segmentation using the CieLab color code to obtain a binary image showing main tree biomass (**B**), and sub-division of 5 × 5 for gap analysis (**C**) and application of image analysis algorithms (Equations (1)–(6)).

2.6. Cocoa Bean Fermentation and Drying Process

The harvested cocoa pods were transported to The University of Melbourne, Australia, and stored for ten days at room temperature between 20 and 25 °C. The storage time allowed a significant increase of the pH related to cocoa flavor potential [26,27]. The seeds were taken from inside each pod sample and mixed with a solution of a commercial yeast *Saccharomyces cerevisiae* (Lowan Whole Foods, Glendenning, NSW, Australia) and water in a plastic container. For the fermentation process, the labeled plastic containers were placed into an incubator I100-G300-D (Thermoline Scientific Equipment Pty Ltd., Wetherill Park NSW, Australia), which was set up at 25 °C (0–12 h), 30 °C (12–24 h), 35 °C (24–36 h), 40 °C (36–48 h), 45 °C (48–72 h), and 48 °C (72–144 h). The temperature was monitored daily, and the seeds were manually flipped every 48 h to ensure aeration and uniform fermentation. After six days of fermentation, a cut test was performed to assess the quality of the process. Subsequently, for the drying process, the beans were set up in a Qualtex Solid State oven, Series 2000 (Qualtex Australasia P/L, Murarrie QLD, Australia) at 45 ± 2 °C, and mixed manually and daily for three days to get a uniform drying.

2.7. Analysis of Volatile Compounds Using Gas Chromatography-Mass Spectroscopy

The cocoa beans from each pod sample were ground in a commercial coffee blender Sunbeam Multigrinder II EM0405 (Sunbeam Products, Boca Raton, FL, USA) for 60 s. One gram of the ground beans was placed into a 15 mL tightly capped vial with 5 mL of saturated salt solution and 100 µL of pyrazine dilution. According to the procedure of De Brito et al. [5], the vials were mixed in a vortex (Vor-Mix VM80) for 1 min and preheated for 45 min at 60 °C, and then placed in a dry block heater and mixed in an orbital mixer OM1 (Ratek Instruments Pty. Ltd., Boronia, Victoria, Australia).

The volatile compounds were extracted by solid phase microextraction in the headspace (SPME-HS) with Supelco fibers 50/30 µm divinylbenzene/carboxene/polydimethylsiloxane (DVB/CAR/PDMS). The extraction of the volatile compounds of each sample took place for 15 min, with 30 min at 60 °C of fiber exposition to the cocoa aromas in the headspace [28]. The fibers were then analyzed by gas chromatography (GC; Agilent Technologies 6850 Series II, Network GC System) coupled to mass spectrometry (MS; Agilent Technologies 5973 Network Mass Selective Detector). The GC–MS was equipped with an Agilent J&W DB-Wax column (30 m × 0.25 mm, 0.25 µm film thickness). The oven temperature was set at 40 °C for 5 min, and then increased to 200 °C at a rate of 10 °C min^{-1}, and finally maintained at 200 °C for 30 min. The carrier gas was helium (Air Liquide or BOC, Ultra-High Purity), with a linear velocity of 36 cm/s and a flow rate of 2.0 mL/min in constant flow mode. The SPME fiber was desorbed in the pulsed splitless injection mode for 5 min at 240 °C and opened after 0.5 min. The temperatures of the selective mass detector quadrupole were MS source 230 °C and MS Quad 180 °C, with an electronic impact ionization system at 70 eV. Compounds identification was performed by comparing the mass spectra of each compound with the National Institute of Standards and Technology MS NIST02.L (National Institute of Standards and Technology, Gaithersburg, MD, USA) library of mass spectra.

2.8. Statistical Analysis and Machine Learning Modeling

An analysis of variance (ANOVA) was conducted to assess significant differences in the LAI values and canopy architectures between the sampled trees, using the SAS® 9.4 software (SAS Institute Inc., Cary, NC, USA). A post-hoc analysis with the Tukey studentized method ($\alpha = 0.05$) was used. Means and standard deviation (SD) were calculated for each parameter.

A machine learning model was developed using the Bayesian Regularization training algorithm after testing 17 different artificial neural network (ANN) algorithms (two backpropagation using Jacobian derivatives, 11 backpropagation using gradient derivatives, and four supervised weight/bias training) using a customized and automatic Matlab® R2018b developed code (unpublished). Six canopy architecture parameters from the canopy measurements ((i) LAI, (ii) LAI_e, (iii) f_f, (iv) f_c, (v) ϕ, and (vi) Ω) were used as inputs to predict six main aroma compounds: (i) Phenethyl acetate, (ii) 2-phenyl-2-butenal, (iii) isoamyl acetate, (iv) tetramethylpyrazine, (v) phenylacetaldehyde, and (vi) 2,3-butanediol. All inputs and target values were normalized from −1 to 1. To develop the model, a random data division was used with 85% (N = 147) of the samples used for training, 15% (N = 26) for testing with a default derivative function, and a performance algorithm based on mean squared error (MSE). The model (Figure 4) was built with a two-layer feedforward network using a tan-sigmoid function in the hidden layer and a linear transfer function in the output layer. A trimming assessment was performed using 3, 5, 7, and 10 neurons to find the model with the best performance and less chance of overfitting. Statistical results are presented for the correlation coefficient (R) and mean squared error values (MSE).

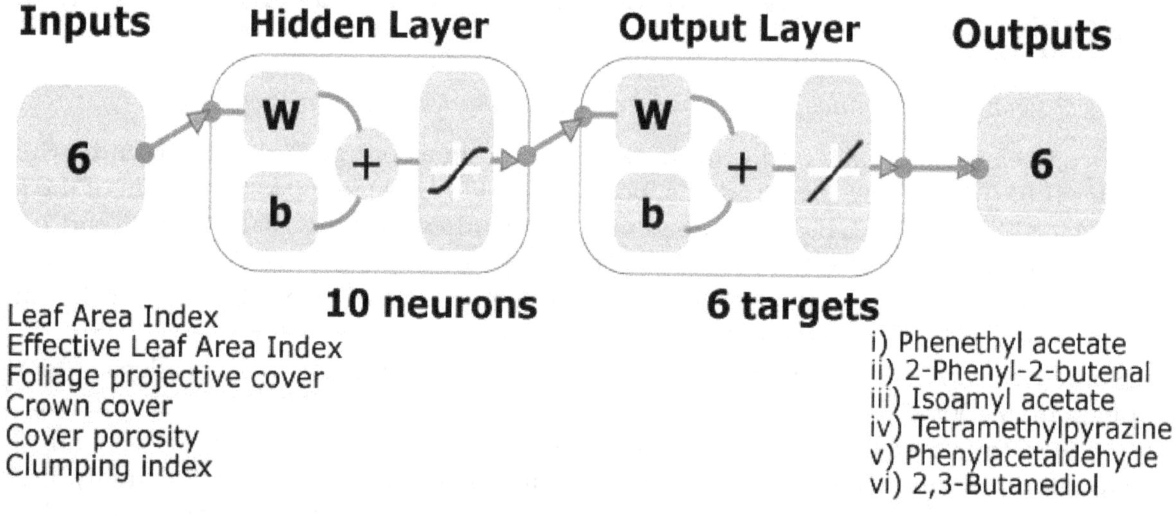

Figure 4. Model diagram showing the two-layer feedforward network with six inputs from the canopy architecture, 10 neurons in the hidden layer with a tan-sigmoid function, and a linear transfer function in the output layer. Six targets/outputs from the volatile compounds were used for the model.

Once developed, the machine learning model was run using the canopy architecture data extracted from the aerial imagery as inputs to obtain the six aromas per tree from the three blocks. Geo-location per tree was obtained from the GPS data from block corners and considering the plantation density through a customized code written in Matlab® R2019a (Fuentes, unpublished). With this information, two maps of LAI_e and the aroma profiles were generated in Matlab, with the first using kriging interpolation algorithms and the second using scattered data representing each tree. A graphical representation of the complete methodology used in this study can be found in the supplementary material (Figure S2).

3. Results

3.1. Canopy Architecture and Volatile Compounds

Table 1 shows the results of the means and SD from the canopy architectures of each of the 24 sampled trees. There were significant differences ($p < 0.05$) between samples for the six parameters. However, for the clumping index, only one sample (B1C4A) was significantly different.

Table 2 shows the means and SD from the results of both blocks, in which significant differences ($p < 0.05$) can be observed for LAI, LAI_e, and crown porosity (Φ). Block 1 had higher mean values of LAI and, although not significantly different, crown cover results compared to Block 2. On the contrary, Block 1 presented lower and significant values of clumping index than Block 2.

Table 3 presents the volatile compounds found in the fermented cocoa beans and the aromas associated with them. It can be observed that most of the associated aromas are related to cocoa, butter, honey, fruity, floral, nutty, and roasted. Figure 5 shows the peak area of each volatile compound found in every sample. It can be seen that most of the samples had phenylethyl alcohol as the most abundant volatile compound, except for the beans of five trees (B1C1A, B1C18A, B1C26B, B2C14A, and B2C16A). Tetramethylpyrazine was present in most of the samples and is higher in those from Block 1 than Block 2. Benzaldehyde and 2,3-butanediol were present in all samples. However, the peak area varied between samples. Pehenthyl acetate and isoamyl acetate were present in all except for two (B1C25A and B2C16A) and one (B1C25A) sample(s), respectively. Other compounds, such as 2-pentanone (fruity) and ethyl acetate (pineapple/fruity/green) were only detected in four (B1C1A, B1C27A, B2C16A, and B2C24A) and six samples (B1C16A, B1C26B, B1C27A, B1C36A, B1C45A, and B2C16A), respectively.

Table 1. Means and standard deviation (SD) of the parameters obtained from the canopy architecture for each sampled tree. Different letters (superscripts) represent statistically significant differences between samples using the Tukey studentized test with $\alpha = 0.05$.

Sample	LAI		LAI_e		Canopy Cover (f_c)		Crown Cover (f_f)		Crown Porosity (Φ)		Clumping Index (Ω)	
	Mean	SD	Mean	SD	Mean	SD	Mean	SD	Mean	SD	Mean	SD
B1C15A	3.08 ^{b}	0.68	2.79 ^{a}	0.76	0.836 ^{a}	0.094	0.962 ^{a}	0.061	0.134 ^{a}	0.050	0.951 ^{a}	0.067
B1C16A	2.78 ^{b}	0.96	2.55 ^{a}	1.12	0.755 ^{a}	0.116	0.879 ^{a}	0.107	0.142 ^{a}	0.050	0.851 ^{a}	0.131
B1C17A	3.60 ^{a}	0.69	3.19 ^{a}	0.36	0.868 ^{a}	0.063	0.957 ^{a}	0.060	0.094 ^{b}	0.035	0.904 ^{a}	0.122
B1C18A	3.47 ^{a}	0.82	3.18 ^{a}	1.08	0.843 ^{a}	0.094	0.932 ^{a}	0.077	0.097 ^{b}	0.036	0.887 ^{a}	0.127
B1C1A	3.62 ^{a}	0.68	3.32 ^{a}	0.98	0.863 ^{a}	0.069	0.945 ^{a}	0.050	0.088 ^{b}	0.032	0.888 ^{a}	0.100
B1C25A	4.56 ^{a}	0.73	4.08 ^{a}	0.73	0.939 ^{a}	0.037	0.994 ^{a}	0.014	0.056 ^{b}	0.026	0.986 ^{a}	0.034
B1C26B	2.55 ^{b}	0.47	2.33 ^{a}	0.69	0.730 ^{a}	0.074	0.863 ^{a}	0.086	0.153 ^{a}	0.050	0.814 ^{a}	0.112
B1C27A	3.31 ^{a}	0.70	3.04 ^{a}	0.97	0.834 ^{a}	0.105	0.928 ^{a}	0.087	0.104 ^{b}	0.030	0.894 ^{a}	0.121
B1C2A	3.50 ^{a}	1.16	3.20 ^{a}	1.33	0.827 ^{a}	0.129	0.915 ^{a}	0.111	0.099 ^{b}	0.051	0.875 ^{a}	0.146
B1C36A	3.68 ^{a}	0.20	3.32 ^{a}	0.56	0.867 ^{a}	0.046	0.946 ^{a}	0.054	0.080 ^{b}	0.011	0.872 ^{a}	0.117
B1C3A	4.29 ^{a}	1.20	3.92 ^{a}	1.43	0.881 ^{a}	0.139	0.935 ^{a}	0.141	0.059 ^{b}	0.031	0.905 ^{a}	0.195
B1C45A	2.65 ^{b}	0.45	2.41 ^{a}	0.67	0.772 ^{a}	0.079	0.913 ^{a}	0.070	0.156 ^{a}	0.030	0.886 ^{a}	0.092
B1C4A	2.40 ^{b}	0.29	2.18 ^{b}	0.51	0.668 ^{b}	0.085	0.775 ^{b}	0.112	0.135 ^{a}	0.036	0.723 ^{b}	0.116
B2C12A	3.06 ^{b}	0.48	2.84 ^{a}	0.75	0.847 ^{a}	0.061	0.980 ^{a}	0.041	0.136 ^{a}	0.035	0.972 ^{a}	0.056
B2C13A	3.33 ^{a}	0.49	3.03 ^{a}	0.56	0.875 ^{a}	0.046	0.991 ^{a}	0.017	0.117 ^{b}	0.035	0.986 ^{a}	0.028
B2C14A	2.61 ^{b}	0.98	2.46 ^{b}	1.13	0.727 ^{b}	0.196	0.853 ^{a}	0.209	0.152 ^{a}	0.053	0.863 ^{a}	0.158
B2C15A	2.30 ^{b}	0.29	2.12 ^{b}	0.48	0.720 ^{b}	0.059	0.888 ^{a}	0.068	0.188 ^{a}	0.035	0.864 ^{a}	0.073
B2C16A	2.81 ^{b}	0.51	2.59 ^{a}	0.66	0.791 ^{a}	0.114	0.919 ^{a}	0.122	0.140 ^{a}	0.021	0.902 ^{a}	0.132
B2C17A	2.88 ^{b}	0.89	2.63 ^{a}	0.89	0.770 ^{a}	0.145	0.885 ^{a}	0.126	0.134 ^{a}	0.053	0.871 ^{a}	0.139
B2C1A	3.27 ^{a}	0.49	2.99 ^{a}	0.55	0.867 ^{a}	0.048	0.984 ^{a}	0.031	0.119 ^{b}	0.028	0.974 ^{a}	0.052
B2C22A	2.91 ^{b}	0.36	2.69 ^{a}	0.60	0.816 ^{a}	0.059	0.947 ^{a}	0.058	0.138 ^{a}	0.027	0.913 ^{a}	0.085
B2C24A	2.56 ^{b}	0.76	2.36 ^{b}	0.81	0.719 ^{b}	0.150	0.849 ^{a}	0.159	0.153 ^{a}	0.063	0.828 ^{a}	0.154
B2C25A	3.02 ^{b}	0.63	2.80 ^{a}	0.83	0.796 ^{a}	0.120	0.901 ^{a}	0.122	0.118 ^{b}	0.030	0.865 ^{a}	0.140
B2C27A	3.13 ^{a}	0.64	3.13 ^{a}	0.64	0.849 ^{a}	0.060	0.976 ^{a}	0.036	0.131 ^{a}	0.038	0.962 ^{a}	0.055

Abbreviations: LAI = Leaf Area Index, LAI_e = effective LAI; sample codes: B1 = Block 1, B2 = Block 2, C# = number of the tree, and A or B = position of the tree (row). Means were taken from six to nine images per tree.

Table 2. Means and standard deviation (SD) of canopy architecture parameters for the two blocks. Different letters represent statistically significant differences between samples using the Tukey studentized test with $\alpha = 0.05$.

Block	LAI		LAI_e		Canopy Cover (f_c)		Crown Cover (f_f)		Crown Porosity (Φ)		Clumping Index (Ω)	
	Mean	SD	Mean	SD	Mean	SD	Mean	SD	Mean	SD	Mean	SD
B1	3.35 [a]	0.93	3.04 [a]	1.01	0.822 [a]	0.11	0.919 [a]	0.095	0.107 [b]	0.048	0.880 [a]	0.126
B2	2.89 [b]	0.67	2.68 [b]	0.76	0.796 [a]	0.12	0.923 [a]	0.116	0.139 [a]	0.043	0.907 [a]	0.116

Abbreviations: LAI = Leaf Area Index, LAI_e = effective LAI, B1 = Block 1, B2 = Block 2.

Table 3. Major volatile compounds found in the fermented cocoa beans, and the associated aromas.

Volatile Compounds	Compound Group	Aroma Associated	References
Phenethyl acetate	*Ester*	*Honey/floral/yeasty/cocoa*	[28,29]
2-Phenyl-2-butenal	*Aldehyde*	*Sweet/cocoa/nutty/beany*	[28,29]
Isoamyl acetate	*Ester*	*Banana/fruity*	[29,30]
Tetramethylpyrazine	*Pyrazine*	*Roasted/cocoa/coffee/nutty/earthy*	[7,29,31]
Phenylacetaldehyde	*Aldehyde*	*Honey/rose/cocoa*	[29,31]
2,3-Butanediol	*Alcohol*	*Cocoa butter/creamy*	[29,30]
Isovaleraldehyde	Aldehyde	Chocolate/nutty/cocoa	[28,29]
2-Pentanone	Ketone	Fruity	[29,32]
Diacetyl/2,3-Butanedione	Ketone	Butter/creamy/caramel	[28,29]
Ethyl acetate	Ester	Pineapple/fruity/green	[28,29]
Acetoin	Ketone	Butter/creamy/dairy/fatty	[28,29]
2-Heptanol	Alcohol	Citrus/fruity	[28,29]
3-Methyl-2-butanol	Alcohol	Cocoa/alcoholic/musty	[29]
Isoamyl alcohol	Alcohol	Malty/banana/pungent	[29,32]
2,3-Dimethylpyrazine	Pyrazine	Caramel/cocoa/coffee/nut skin	[29,33]
2,3-Dimethyl-5-ethylpyrazine	Pyrazine	Burnt/roasted/cocoa	[29]
Benzaldehyde	Aldehyde	Cherry/almond/bitter	[29,33]
Ethyl Phenylacetate	Esters	Honey/rose/dark chocolate	[29]
Benzyl alcohol	Alcohol	Sweet/floral/rose/phenolic	[28]
Phenylethyl alcohol	Alcohol	Honey/bready/floral	[29]

Compounds in italics were used as targets for machine learning modeling.

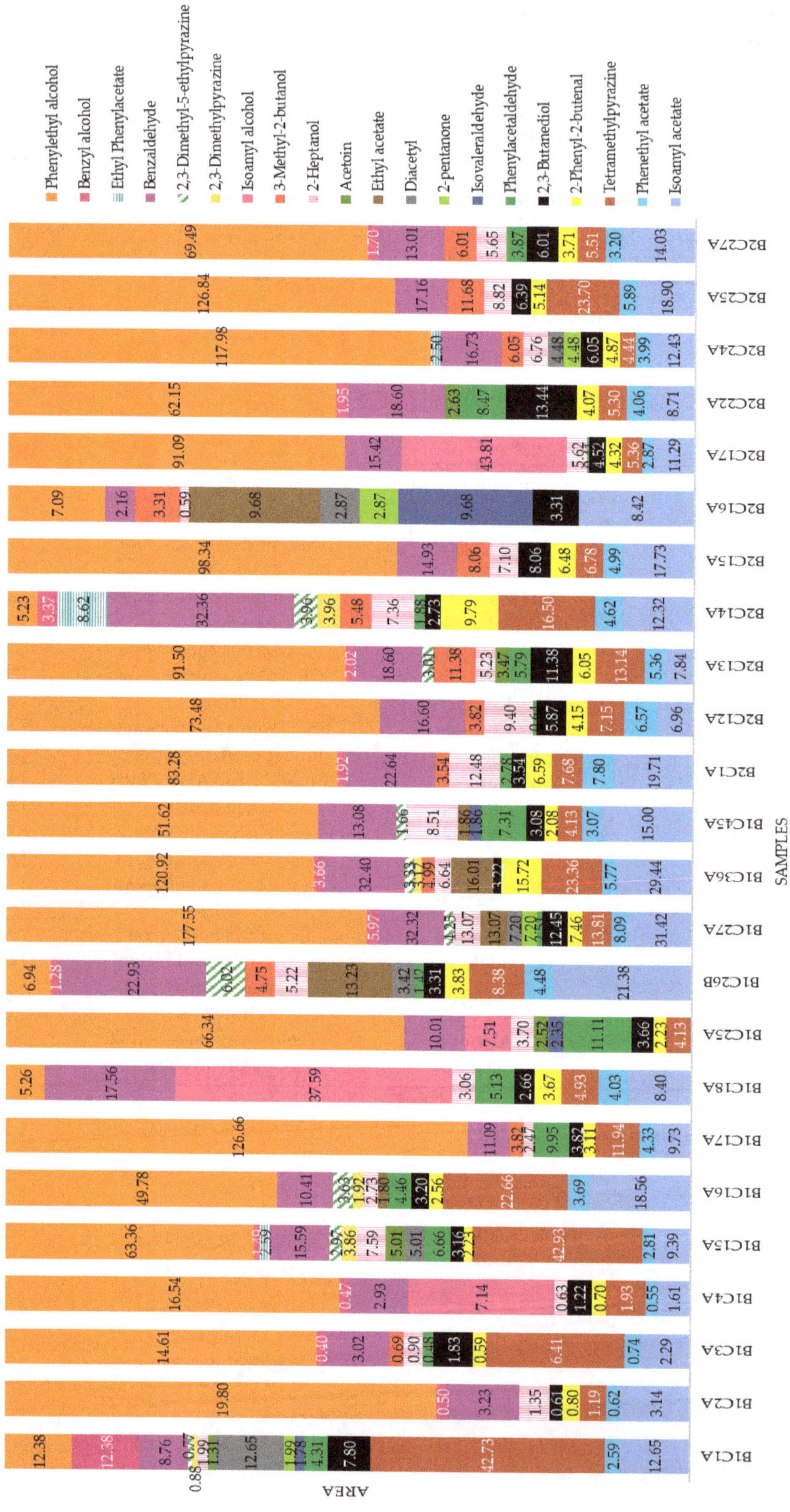

Figure 5. Peak areas of the volatile compounds found in the fermented cocoa bean samples. All values are reported in scientific notation 10^3. Sample codes: B1 = Block 1, B2 = Block 2, C# = number of the tree, and A or B = position of tree (row).

3.2. *Machine Learning Modeling*

Figure 6 shows the machine learning model to predict six of the main volatile compounds (Figure S3) in cacao beans (phenethyl acetate, 2-phenyl-2-butenal, isoamyl acetate, tetramethylpyrazine, phenylacetaldehyde, and 2,3-butanediol) using the LAI values and canopy architecture data as inputs. The overall model had 4.8% of outliers according to the 95% confidence bounds. In Table 4, it can be observed that the training stage had a correlation coefficient of R = 0.82, while the testing stage and the overall model had correlation coefficients of R = 0.81 and 0.82, respectively. Furthermore, the model had a high performance MSE = 0.09 for the training stage and MSE = 0.11 for the testing stage, which is indicative of an adequate model.

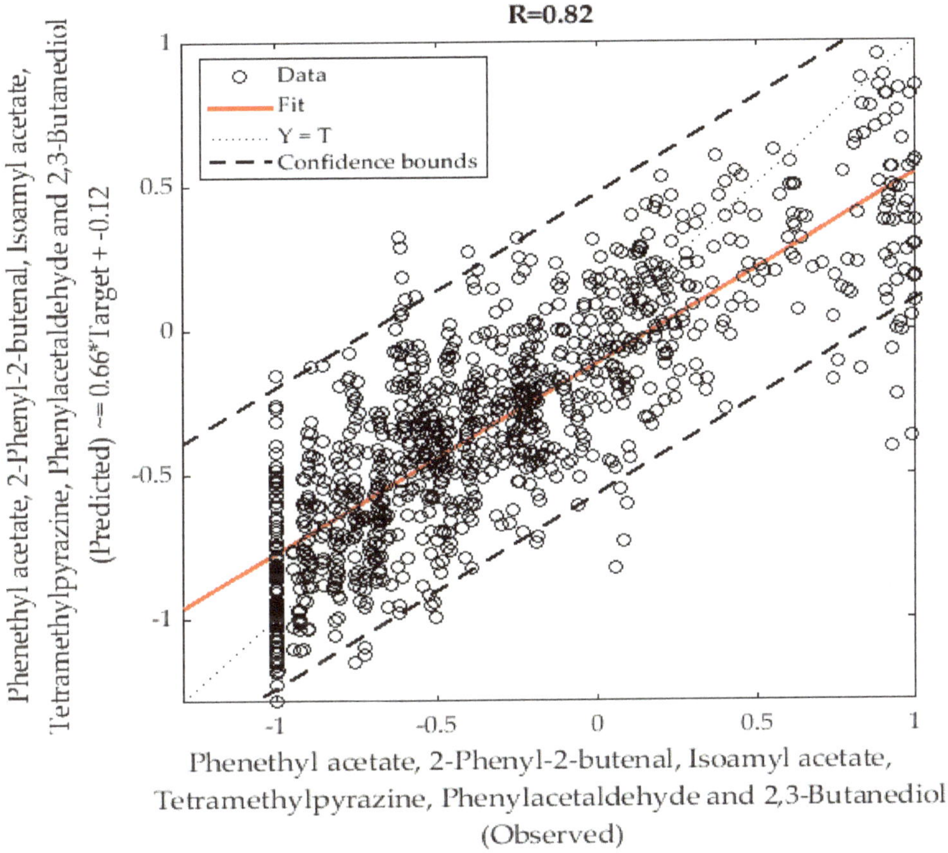

Figure 6. Artificial neural network overall model showing the observed responses (targets) in the x-axis and the predicted values (outputs) in the y-axis of the six volatile compounds. R = correlation coefficient.

Table 4. Results from the artificial neural network model showing the numbers of samples and observations, correlation coefficient (R), and performance based on mean squared error (MSE) for each stage.

Stage	Samples	Observations (Samples × 6 Targets)	R	MSE
Training	147	882	0.82	0.09
Testing	26	156	0.81	0.11
Overall	173	1038	0.82	-

3.3. *Implementation of the Machine Learning Model Developed to Map Aroma Compounds from Aerial Imagery*

Aerial imagery allowed the extraction of canopy architecture parameters from sub-images per plant that are effectively comparable to upward-looking images from the VitiCanopy App. This was validated using a simple linear regression between the averaged LAI_e parameters from the 24 sampled trees in the field using VitiCanopy (observed) and those obtained from the same plants extracted from the aerial image (correlation coefficient (R) = 0.91; data not shown).

The machine learning model developed and described in Figure 6 and Table 3 was implemented using the canopy architecture parameters obtained from the aerial imagery. Figure 7 shows the LAI_e and aroma profile data from each of the three blocks studied, using a kriging interpolation technique. The denormalized GC area values obtained from the machine learning model implementation were from 0 to 60×10^3, which was used as a scale for all the aroma profiles. Since LAI_e is one of the most important canopy architectural parameters, it was individually mapped with values ranging from 0 (missing plants) to 6. Figure 8 shows the scattered values, which represent the LAI_e and aroma profile for individual plants from the field studied.

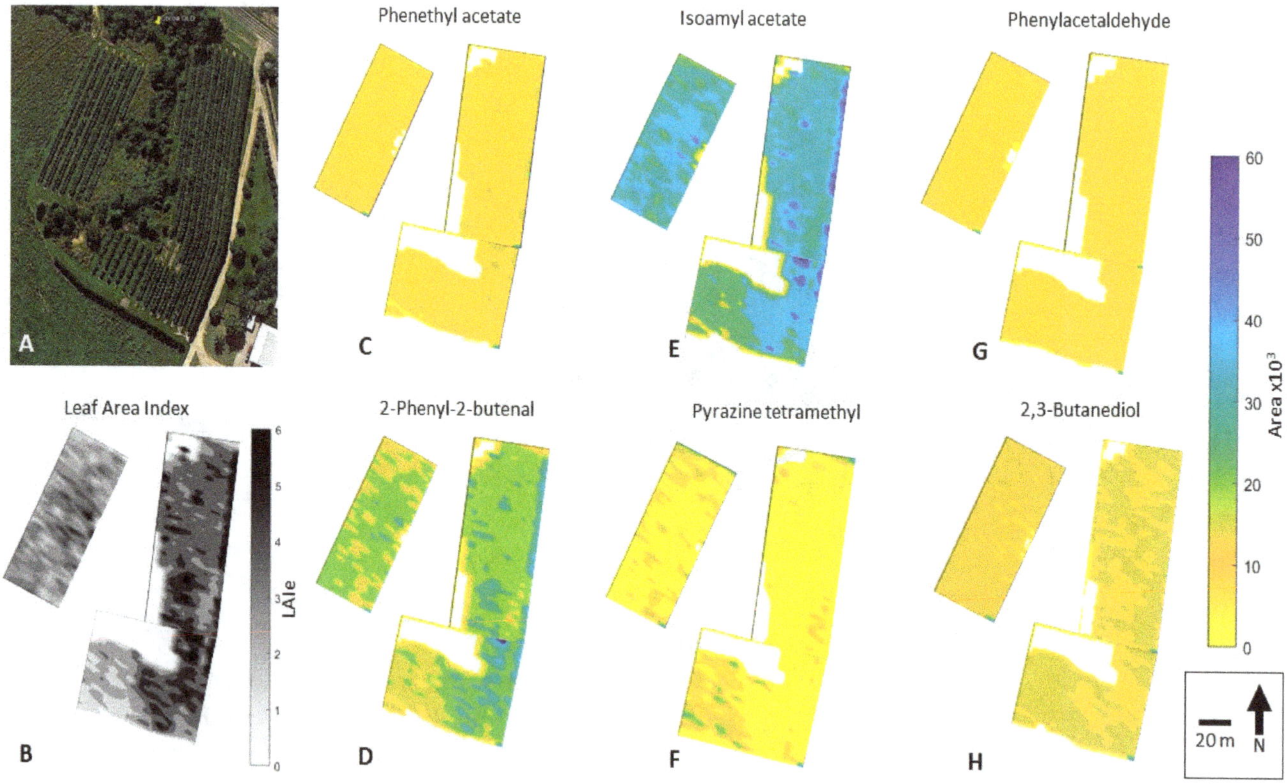

Figure 7. Original aerial image obtained from Google Maps Pro (**A**), and mapping of effective leaf area indexes (LAI_e) (**B**) and aroma profiles for phenethyl acetate (honey/floral/yeasty/cocoa) (**C**); 2-phenyl-2-butenal (sweet/cocoa/nutty/beany) (**D**); isoamyl acetate (banana/fruity) (**E**); pyrazine tetramethyl (roasted/cocoa/coffee/nutty/earthy) (**F**); phenylacetaldehyde (honey/rose/cocoa) (**G**); and 2,3-butanediol (cocoa butter/creamy) (**H**).

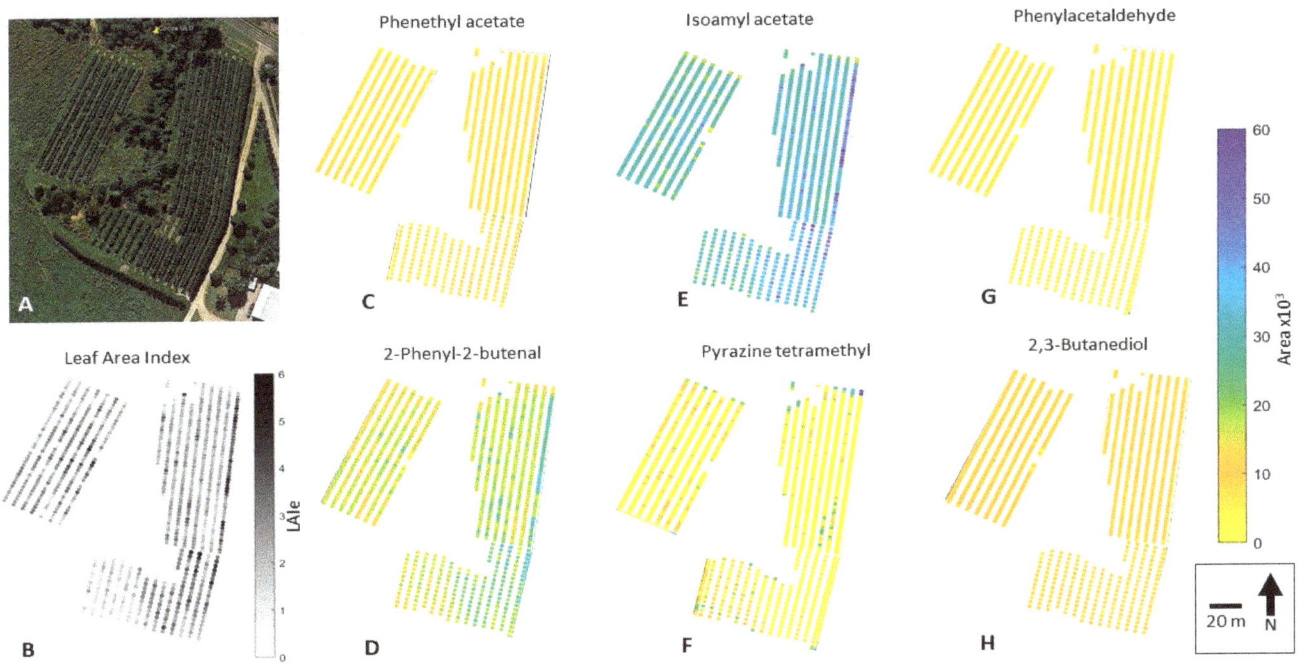

Figure 8. Original aerial image obtained from Google Map Pro (**A**), and plant-by-plant mapping of effective leaf area indexes (LAI_e) (**B**) and aroma profiles for phenethyl acetate (honey/floral/yeasty/cocoa) (**C**); 2-phenyl-2-butenal (sweet/cocoa/nutty/beany) (**D**); isoamyl acetate (banana/fruity) (**E**); pyrazine tetramethyl (roasted/cocoa/coffee/nutty/earthy) (**F**); phenylacetaldehyde (honey/rose/cocoa) (**G**); and 2,3-butanediol (cocoa butter/creamy) (**H**).

4. Discussion

4.1. Canopy Architecture and Volatile Compounds

The mean values of LAI and LAI_e for all trees fall within those reported by Daymond et al. [23] (2.2–4.5) for different clones of cacao in Brazil. The range of the mean values of canopy cover found in the present study was between 67% and 94%, which is similar to the range (72%–91%) obtained by Bos et al. [34] in Criollo and Forastero hybrids of cacao trees from Indonesia. The significant differences found between the two blocks for LAI, LAI_e, and crown porosity may be due to the fact that Block 1 was planted in March 2013, while Block 2 was planted in October 2013, and therefore, their maturity stages were different at the time of measurements, giving an average LAI of 2.89 for Block 2, which is lower than that of Block 1 (3.35). High-resolution aerial images could be obtained using Unmanned Aerial Vehicle (UAV) platforms and high-resolution visible cameras. It is expected that by using the latter type of high-resolution images, the correlation between VitiCanopy and the aerial extraction of canopy architecture features will improve due to a more accurate gap analysis.

All volatile compounds found in the cocoa beans have been previously reported in other studies. Ten out of 20 of these compounds provide aromas related to nuts, roasted, cocoa, and chocolate, while six compounds provide fruity aromas including banana, cherry, and pineapple. Acetoin and 2,3-butanediol, which elicit creamy and butter aromas, were also found. According to Aprotosoaie et al. [35], the alcohols found in cocoa are due to fermentation and decrease with drying and roasting. However, 2-heptanol and phenylethyl alcohol, which give fruity and floral aromas, are commonly found in cacao. In the present study, phenylethyl alcohol was the most abundant compound in most of the samples. Furthermore, aldehydes and ketones, followed by esters and pyrazines, are the most important to determine cocoa quality [28,35], and in this study, seven compounds from aldehydes and ketones, four esters, and three pyrazines were found (Table 2). According to Rodriguez-Campos et al. [28], phenethyl acetate is an important ester in fermented cocoa beans, and higher concentrations of this compound are an indicator of high cocoa quality. In the present study, this volatile compound was identified in 22 out of 24 samples. On the other hand, similar to the present study, tetramethylpyrazine has been reported in other studies as the most abundant pyrazine and is one of the most important volatiles for cocoa flavor [36]. This was found in all except one sample from this study (Figure 5).

4.2. Machine Learning Modeling to Predict the Aroma Profile

Besides fermentation and drying, other factors that contribute to differences in aromas and flavors in cacao are the genotype, location and growth conditions, ripening, and amount of shade [37]. Therefore, the machine learning model developed in this study based on canopy architecture parameters has a strong basis on the physiological factors affecting the aroma profile. Hence, the high accuracy and performance obtained (R = 0.82; MSE = 0.09) using the canopy architecture information, which is related to the amount of shade, accurately represents the six volatile compounds commonly found in cocoa and that are determinant for its quality (Table 2). Up to date, there are no known studies that have managed to predict the volatile compounds in cacao using the canopy architecture. However, there was a study in grapevines in which flavonol contents were predicted using crown porosity and LAI using partial least squares (PLS) regression [38].

The model presented in this paper may be useful by producers to modify the canopy architecture of the cocoa trees according to the desired aroma characteristics on the final product. This would allow them to avoid financial loses or predict the end use of the final product by anticipating the quality based on the aroma profile. This model may be fed with more data from samples of other regions and genotypes to create a general predictive model to be used by any grower. On the other hand, similar models may be developed for other crops to predict and improve the quality at early stages.

4.3. Implementation of the Machine Learning Model Developed to Map Aroma Compounds from Aerial Imagery

Spatial distribution of LAI_e and aroma profiles were possible due to high-resolution aerial images, which can be effectively used to implement the canopy cover photography algorithms described (Equations (1)–(6)). A similar approach was implemented for grapevines using f_c, with similar results from imagery obtained using unmanned aerial vehicles [24]. From the kriging interpolated maps (Figure 7) it can be seen that some compounds are highly dependent on vigor characteristics of plants, such as isoamyl acetate, with clear differences of higher areas found for the far-right row in Block 1 (Figure 7E). These trees corresponded to almost double LAI_e values (~5) compared to the average of trees in Blocks 1 and 2 (Figure 7A,B). Other compounds, such as phenethyl acetate, phenylacetaldehyde, and pyrazine tetramethyl, do not have many spatial differences among the three blocks (Figure 7C,G,F, respectively). The latter results confirm that the pattern of the machine learning models was effectively identified, and they are independent for each aroma related to canopy architecture parameters.

The differences in the spatial distribution of growth and aroma profiles can be more clearly seen using a scattering plot, which represents the tree-by-tree scale (Figure 8). This spatial distribution map could offer growers a powerful tool to perform tailored canopy management strategies to manage aroma profile outputs that could be considered desirable or to obtain a uniform quality of pods and, finally, the chocolate produced.

5. Conclusions

The data analysis tools presented in this paper could contribute significantly to adapting management strategies from growers, specifically in relation to canopy management, to obtain desired aroma profiles at the tree-by-tree scale. Specific practical applications could be related to homogenize the aroma quality profile within plants and the entire plantation or to increase desired and specific aromas to change the style of the cocoa volatiles. These management strategies can be implemented within the critical stages of the growing period of cocoa trees, or close to harvest. By implementing VitiCanopy and the machine learning modeling strategies presented, these potential changes in the volatile composition of beans related to canopy architecture changes due to canopy management can be obtained in real time at the fingertips of growers through their portable electronic devices (smartphones and tablet PCs). The latter will enhance crop management capabilities from major industries, but more importantly, by growers in developing countries.

Supplementary Materials: Figure S1. Map showing the site location where the study was conducted. The image was obtained from Google Earth (Google Technology company, Menlo Park, CA, USA); Figure S2. Diagram representing the methods used in the study, where (A) represents the canopy architecture assessment at ground level using the VitiCanopy app,(B) depicts the cacao pods processing and assessment of volatile compounds using gas chromatography, mass spectroscopy, (C) the development of the machine learning modelling using (A) as inputs and (B) as targets/outputs,
(D) represents the canopy architecture assessment from aerial images using a customized Matlab® code, and (E) the development of spatial distribution maps using the outputs from (C) using the data from (D) as inputs; Figure S3. Peak areas of the six volatile compounds found in the fermented cocoa bean samples used to develop the machine learning model. All values are reported in scientific notation 10^3. Sample codes: B1 = Block 1, B2 = Block 2, C# = number of the tree, and A or B = position of the tree (row).

Author Contributions: Conceptualization, S.F.; data curation, S.F. and C.G.V.; formal analysis, S.F. and C.G.V.; investigation, S.F., G.C., D.D.T., and A.Z.; methodology, S.F., G.C., D.D.T., A.Z., and C.G.V.; project administration, S.F.; software, S.F.; supervision, S.F.; validation, C.G.V.; visualization, S.F. and C.G.V.; writing—original draft, S.F., G.C., D.D.T., A.Z., and C.G.V.; writing—review & editing, S.F. and C.G.V.

Acknowledgments: The authors would like to acknowledge the support from Yan Diczbalis, Principal Horticulturist from the Department of Agriculture and Fisheries, Queensland, Australia. Furthermore, the authors acknowledge Pangzhen Zhang for his support in the use of GC-MS in the chemistry laboratory of The University of Melbourne, Australia.

References

1. Dand, R. Chapter 9: Cocoa Bean Processing and the Manufacture of Chocolate. In *The International Cocoa Trade*; MacFadyen, H., Ed.; Woodhead Publishing: Cambridge, MA, USA, 2011.
2. ICCO. *Production of Cocoa Beans*; International Cocoa Organization: London, UK, 2017.
3. Astika, I.; Solahudin, M.; Kurniawan, A.; Wulandari, Y. Determination of cocoa bean quality with image processing and artificial neural network. In *The Quality Information for Competitive Agricultural Based Production System and Commerce, Proceedings of the AFITA 2010 International Conference, Bogor, Indonesia, 4–7 October 2010*; Asian Federation for Information Technology in Agriculture: Bogor, Indonesia, 2010.
4. Beckett, S.T.; Fowler, M.S.; Ziegler, G.R. *Beckett's Industrial Chocolate Manufacture and Use*; John Wiley & Sons: Hoboken, NJ, USA, 2017.
5. De Brito, E.S.; García, N.H.P.; Amancio, A.C.; Valente, A.L.; Pini, G.F.; Augusto, F. Effect of autoclaving cocoa nibs before roasting on the precursors of the Maillard reaction and pyrazines. *Int. J. Food Sci. Technol.* **2001**, *36*, 625–630. [CrossRef]
6. Ilangantileke, S.; Wahyudi, T.; Bailon, M.G. Assessment methodology to predict quality of cocoa beans for export. *J. Food Qual.* **1991**, *14*, 481–496. [CrossRef]
7. Afoakwa, E.O.; Paterson, A.; Fowler, M.; Ryan, A. Flavor formation and character in cocoa and chocolate: A critical review. *Crit. Rev. Food Sci. Nutr.* **2008**, *48*, 840–857. [CrossRef] [PubMed]
8. Elwers, S.; Zambrano, A.; Rohsius, C.; Lieberei, R. Differences between the content of phenolic compounds in Criollo, Forastero and Trinitario cocoa seed (*Theobroma cacao* L.). *Eur. Food Res. Technol.* **2009**, *229*, 937–948. [CrossRef]
9. Hansen, C.E.; Mañez, A.; Burri, C.; Bousbaine, A. Comparison of enzyme activities involved in flavour precursor formation in unfermented beans of different cocoa genotypes. *J. Sci. Food Agric.* **2000**, *80*, 1193–1198. [CrossRef]
10. Saravia-Matus, S.L.; Rodríguez, A.G.; Saravia, J.A. Determinants of certified organic cocoa production: Evidence from the province of Guayas, Ecuador. *Org. Agric.* **2019**, 1–12. [CrossRef]
11. Department of Agriculture and Fisheries. *Growing Cocoa*; Government of Queensland: Brisbane, Australia, 2015.

12. Trognitz, B.; Cros, E.; Assemat, S.; Davrieux, F.; Forestier-Chiron, N.; Ayestas, E.; Kuant, A.; Scheldeman, X.; Hermann, M. Diversity of cacao trees in Waslala, Nicaragua: Associations between genotype spectra, product quality and yield potential. *PLoS ONE* **2013**, *8*, e54079. [CrossRef]
13. Smulders, M.; Esselink, D.; Amores, F.; Ramos, G.; Sukha, D.; Butler, D.; Vosman, B.; Van Loo, E. Identification of cocoa (*Theobroma cacao* L.) varieties with different quality attributes and parentage analysis of their beans. *IGENIC Newsl.* **2008**, *12*, 1–13.
14. Olujide, M.; Adeogun, S. Assessment of cocoa growers' farm management practices in Ondo State, Nigeria. *Span. J. Agric. Res.* **2006**, *4*, 173–179. [CrossRef]
15. Daymond, A.; Hadley, P. Differential effects of temperature on fruit development and bean quality of contrasting genotypes of cacao (*Theobroma cacao*). *Ann. Appl. Biol.* **2008**, *153*, 175–185. [CrossRef]
16. Saunders, J.A.; Mischke, S.; Leamy, E.A.; Hemeida, A.A. Selection of international molecular standards for DNA fingerprinting of *Theobroma cacao*. *Theor. Appl. Genet.* **2004**, *110*, 41–47. [CrossRef] [PubMed]
17. De Bei, R.; Fuentes, S.; Gilliham, M.; Tyerman, S.; Edwards, E.; Bianchini, N.; Smith, J.; Collins, C. VitiCanopy: A free computer app to estimate canopy vigor and porosity for grapevine. *Sensors* **2016**, *16*, 585. [CrossRef]
18. Poblete-Echeverría, C.; Fuentes, S.; Ortega-Farias, S.; Gonzalez-Talice, J.; Yuri, J.A. Digital cover photography for estimating leaf area index (LAI) in apple trees using a variable light extinction coefficient. *Sensors* **2015**, *15*, 2860–2872. [CrossRef] [PubMed]
19. Macfarlane, C.; Hoffman, M.; Eamus, D.; Kerp, N.; Higginson, S.; McMurtrie, R.; Adams, M. Estimation of leaf area index in eucalypt forest using digital photography. *Agric. For. Meteorol.* **2007**, *143*, 176–188. [CrossRef]
20. Miyaji, K.-I.; Da Silva, W.S.; Alvim, P.D.T. Longevity of leaves of a tropical tree, *Theobroma cacao*, grown under shading, in relation to position within the canopy and time of emergence. *New Phytol.* **1997**, *135*, 445–454. [CrossRef]
21. Fuentes, S.; Palmer, A.R.; Taylor, D.; Zeppel, M.; Whitley, R.; Eamus, D. An automated procedure for estimating the leaf area index (LAI) of woodland ecosystems using digital imagery, MATLAB programming and its application to an examination of the relationship between remotely sensed and field measurements of LAI. *Funct. Plant Biol.* **2008**, *35*, 1070–1079. [CrossRef]
22. Yapp, J.H.; Hong, H. *A Study into the Potential for Enhancing Productivity in Cocoa (Theobroma cacao L.) through Exploitation of Physiological and Genetic Variation*; University of Reading: Reading, UK, 1992.
23. Daymond, A.; Hadley, P.; Machado, R.; Ng, E. Canopy characteristics of contrasting clones of cacao (*Theobroma cacao*). *Exp. Agric.* **2002**, *38*, 359–367. [CrossRef]
24. Baofeng, S.; Jinru, X.; Chunyu, X.; Yuyang, S.; Fuentes, S. Digital surface model applied to unmanned aerial vehicle based photogrammetry to assess potential biotic or abiotic effects on grapevine canopies. *Int. J. Agric. Biol. Eng.* **2016**, *9*, 119–130.
25. Xue, J.; Fan, Y.; Su, B.; Fuentes, S. Assessment of canopy vigor information from kiwifruit plants based on a digital surface model from unmanned aerial vehicle imagery. *Int. J. Agric. Biol. Eng.* **2019**, *12*, 165–171. [CrossRef]
26. Biehl, B.; Meyer, B.; Crone, G.; Pollmann, L.; Said, M.B. Chemical and physical changes in the pulp during ripening and post-harvest storage of cocoa pods. *J. Sci. Food Agric.* **1989**, *48*, 189–208. [CrossRef]
27. Selamat, J.; Mordingah Harun, S. Formation of methyl pyrazine during cocoa bean fermentation. *Pertanika* **1994**, *17*, 27.
28. Rodriguez-Campos, J.; Escalona-Buendía, H.; Contreras-Ramos, S.; Orozco-Avila, I.; Jaramillo-Flores, E.; Lugo-Cervantes, E. Effect of fermentation time and drying temperature on volatile compounds in cocoa. *Food Chem.* **2012**, *132*, 277–288. [CrossRef] [PubMed]
29. The Good Scents Company. The Good Scents Company Information System. Available online: http://www.thegoodscentscompany.com (accessed on 3 March 2019).
30. Ramos, C.L.; Dias, D.R.; PedrozoMiguel, M.G.D.; FreitasSchwan, R. Impact of different cocoa hybrids (*Theobroma cacao* L.) and S. cerevisiae UFLA CA11 inoculation on microbial communities and volatile compounds of cocoa fermentation. *Food Res. Int.* **2014**, *64*, 908–918. [CrossRef] [PubMed]
31. Owusu, M.; Petersen, M.A.; Heimdal, H. Assessment of aroma of chocolate produced from two Ghanaian cocoa fermentation types. In Proceedings of the 12th International Weurman Flavour Research Symposium, Interlaken, Switzerland, 1–4 July 2008.

32. Rodriguez-Campos, J.; Escalona-Buendía, H.; Orozco-Avila, I.; Lugo-Cervantes, E.; Jaramillo-Flores, M.E. Dynamics of volatile and non-volatile compounds in cocoa (*Theobroma cacao* L.) during fermentation and drying processes using principal components analysis. *Food Res. Int.* **2011**, *44*, 250–258. [CrossRef]
33. Bonvehí, J.S. Technology. Investigation of aromatic compounds in roasted cocoa powder. *Eur. Food Res. Technol.* **2005**, *221*, 19–29. [CrossRef]
34. Bos, M.M.; Steffan-Dewenter, I.; Tscharntke, T. Shade tree management affects fruit abortion, insect pests and pathogens of cacao. *Agric. Ecosyst. Environ.* **2007**, *120*, 201–205. [CrossRef]
35. Aprotosoaie, A.C.; Luca, S.V.; Miron, A. Flavor chemistry of cocoa and cocoa products—An overview. *Compr. Rev. Food Sci. Food Saf.* **2016**, *15*, 73–91. [CrossRef]
36. Ramli, N.; Hassan, O.; Said, M.; Samsudin, W.; Idris, N.A. Influence of roasting conditions on volatile flavor of roasted Malaysian cocoa beans. *J. Food Process. Preserv.* **2006**, *30*, 280–298. [CrossRef]
37. Kongor, J.E.; Hinneh, M.; Van de Walle, D.; Afoakwa, E.O.; Boeckx, P.; Dewettinck, K. Factors influencing quality variation in cocoa (*Theobroma cacao*) bean flavour profile—A review. *Food Res. Int.* **2016**, *82*, 44–52. [CrossRef]
38. Martínez-Lüscher, J.; Brillante, L.; Kurtural, S.K. Flavonol profile is a reliable indicator to assess canopy architecture and the exposure of red wine grapes to solar radiation. *Front. Plant Sci.* **2019**, *10*, 10. [CrossRef]

7

A Novel Cosmic-Ray Neutron Sensor for Soil Moisture Estimation over Large Areas

Luca Stevanato [1,*], Gabriele Baroni [2], Yafit Cohen [3], Cristiano Lino Fontana [1], Simone Gatto [4], Marcello Lunardon [1], Francesco Marinello [4], Sandra Moretto [1] and Luca Morselli [1]

1 Department of Physics and Astronomy, University of Padova, Via Marzolo 8, 35131 Padova, Italy; cristiano.fontana@unipd.it (C.L.F.); marcello.lunardon@unipd.it (M.L.); sandra.moretto@unipd.it (S.M.); luca.morselli.1@studenti.unipd.it (L.M.)

2 Department of Agricultural and Food Sciences, University of Bologna, Viale Fanin 50, 40127 Bologna, Italy; g.baroni@unibo.it

3 Department of Sensing, Information and Mechanization Engineering, Agricultural Research Organization (ARO), Volcani Center, Rishon Lezion 7505101, Israel; yafitush@volcani.agri.gov.il

4 Department of Land, Environment, Agriculture and Forestry, University of Padova, Viale dell'Università 16, 35020 Legnaro, Italy; simonegatto@portofelloni.com (S.G.); francesco.marinello@unipd.it (F.M.)

* Correspondence: luca.stevanato@unipd.it

Abstract: A correct soil moisture estimation is a fundamental prerequisite for many applications: agriculture, meteorological forecast, flood and drought prediction, and, in general, water accounting and management. Traditional methods typically provide point-like measurements, but suffer from soil heterogeneity, which can produce significant misinterpretation of the hydrological scenarios. In the last decade, cosmic-ray neutron sensing (CRNS) has emerged as a promising approach for the detection of soil moisture content. CRNS can average soil moisture over a large volume (up to tens of hectares) of terrain with only one probe, thus overcoming limitations arising from the heterogeneity of the soil. The present paper introduces the development of a new CRNS instrument designed for agricultural applications and based on an innovative neutron detector. The new instrument was applied and tested in two experimental fields located in Potsdam (DE, Germany) and Lagosanto (IT, Italy). The results highlight how the new detector could be a valid alternative and robust solution for the application of the CRNS technique for soil moisture measurements in agriculture.

Keywords: CRNS; neutron; cosmic-ray; soil moisture; water; precision farming

---

1. Introduction

Water scarcity and drought problems in several parts of the world highlight the necessity for new solutions for better management of water resources. The Food and Agriculture Organization of the United Nations (FAO) calculates that 70% of employed water resources are dedicated to agriculture on the global scale [1]. Sustaining agricultural productivity requires an efficient management of agricultural water resources that involves a clear understanding of the temporal dynamics and spatial variability of soil moisture. Such dynamics are essential to optimize irrigation, preserve water for drought periods, and to optimize other production inputs, such as fertilizer application and water pumping power. The prerequisite is reliable soil moisture data, measured over large-scales and in real-time.

Due to this crucial role, many devices have been developed to measure soil moisture at different spatial and temporal scales [2–4]. Available technologies range from point-scale invasive approaches, like time domain reflectometry (TDR) probes [5], to satellite remote sensing approaches [6]. The former technology can achieve accurate measurements (RMSE, root-mean-square error <2%) at high temporal resolution (minutes) and at different soil depths [5]. However, difficulties arise for the coverage of large areas (e.g., >500 m^2), since soil heterogeneity can produce significant misinterpretation of hydrological conditions [4,7]. To monitor a large area, several point-scale probes must be installed, but this might generate technical problems (such as the definition of statistically relevant positions, electrical power supply, data transmission) and high costs, due to multiple instruments and to low accessibility of specific locations (especially in the case of extensive crops) [8–10]. Furthermore, their set-up in agricultural fields is limited by tillage and other land management operations and farming practices. In addition, the detectors are invasive (buried in the soil) and they require high maintenance. For these reasons, they are not suitable for covering heterogeneous and inaccessible sites (mountain sides and cropped fields) and they are expensive for long-term monitoring observatories.

A completely different alternative is represented by remote sensing approaches typically based on microwaves (1 mm–1 m). Compared to point-scale methods, satellite remote sensing provides soil moisture observations at a large scale (>km^2) and covers global areas, so it is more suitable for hydrological applications [11]. However, the signal is sensitive only to the very first centimeters of soil interface [12] and the temporal resolution (e.g., weekly measurements) is not always suitable for many applications. Large-scale satellite remote sensing methods have other limitations [13], including limited capability to penetrate vegetation, inability to measure soil ice, and sensitivity to surface roughness [14]. For these reasons, despite the progress over the last decades, accuracy of remote sensing estimation is still too high for several applications (e.g., RMSE >4%) [15] and many studies are focusing on possible improvements [6].

In the last decade, to overcome the aforementioned operational challenges, a proximal geophysical method has been developed in order to fill the gap between point-scale and remote sensing approaches: cosmic-ray neutron sensing (CRNS) [16–18]. CRNS is a valid and robust alternative, offering many advantages: it is contactless, allows quantification of soil moisture averaged over large areas with only one probe, and is not invasive for field agricultural operations. The major advantages of CRNS are its large horizontal footprint (up to tens of hectares) and the penetration depth of tens of centimeters, enough to reach typical roots' depth [19,20].

The technique is based on the natural neutrons detected on the earth's surface, that are mostly generated by cosmic rays, according to various processes. The high energy protons component of cosmic rays, produced by galactic sources, interact with atomic nuclei in the upper atmosphere and produce secondary high-energy particles cascades. Muons and fast neutrons are constituents of such cascades. The secondary neutrons reach the ground level and interact with soil atoms. Another mechanism of neutron production is the so-called spallation effect, due to high-energy secondary muons interacting with ground atoms.

Hydrogen in water molecules becomes the dominant factor for slowing down and absorbing neutrons (also known as neutrons moderation). The fast neutrons, produced in air and soil, travel in all directions within the air–soil–vegetation continuum and therefore an equilibrium concentration of neutrons is established. The equilibrium is shifted in response to changes in the water presence above and below the land surface. For example, a drier soil, having a lower moderation capacity, reflects a greater number of neutrons compared to a more humid soil. In the latter, neutrons are more easily moderated, thus slowed down and partially absorbed; the net effect is an increase of the slow population, in respect to the fast one. The resultant neutron intensity above the land surface is inversely proportional to soil water content.

The portion of the neutrons energy spectrum that is most sensitive to soil moisture is the epithermal/fast region from energies of 0.25 eV to 100 keV [20]. Evaporation neutrons from 100 keV to 10 MeV give additional information especially for snow measurements [21]. Finally, high-energy neutrons over 10 MeV are not dependent from local conditions and are directly proportional to the primary incoming flux. Many studies relied on the performance of a set of CRNS probes for monitoring [22,23], modeling [24,25], or remote-sensing validation purposes [26,27].

It is important to underline that the measured intensity of environmental neutrons depends not only on the water in the soil but also on the incoming cosmic-ray neutrons flux [28]. This component changes with changing atmospheric conditions and also with variation of the incoming flux of galactic cosmic rays [17]. For this reason, on one hand CRNS is typically equipped with sensors for air pressure p, air temperature T, and relative humidity h_{rel}. On the other hand, it is worth noting that normally the correction by the incoming-flux is not directly measured in situ, but it is extrapolated offline using the Neutron Monitor Database (NMDB) [29], which provides data from several stations around the world.

The aim of the present study was to present and assess a new sensor for detecting soil moisture based on the CRNS approach. For this reason, several tests have been conducted in different agro-environmental conditions in comparison to current commercial CRNS probes and point-scale soil moisture measurements. The discussion focuses particularly on the applicability of the new sensor for agricultural applications.

2. Materials and Methods

2.1. Instrumentation

For many years neutrons have been detected using ^{3}He proportional counter tubes (for the thermal component) and liquid/plastic scintillators (for the fast component); these two well-established technologies show some important limitations in practical applications. ^{3}He is a nuclide produced almost entirely in artificial contexts, as the product of the tritium decay. The current storage is depleting, and the price is high and rising, since it comes mainly from the production or dismantling of nuclear weapons of past decades [30]. Liquid/plastic scintillators are often made of toxic or hazardous materials, safely used in research contexts, but not suitable for agricultural, civil, or industrial applications. The interest in neutron detection for homeland security applications has triggered, in the last decade, the development of new detectors made up of liquid or plastic scintillation materials with low toxicity, that are safe and easy to use.

We studied a new solution [31], namely Finapp, based on a composite detector made of commercial detectors: EJ-299-33A and EJ-420(6), both manufactured by Eljen Technology (Sweetwater, TX, USA). EJ-420 and EJ-426 are inorganic scintillators, that have proven to have a good response to thermal neutrons [32]. EJ-299-33A was the first plastic detector to become commercially available for gamma/fast-neutron discrimination [33]. The discrimination capability of neutrons from gamma-rays is a fundamental prerequisite, in order to discriminate neutrons from the naturally occurring gamma-ray environmental background, that is not correlated to soil moisture in the same way as neutrons.

We assembled the composite detector by wrapping the plastic scintillator EJ-299-33A (a cylinder of $3'' \times 3''$) with thermal neutron detectors. In the plane face we used an EJ420 detector with a diameter of $3''$. On the lateral face we used EJ426, which has a lower efficiency but is flexible. The optical-grade rubber EJ-560 ensures proper optical contact between the different detectors. A single photomultiplier (PMT) Mod. H6553 (Hamamatsu Photonics, Hamamatsu, Japan) was used to collect light emissions of the scintillators. Figure 1 shows an exploded view of the assembly.

Normally EJ420(6) and ^{3}He tubes are able to detect only thermal neutrons with energy below 0.025 eV but, according to Kohli et al. [20], the most sensitive part to soil moisture of the neutron spectra is the epithermal region between 0.025 eV and 100 keV. In order to collect these neutrons, detectors are normally equipped with a few cm of polyethylene, a material enriched with hydrogen that acts as a moderator slowing down neutron energy from the epithermal/fast region to the thermal region.

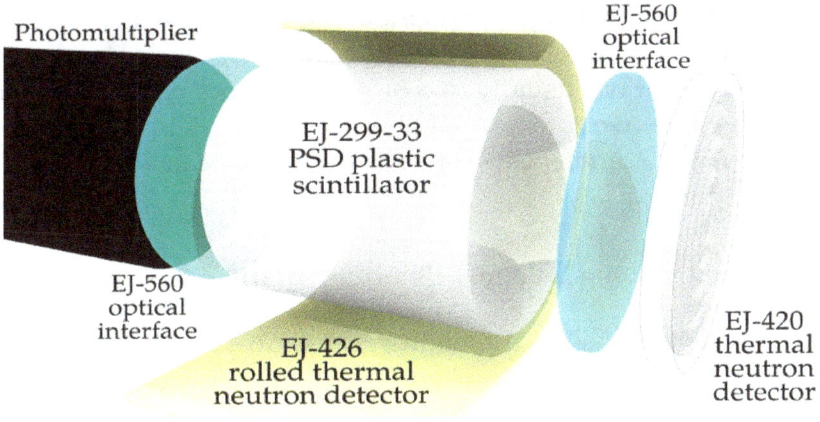

Figure 1. Assembly of the detector, Finapp, to employ in cosmic-ray neutron sensing (CRNS) technique. Abbreviation: PSD, pulse shape discrimination.

2.2. Data Acquisition

The data acquisition system (DAQ) was composed of an electronic signal digitizer model DT5790 (CAEN Spa, Viareggio, Italy) featuring: two input channels, that generate digital waveforms from the analog signals, with a 12 bit resolution and a sampling rate of 250 MS/s (samples per second); and two channels of high voltage power supply, for the photomultiplier. The digitizer is interfaced with a low-cost, low-power, embedded computer (Beaglebone Black). The software controlling the digitizer is an open-source, distributed data acquisition system, called ABCD [34,35] developed for the H2020 C-BORD project [36,37]. ABCD is employed to provide a continuous stream of data to a custom analysis system, based on the group's experience developed during the FP7 TAWARA_RTM project [38]. The analysis is fully automatic and performed online. The probe was equipped with internal temperature sensors and was configured to gather weather data from a local weather station installed near the probe. The operating negative voltage for the photomultiplier was set at 1600 V. Data transmission was ensured by a cellphone modem and data were stored locally in a Secure Digital (SD) memory. Figure 2 shows a block diagram of the probe functionality.

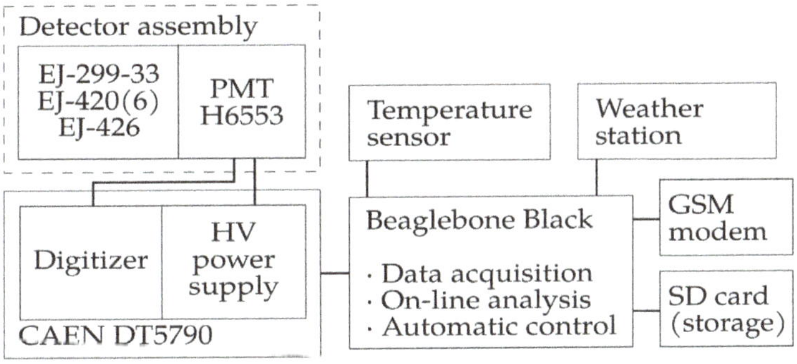

Figure 2. Block diagram of the data acquisition chain of the probe. Abbreviations: PMT, photomultiplier; HV, high voltage; GSM, protocol for data trasmission; SD, Secure Digital.

2.3. Data Processing and Analysis

2.3.1. Particle Discrimination

The significant parameters (e.g., signal integrals), needed for the signals' identification and discrimination, are extracted from the digital waveforms acquired by the digitizer. Together with the auxiliary sensors' data (e.g., temperature, pressure, etc.), the analysis software performs a data merge of the information, in order to correct the parameters and improve the signals identification and discrimination capabilities. Particles are identified and discriminated according to the generated signals, with an algorithm based on the most popular method of pulse shape discrimination (PSD) [39]. This technique exploits the different processes activated by different particles interacting in the scintillator; in particular, the produced light has usually two or more components, characterized by their decay time τ_i. The various light components are excited with different yields, depending on the interacting particle. These processes lead to different waveform shapes, that can be distinguished calculating the so-called PSD parameter, defined as:

$$\text{PSD} = (\text{Long Integral} - \text{Short Integral})/\text{Long Integral} \tag{1}$$

where "Long Integral" is the PMT signal charge-integrated over a long integration gate, while "Short Integral" is calculated on a short integration gate. The "Long Integral" is associated with the light coming from all the components. The "Short Integral" accounts only for the light associated with the fastest component.

Figure 3 shows the typical sampled signals from gamma-rays, fast neutrons, and thermal neutrons (Figure 3a) and PSD parameter versus energy (Figure 3b). Discrimination between particles can be made simply by selecting the significant regions, on the PSD plot, in order to separate the three contributes (Figure 3b).

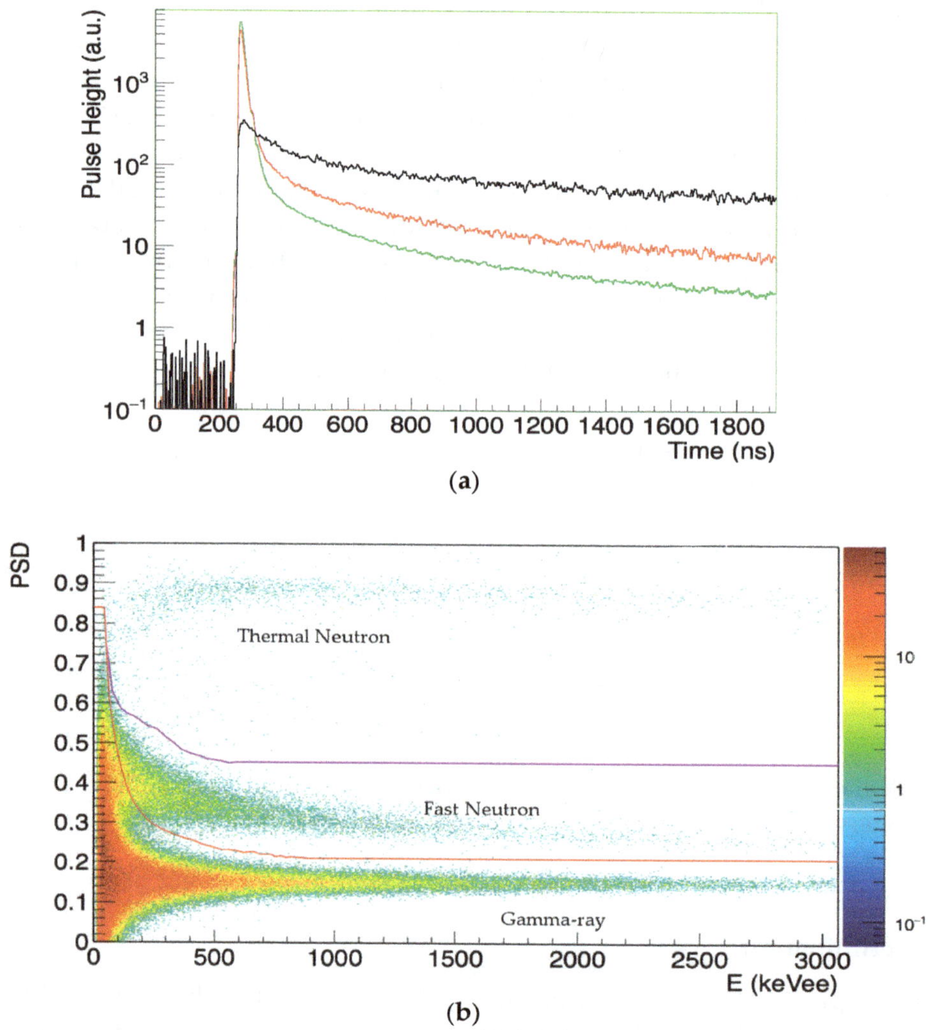

Figure 3. (**a**) Characteristic signals from gamma-ray (green), fast neutron (red), and thermal neutron (black). (**b**) PSD parameter versus energy; different zones are identified by a polynomial line.

2.3.2. From Raw Neutron Counts to Soil Moisture Estimation

The number of neutrons depends not only on the water content in the environment but also on atmospheric conditions and incoming galactic cosmic rays [17]. For this reason, data from a close by weather station was used to normalize neutron counts.

The standard procedure [40] to correct neutron counts (N_{raw}) by atmospheric variations and incoming fluctuations follows:

$$N = N_{raw} \cdot exp\Big(\beta\big(\langle p\rangle - p_{ref}\big)\Big) \cdot \Big(1 - \alpha\big(\langle h\rangle - h_{ref}\big)\Big) \cdot \left(1 + \gamma\left(\frac{I_{ref}}{I} - 1\right)\right) \tag{2}$$

where, h is the absolute humidity in g·m^{-3}, I the incoming flux of galactic cosmic-ray, $\beta = 0.0076$, $\alpha = 0.0054$, $\gamma = 1$, and h_{ref}, p_{ref} are the mean value of humidity and pressure during the measuring period, respectively. I_{ref} is the average value of the incoming fluctuation over a long period and depends on the efficiency of the station used for correction.

The corrected environmental neutrons at ground level are converted into (soil) water equivalent θ with the following empirical formula:

$$\theta(N) = \left(\frac{0.0808}{\frac{N}{N_0} - 0.372} - 0.115 - \theta_{offset}\right) \cdot \rho_{bulk} \tag{3}$$

where, ρ_{bulk} is the soil bulk density (kg·m^{-3}), N is the corrected neutron flux, θ_{offset} is the gravimetric water equivalent of additional hydrogen pools (e.g., lattice water, soil organic carbon), and N_0 is the counting rate over dry soil [21,41]. N_0 could be calibrated based on independent soil sampling campaigns as suggested in different studies [18]. Since the aim of the present study was to assess the capability of the Finapp probe to provide the correct hydrological behavior, N_0 was used as a tuning parameter to fit the soil moisture dynamics measured by point-scale measurements.

Concerning the incoming corrections, the standard procedure foresees the use of the nearest Neutron Monitor Database (NMBD) station close to the site measurement, but this this could introduce problems: (i) the distance between the CNRS probe and NMDB stations may be of the order of several hundreds of km; (ii) delays in collecting data from the NMDB can create problems for online monitoring; (iii) dependence from an external source of information, the availability of NMBD data is at the discretion of the institution that maintains the station. To overcome these problems, our probe measured the incoming fluctuations directly in situ through the measurement of muons. Their flux at sea level depends on atmospheric conditions and incoming cosmic-ray fluctuations, in the same way as high-energy neutrons. Muons release a great amount of energy in plastic scintillators; thus, it is possible to identify these particles by putting an appropriate energy threshold to exclude completely the contribution of gamma-rays from the environmental background.

3. Experimental Sites

3.1. Potsdam, Germany

The first experimental site was at the campus of Potsdam University, in the Brandenburg region near Berlin, Germany. This experimental site was mainly dedicated to the assessment of the capabilities of the Finapp probe, based on the comparison with two ^{3}He tubes installed in the field. The two tubes were from Hydroinnova LLC (Albuquerque, NM, USA) and Canberra Industries, (Meriden, CT, USA): two commercial CRNS probes used as a standard for cosmic-rays neutron sensing for many years.

Figure 4 shows test-site pictures. Our innovative detector Finapp measures neutrons in the same energy range of the commercial tubes without the use of ^{3}He gas. In a two-month period of outdoor field tests (from 29 May 2018 to 17 July 2018), Finapp was compared with the Hydroinnova CRS-1000 and the CANBERRA tube. The experimental site is equipped with a standard weather station that was extended to monitor four soil moisture profiles based on point-scale soil moisture sensors (5TE Meter group) installed at 5, 15, 25, and 35 cm depth that are used for comparison.

The climate in Potsdam is warm and temperate, with an average temperature of 9.2 °C and an annual rainfall of 600 mm evenly distributed throughout the year. During summer months, thunderstorms are frequent. The WGS84 coordinates of the installation site are N 52.410087, E 12.978808. There are no cultivated fields nearby; the probes have been installed on an uncultivated grassy lawn.

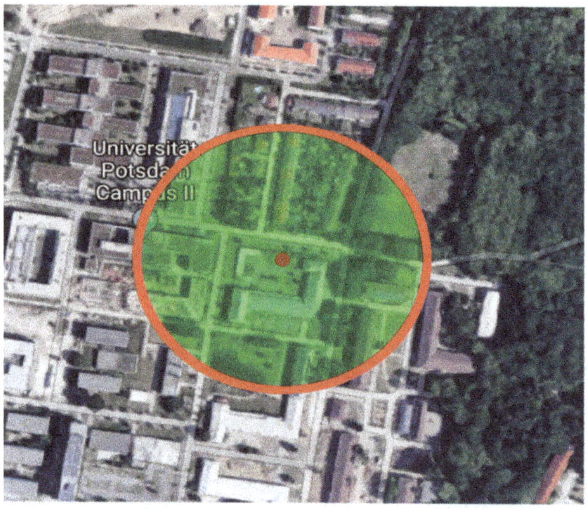

(a) (b)

Figure 4. (**a**) Campus of Potsdam University, red point is where CRNS probes were installed and the red circle is the footprint. (**b**) Picture of the installation.

3.2. *Lagosanto, Italy*

The second experimental site was located in Lagosanto, Italy, in an experimental field in the Porto Felloni agricultural company. This company is renowned as being one of the most technologically advanced in Italy, with a continuous and effective implementation of precision farming practices. The area is located in Emilia Romagna, a few km from the sea and 50 km from Ferrara. The climate is warm and temperate with an average temperature of 14 °C and an annual rainfall of 600 mm, with a dry period during the summer months. The probe was installed in a recent orchard with walnut trees, characterized by a small trunk. During summer months water is provided daily by drip irrigation. The probe was positioned as shown in Figure 5a at WGS84 coordinates N 44.752756, E 12.134761. The probe's footprint is an area with a radius of about 150 m, as highlighted by the red circle. Figure 5b shows a picture of the installation; the probe was installed 1.8 m from the ground. At about 90 m from the probe, five classical Sentek sensors for soil moisture were installed at 10, 20, 30, 40, and 50 cm depth, while a weather station was installed less than 1 km from the probe. Data were collected from August 8, 2018 to November 11, 2018.

In the test area, soil is in general homogenous, and characterized by a sandy loamy texture, with a 2% organic matter, as reported in Table 1.

Table 1. Texture of the soil where the Sentek probes were installed.

Loam	Sand	Clay	Organic Matter	ρ_{bulk}
42.1%	38.5%	19.4%	2.1%	1.4 g/cm^3

(a) (b)

Figure 5. (**a**) Field site, red point is where the Finapp probe was installed, the red circle is the footprint, the orange point represents where the five Sentek probes are installed at different depths, and the light blue line is the drainage canal. (**b**) Picture of the installation.

The drainage of the soil favors infiltration, calling for the need of frequent irrigation. The plant is vulnerable to dehydration and the soil must therefore always be kept damp, but waterlogging must be avoided, as this can cause asphyxia and blossom-end rot of the fruit.

A particularity of this site is the presence of a very shallow saturated zone. The territory is a reclamation area, kept dry thanks to water pumps that operate 24 h a day. This creates a very shallow saturated zone that reaches 50 cm deep. This is clearly visible from Figure 6, where the soil moisture sensors from the Sentek probes reaches 50% of the water equivalent at a depth of 50 cm. Furthermore, the owner of the property has underlined how the nearby reclamation canal (identified in Figure 5a with the light blue line) creates infiltrations in the field and this creates problems of asphyxiation of the plants due to too much water in the soil.

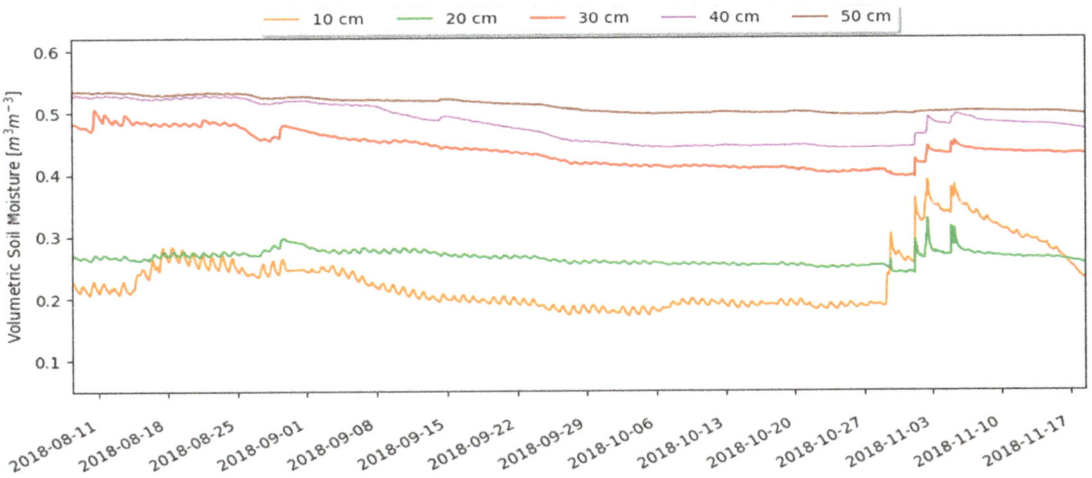

Figure 6. Volumetric soil moisture from Sentek sensor in Lagosanto for the field test period, from from August 8, 2018 to November 11, 2018.

4. Results

4.1. Potsdam Results

The Finapp probe was compared with two commercial CRNS probes, one from Hydroinnova (CRS-1000) and another from CANBERRA during the two months of outdoor testing at Potsdam University. Figure 7 shows the time series of epithermal neutron counts corrected for air pressure, for the whole study period. The neutron counts are averaged over 12 h intervals. The y-axis reports the variation in respect to the average counts of the single probes in the whole period.

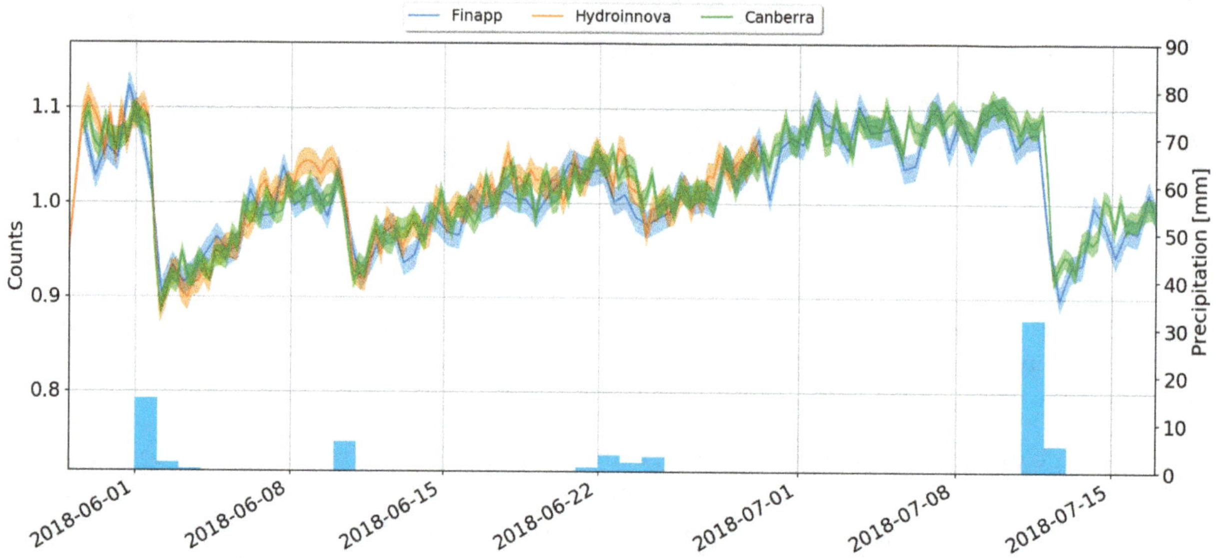

Figure 7. Epithermal counts: Finapp (blue), CRS-1000 from Hydroinnova (orange), and He-3 tube from CANBERRA (green). Data is corrected for atmospheric pressure variations. The colored zone indicates statistic uncertainty. Daily rain is reported in the bottom part of the plot.

Results show very good agreement between epithermal counts. The epithermal neutron counts drop during rain periods when the soil moisture increases. The efficiency of Finapp in respect to CRS-1000 from Hydroinnova, quantified in terms of epithermal counts, was 55%. This aspect means that the Finapp probe needs about twice the time, needed by the Hydroinnova probe, to have the same accuracy in measurements.

Figure 8 shows soil moisture from neutrons count by Equation (3) using data from the Finapp probe. The same figure reports the measurements of classical TDR probes at different depths. It is possible to note that the soil moisture in Potsdam is more uniform at different depths, compared to the Lagosanto data (Figure 6) where the soil moisture increases sensibly with depth. At Potsdam, the soil is much drier and the groundwater system is much deeper, thus it does not affect the conditions in the first 50 cm of the soil.

There is very good agreement between the three rain events (stormy) on June 1, June 11, and July 12. It is worth pointing out the rain event between June 21–26, where light rain increased the soil moisture according to CRNS, but it was not seen from 5TE probes. The behavior could be explained by the small amount of rain which did not infiltrate into the soil but remained on the land surface. For this reason, the 5TE did not record the event while the Finapp detected the water canopy interception or ponding water in the surrounding urban areas. Similar behaviors have been identified in previous studies [40,42].

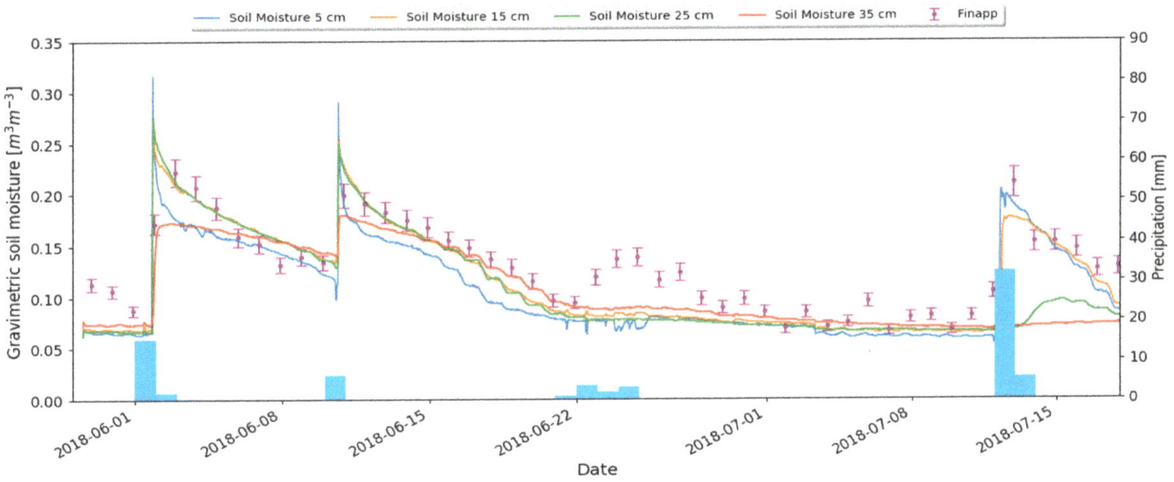

Figure 8. Gravimetric soil moisture from classical time domain reflectometry (TDR) probes (blue, orange, green, red) at different depths. The magenta points are CRNS gravimetric soil moisture using Desilet formula [19] and epithermal counts from Finapp. Daily rain is reported in the bottom part of the plot.

4.2. Lagosanto Results

Raw thermal counts were corrected hourly by absolute humidity and pressure from the near weather station. The reference values were set to $h_{ref} = 12$ g/m^3 and $p_{ref} = 1016.4$ hPa. The incoming variation was corrected by the muons measured by Finapp. The flux is mostly stable during the whole period and fluctuation is of the order of 2%. The reference value was set to $I_{ref} = 4667$.

Furthermore, to improve the comparison with point-scale probes, we computed a weighted average of the measurements (Figure 6) at different depths as described in [43]. CRNS measures the water content in the first 20–70 cm of the soil depending on soil moisture; the higher the water content, less the depth sensitivity, because wet soil has a larger moderation power for neutrons. Finapp acquired data during the second part of summer and first part of autumn.

The end of the vegetative season for the orchard is in September and the irrigation drip was stopped on September 1. September and October were relatively dry months with a rainfall of 60 mm. From October 27 a rainy period significantly increased the soil moisture. Figure 9 shows the averaged soil moisture from the Sentek probe and the soil moisture from Finapp; the lower part of the figure shows the cumulated precipitation with a 24 h interval. Focusing attention on the Finapp data, it is possible to notice a very good agreement between precipitation and soil moisture. Every rain event corresponds to an increase in volumetric soil moisture. Only the events on October 13 report an increase of soil moisture without any precipitation. This increase is probably due to infiltration from the near reclamation canal, that increased soil moisture in the footprint of our probe, as suggested by the property owner. On the contrary, there are some differences if we compare Finapp with point-scale probes. First of all, during the irrigation period the point-scale measurements are less sensitive to precipitation because the drip irrigation biases the point measurement due to a heterogeneous water distribution. In September there is good agreement between the Finapp and Sentek probes; the soil moisture decreases in both cases due to the dry period. In October there is again some discrepancy, probably due to a malfunction of the Sentek probe. The property owner notes that in October the probes are no longer checked daily because the data is no longer used, and periods of malfunctioning occur due to various technical problems (power, batteries, data transmission, etc.). In November the agreement is good again and both the Sentek and CRNS probes respond to precipitation in the same way.

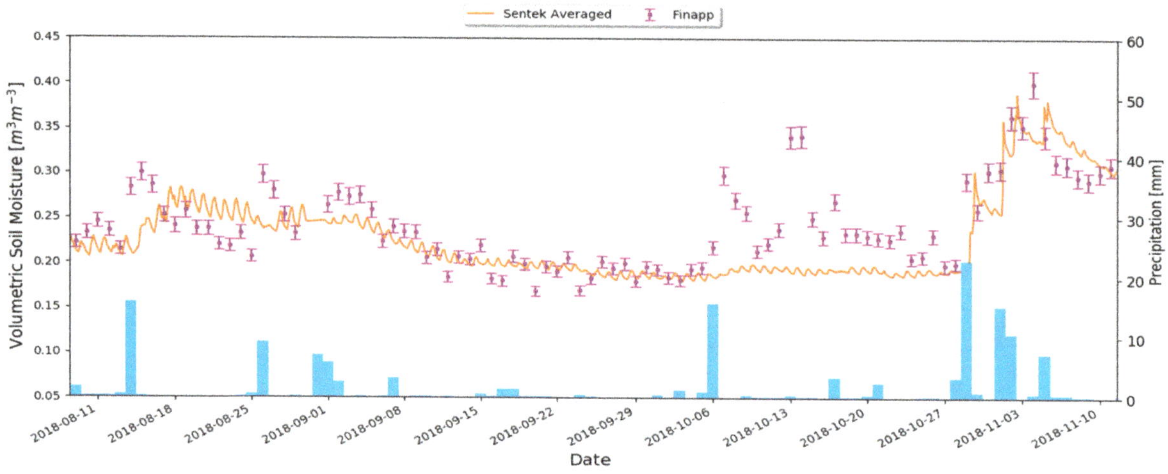

Figure 9. Volumetric soil moisture from Finapp (magenta) and average soil moisture from Sentek (orange); for details see the text. Daily rain is reported in the bottom part of the plot.

5. Discussion and Conclusions

The Finapp probe was tested as a novel sensor for soil moisture estimation over a large area. The sensor makes use of the CRNS technique to reconstruct soil moisture from environmental epithermal neutron counts. The results show that Finapp measures the neutrons in the epithermal region like the well-tested ^{3}He tubes. Efficiency as an epithermal neutrons counter is 55%, in respect to the commercial CRS-1000 from Hydroinnova. Efficiency could be increased, in order to reduce statistic uncertainty, by growing the size of the detector and/or using new materials for thermal neutron detection. In particular, good results were achieved by our group [44] and more research is currently underway.

We reconstructed soil moisture in Figures 8 and 9 for the Potsdam and Lagosanto field trials, respectively. Data is well correlated with precipitations, and it shows the limit of classical point-like measurements, that can be influenced by the heterogeneity of the soil and/or the irrigation distributions. Averaging soil moisture over a large area overcomes the problem of these heterogeneities, making the measurement better suitable for supporting agricultural management conducted at field scale (e.g., irrigation). The soil moisture in Figures 8 and 9 have a time-interval of 24 h between data points. We use such integration time in order to reduce statistic uncertainty on the final measurements. Typical integration values for CRNS commercial probes are 6 or 12 h.

In respect to the currently-employed technology with ^{3}He tubes, Finapp offers more measured parameters, like gamma-rays and fast neutrons. Future development foresees the analysis of this information. In particular, gamma-rays were found to be correlated to soil moisture as well [45,46], but with a smaller footprint. The fast neutrons and high energy events detected can be used to normalize the incoming galactic fluctuations directly in-situ with the data from the sensor.

Overall, Finapp uses non-toxic and non-carcinogenic plastic materials, is easily recyclable, and its technology can be easily scaled to market size. Scalability will be the key point for future development, in order to reduce the price to the same level as professional point-scale soil moisture probes currently on the market.

Author Contributions: Conceptualization, L.S., M.L., S.M. and C.L.F.; methodology, L.S., M.L., G.B. and F.M.; software, L.S., C.L.F. and L.M.; validation, L.S., M.L., G.B. and F.M.; formal analysis, L.S. and L.M.; investigation, L.S., M.L., G.B. and F.M.; resources, L.S., G.B., F.M. and S.G.; data curation, L.S., G.B. and S.G.; writing—original draft preparation, L.S.; writing—review and editing, L.S., G.B., F.M., M.L., S.M., C.L.F. and Y.C.; visualization, L.S., M.L., L.M., G.B. and C.L.F.; supervision, L.S.; project administration, L.S.; funding acquisition, L.S.

Acknowledgments: The authors thank the property of the Porto Felloni agricultural company for hosting the Finapp instrumentation for three months of outdoor tests. We are also grateful to Potsdam University for giving us the opportunity to compare the Finapp probe with commercial CRNS ^{3}He tubes, and to Peter Bíró and Andreas Bauer for their technical support.

References

1. Food and Agriculture Organization (FAO). Introduction. In *Water for Sustainable Food and Agriculture*; FAO: Rome, Italy, 2017; p. 1. ISBN 978-92-5-109977-3.
2. Corradini, C. Soil moisture in the development of hydrological processes and its determination at different spatial scales. *J. Hydrol.* **2014**, *516*, 1–5. [CrossRef]
3. Ochsner, T.E.; Cosh, M.H.; Cuenca, R.H.; Dorigo, W.A.; Draper, C.S.; Hagimoto, Y.; Kerr, Y.H.; Njoku, E.G.; Small, E.E.; Zreda, M. State of the Art in Large-Scale Soil Moisture Monitoring. *Soil Sci. Soc. Am. J.* **2013**, *77*, 1888–1919. [CrossRef]
4. Robinson, D.A.; Campbell, C.S.; Hopmans, J.W.; Hornbuckle, B.K.; Jones, S.B.; Knight, R.; Ogden, F.; Selker, J.; Wendroth, O. Soil Moisture Measurement for Ecological and Hydrological Watershed-Scale Observatories: A Review. *Vadose Zone J.* **2008**, *7*, 358–389. [CrossRef]
5. Romano, N. Soil moisture at local scale: Measurements and simulations. *J. Hydrol.* **2014**, *516*, 6–20. [CrossRef]
6. Bauer-Marschallinger, B.; Freeman, V.; Cao, S.; Paulik, C.; Schaufler, S.; Stachl, T.; Modanesi, S.; Massari, C.; Ciabatta, L.; Brocca, L.; et al. Toward Global Soil Moisture Monitoring with Sentinel-1: Harnessing Assets and Overcoming Obstacles. *IEEE Trans. Geosci. Remote Sens.* **2018**, *57*, 520–539. [CrossRef]
7. Susha Lekshmi, S.U.; Singh, D.N.; Baghini, M.S. A critical review of soil moisture measurement. *Measurement* **2014**, *54*, 92–105. [CrossRef]
8. Famiglietti, J.S.; Deveraux, J.A.; Laymon, C.A.; Tsegaye, T.; Houser, P.R.; Jackson, T.J.; Graham, S.T.; Rodell, M.; van Oevelen, P.J. Ground-based investigation of soil moisture variability within remote sensing footprints During the Southern Great Plains 1997 (SGP97) Hydrology Experiment. *Water Resour. Res.* **1998**, *35*, 1839–1852. [CrossRef]
9. Famiglietti, J.S.; Ryu, D.; Berg, A.A.; Rodell, M.; Jackson, T.J. Field observations of soil moisture variability across scales. *Water Resour. Res.* **2008**, *44*, W01423. [CrossRef]
10. Western, A.W.; Grayson, R.B.; Bioshl, G. Scaling of Soil Moisture: A Hydrologic Perspective. *Annu. Rev. Earth Planet. Sci.* **2002**, *30*, 149–180. [CrossRef]
11. Zhuo, L.; Han, D. The Relevance of Soil Moisture by Remote Sensing and Hydrological Modelling. *Procedia Eng.* **2016**, *154*, 1368–1375. [CrossRef]
12. Petropoulos, G.P.; Ireland, G.; Barrett, B. Surface soil moisture retrievals from remote sensing: Current status, products & future trends. *Phys. Chem. Earth* **2015**, *83*, 35–56. [CrossRef]
13. Entekhabi, D.; Njoku, E.G.; Houser, P.; Spencer, M.; Doiron, T.; Kim, Y.; Smith, J.; Girard, R.; Belair, S.; Crow, W.; et al. The Hydrosphere State (Hydros) Satellite Mission: An Earth System Pathfinder for Global Mapping of Soil Moisture and Land Freeze/Thaw. *IEEE Trans. Geosci. Remote Sens.* **2004**, *42*, 2184–2195. [CrossRef]
14. Wanders, N.; Karssenberg, D.; Bierkens, M.; Parinussa, R.; de Jeu, R.; van Dam, J.; de Jong, S. Observation uncertainty of satellite soil moisture products determined with physically-based modeling. *Remote Sens. Environ.* **2012**, *127*, 341–356. [CrossRef]
15. Fang, B.; Lakshmi, V. Soil moisture at watershed scale: Remote sensing techniques. *J. Hydrol.* **2014**, *516*, 258–272. [CrossRef]
16. Zreda, M.; Desilets, D.; Ferré, T.; Scott, R. Measuring soil moisture content non-invasively at intermediate spatial scale using cosmic-ray neutrons. *Geophys. Res. Lett.* **2008**, *35*, L21402. [CrossRef]
17. Zreda, M.; Shuttleworth, W.; Zeng, X.; Zweck, C.; Desilets, D.; Franz, T.; Rosolem, R. COSMOS: The COsmic-ray Soil Moisture Observing. *Hydrol. Earth Syst. Sci.* **2012**, *16*, 4079–4099. [CrossRef]
18. Franz, T.E.; Zreda, M.; Rosolem, R.; Ferre, T.P.A. Field Validation of a Cosmic-Ray Neutron Sensor Using a Distributed Sensor Network. *Vadose Zone J.* **2012**, *11*. [CrossRef]
19. Desilets, D.; Zreda, M. Footprint diameter for a cosmic-ray soil moisture probe: Theory and Monte Carlo simulations. *Water Resour. Res.* **2013**, *49*, 3566–3575. [CrossRef]

20. Kohli, M.; Schron, M.; Zreda, M.; Schmidt, U.; Dietrich, P.; Zacharias, S. Footprint characteristics revised for field-scale soil moisture monitoring with cosmic-ray neutrons. *Water Resour. Res.* **2015**, *51*, 5772–5790. [CrossRef]
21. Desilets, D.; Zreda, M.; Ferré Ty, P.A. Nature's neutron probe: Land surface hydrology at an elusive scale with cosmic rays. *Water Resour. Res.* **2010**, *46*, W11505. [CrossRef]
22. Villarreyes, R.C.; Baroni, G.; Oswald, S. Integral quantification of seasonal soil moisture changes in farmland by cosmic-ray neutrons. *Hydrol. Earth Syst. Sci.* **2011**, *15*, 3843–3859. [CrossRef]
23. Evans, J.G.; Ward, H.C.; Blake, J.R.; Hewitt, E.J.; Morrison, R.; Fry, M.; Ball, L.A.; Doughty, L.C.; Libre, J.W.; Hitt, O.E.; et al. Soil water content in southern England derived from a cosmic-ray soil moisture observing system—COSMOS-UK. *Hydrol. Process.* **2016**, *30*, 4987–4999. [CrossRef]
24. Baatz, R.; Bogena, H.R.; Franssen, H.-J.H.; Huisman, J.A.; Montzka, C.; Vereecken, H. An empirical vegetation correction for soil water content quantification using cosmic ray probes. *Water Resour. Res.* **2015**, *51*, 2030–2046. [CrossRef]
25. Andreasen, M.; Jensen, H.K.; Zreda, M.; Desilets, D.; Bogena, H.; Looms, C.M. Modeling cosmic ray neutron field measurements. *Water Resour. Res.* **2016**, *52*, 6451–6471. [CrossRef]
26. Holgate, C.M.; De Jeu, R.A.M.; van Dijk, A.I.J.M.; Liu, Y.Y.; Renzullo, L.J.; Kumar, V.; Dharssi, I.; Parinussa, R.M.; Van Der Schalie, R.; Gevaert, A.; et al. Comparison of remotely sensed and modelled soil moisture data sets across Australia. *Remote Sens. Environ.* **2016**, *186*, 497–500. [CrossRef]
27. Montzka, C.; Bogena, H.; Zreda, M.; Monerris, A.; Morrison, R.; Muddu, S.; Vereecken, H. Validation of Spaceborne and Modelled Surface Soil Moisture Products with Cosmic-Ray Neutron Probes. *Remote Sens.* **2017**, *9*, 103. [CrossRef]
28. Schrön, M.; Rosolem, R.; Köhli, M.; Piussi, L.; Schröter, I.; Iwema, J.; Kögler, S.; Oswald, S.E.; Wollschläger, U.; Samaniego, L.; et al. Cosmic-ray Neutron Rover Surveys of Field Soil Moisture and the Influence of Roads. *Water Resour. Res.* **2018**, *54*, 6441–6459. [CrossRef]
29. Mavromichalaki, H.; Papaioannou, A.; Plainaki, C.; Sarlanis, C.; Souvatzoglou, G.; Gerontidou, M.; Papailiou, M.; Eroshenko, E.; Belov, A.; Yanke, V.; et al. Applications and usage of the real-time Neutron Monitor Database. *Adv. Space Res.* **2011**, *47*, 2210–2222. [CrossRef]
30. Kouzes, R. *The ^{3}He Supply Problem*; PNNL-18388; Pacific Northwest National Laboratory: Richland, WA, USA, 2009.
31. Cester, D.; Lunardon, M.; Moretto, S.; Nebbia, G.; Pino, F.; Sajo-Bohus, L.; Stevanato, L.; Bonesso, I.; Turato, F. A novel detector assembly for detecting thermal neutrons, fast neutrons and gamma rays. *Nucl. Instrum. Methods A* **2016**, *830*, 191–196. [CrossRef]
32. Pino, F.; Stevanato, L.; Cester, D.; Nebbia, G.; Sajo-Bohus, L.; Viesti, G. Study of the thermal neutron detector ZnS(Ag)/LiF response using digital pulse processing. *J. Instrum.* **2015**, *10*, T08005. [CrossRef]
33. Cester, D.; Nebbia, G.; Stevanato, L.; Pino, F.; Viesti, G. Experimental tests of the new plastic scintillator with pulse shape discrimination capabilities EJ-299-33. *Nucl. Instrum. Methods A* **2014**, *735*, 202–206. [CrossRef]
34. Fontana, C.L. A distributed data acquisition system for signal digitizers with on-line analysis capabilities. In Proceedings of the IEEE Nuclear Science Symposium and Medical Imaging Conference, Atlanta, GA, USA, 21–28 October 2017; pp. 1–5. [CrossRef]
35. Fontana, C.L. A distributed data acquisition system for nuclear detectors. *Int. J. Mod. Phys. Conf. Ser.* **2018**, *48*, 1860118. [CrossRef]
36. Sardet, A.; Pérot, B.; Carasco, C.; Sannié, G.; Moretto, S.; Nebbia, G.; Fontana, C.; Moszyński, M.; Sibczyński, P.; Grodzicki, K.; et al. Design of the rapidly relocatable tagged neutron inspection system of the C-BORD project. In Proceedings of the IEEE Nuclear Science Symposium, Medical Imaging Conference and Room-Temperature Semiconductor Detector Workshop, Strasbourg, France, 29 October–5 November 2016; pp. 1–5. [CrossRef]
37. Sibczynski, P.; Dziedzic, A.; Grodzicki, K.; Iwanowska-Hanke, J.; Mianowska, Z.; Moszyński, M.; Swiderski, L.; Syntfeld-Każuch, A.; Szawłowski, M.; Wolski, D.; et al. C-BORD—An overview of efficient toolbox for high-volume freight inspection. In Proceedings of the IEEE Nuclear Science Symposium and Medical Imaging Conference (NSS/MIC), Atlanta, GA, USA, 21–28 October 2017; pp. 1–3. [CrossRef]
38. Fontana, C.L. A Distributed Data Acquisition System for the Sensor Network of the TAWARA_RTM Project. *Phys. Procedia* **2017**, *90*, 271–278. [CrossRef]
39. Cester, D.; Lunardon, M.; Nebbia, G.; Stevanato, L.; Viesti, G.; Petrucci, S.; Tintori, C. Pulse shape discrimination with fast digitizers. *Nucl. Instrum. Methods A* **2014**, *748*, 33–38. [CrossRef]

40. Schrön, M.; Zacharias, S.; Womack, G.; Köhli, M.; Desilets, D.; Oswald, S.E.; Bumberger, J.; Mollenhauer, H.; Kögler, S.; Remmler, P.; et al. Intercomparison of cosmic-ray neutron sensors and water balance monitoring in an urban environment. *Geosci. Instrum. Methods Data Syst.* **2018**, *7*, 83–99. [CrossRef]
41. Bogena, H.R.; Huisman, J.A.; Baatz, R.; Hendricks Franssen, H.-J.; Vereecken, H. Accuracy of the cosmic-ray soil water content probe in humid forest ecosystems: The worst case scenario. *Water Resour. Res.* **2013**, *49*, 5778–5791. [CrossRef]
42. Baroni, G.; Oswald, S.E. A scaling approach for the assessment of biomass changes and rainfall interception using cosmic-ray neutron sensing. *J. Hydrol.* **2015**, *525*, 264–276. [CrossRef]
43. Franz, T.E.; Zreda, M.; Ferre, T.P.A.; Rosolem, R.; Zweck, C.; Stillman, S.; Zeng, X.; Shuttleworth, W.J. Measurement depth of the cosmic ray soil moisture probe affected by hydrogen from various sources. *Water Resour. Res.* **2012**, *48*, W08515. [CrossRef]
44. Carturan, S.M.; Vesco, M.; Bonesso, I.; Quaranta, A.; Maggioni, G.; Stevanato, L.; Zanazzi, E.; Marchi, T.; Fabris, D.; Cinausero, M.; et al. Flexible scintillation sensors for the detection of thermal neutrons based on siloxane 6LiF containing composites: Role of 6LiF crystals size and dispersion. *Nucl. Instrum. Methods A* **2019**, *925*, 109–115. [CrossRef]
45. Baldoncini, M.; Albéri, M.; Bottardi, C.; Chiarelli, E.; Raptis, K.G.C.; Strati, V.; Mantovani, F. Investigating the potentialities of Monte Carlo simulation for assessing soil water content via proximal gamma-ray spectroscopy. *J. Environ. Radioact.* **2018**, *192*, 105–116. [CrossRef]
46. Baldoncini, M.; Albéri, M.; Bottardi, C.; Chiarelli, E.; Raptis, K.G.C.; Strati, V.; Mantovani, F. Biomass water content effect on soil moisture assessment via proximal gamma-ray spectroscopy. *Geoderma* **2018**, *335*, 67–77. [CrossRef]

8

Vibration Monitoring of the Mechanical Harvesting of Citrus to Improve Fruit Detachment Efficiency

Sergio Castro-Garcia [1,*], Fernando Aragon-Rodriguez [1], Rafael R. Sola-Guirado [1], Antonio J. Serrano [2], Emilio Soria-Olivas [2] and Jesús A. Gil-Ribes [1]

[1] Department of Rural Engineering, Universidad de Cordoba, 14004 Cordoba, Spain; g92arrof@uco.es (F.A.-R.); ir2sogur@uco.es (R.R.S.-G.); gilribes@uco.es (J.A.G.-R.)

[2] IDAL, Intelligent Data Analysis Laboratory, Universidad de Valencia, 46100 Valencia, Spain; ajserran@uv.es (A.J.S.); emilio.soria@uv.es (E.S.-O.)

* Correspondence: scastro@uco.es.

Abstract: The introduction of a mechanical harvesting process for oranges can contribute to enhancing farm profitability and reducing labour dependency. The objective of this work is to determine the spread of the vibration in citrus tree canopies to establish recommendations to reach high values of fruit detachment efficiency and eliminate the need for subsequent hand-harvesting processes. Field tests were carried out with a lateral tractor-drawn canopy shaker on four commercial plots of sweet oranges. Canopy vibration during the harvesting process was measured with a set of triaxial accelerometer sensors with a datalogger placed on 90 bearing branches. Monitoring of the vibration process, fruit production, and branch properties were analysed. The improvement of fruit detachment efficiency was possible if both the hedge tree and the machinery were mutually adjusted. The hedge should be trained to facilitate access of the rods and to encourage external fructification since the internal canopy branches showed 43% of the acceleration vibration level of the external branches. The machine should be adjusted to vibrate the branches at a vibration time of at least 5.8 s, after the interaction of the rod with the branch, together with a root mean square acceleration value of 23.9 m/s^2 to a complete process of fruit detachment.

Keywords: *Citrus sinensis* L. Osbeck; mechanical harvesting; acceleration sensor; vibration time; logistic regression

---

1. Introduction

Citrus fruits, whether for fresh consumption or industrial processing, are mainly harvested by hand. Worldwide, 147 million tonnes of citrus were produced in 2017, including orange, grapefruit, lemon, mandarin, and other citrus fruits [1]. Spain is the sixth largest producer of citrus fruit in the world, with an approximate production in 2017–2018 of more than seven million tonnes. In Spain, the predominant citrus orchards are trained for manual harvesting, with an orientation towards the fresh market. Manually harvested orchards experience problems due to the availability of labour and the high cost of operation.

Within the citrus production process, harvesting is a phase of enormous economic importance due to its high impact on the final cost of production. The manual harvesting in Southern Spain requires an average of 95 days' work per hectare and represents between 25% and 35% of the final cost of production [2]. Roka and Hyman [3] stated that, under Florida conditions, the application of mechanical harvesting for industrial processing could provide a 50% cost reduction, while increasing labour productivity by ten per cent. With the current approach to citrus production, the high costs of manual harvesting could compromise the profitability of the activity and the future of plantations in the long term [4].

Since the 1970s, the development of mechanical citrus harvesting systems for the juice industry has mainly taken place in Florida. However, none of the mechanised systems have been able to match the flexibility and fruit selection capabilities of manual harvesting [5]. The foremost mechanical harvesting systems are the trunk and canopy shaker systems that were applied to and developed for citrus fruits, and which reach high values of harvesting efficiency ranging between 84–95% and 55–95%, respectively [6]. However, the detachment of immature fruitlets has been identified in both systems; this occurs particularly in the late varieties that are of special interest to the juice industry and, in addition to a reduction in the working capacity (ha/h), it could represent an obstacle to the adoption of these mechanical harvesting systems by farmers. Roka et al. [7] showed yield reduction values for the use of these machines compared to manual harvesting of 20–50% according to their use and regulation. In parallel, the development and testing of abscission agents have permitted an increase in the working capacity and an improvement in the detachment of mature fruit with these harvesting technologies [8]. It was shown that a moderate reduction in fruit detachment force, through the application of an abscission agent, was enough to significantly increase the harvesting efficiency [9]. Subsequently, it was demonstrated that the use of an abscission agent together with an adjustment of the vibration parameters, fundamentally time and frequency of vibration, allowed high percentages of mature fruit detachment to be achieved without significantly affecting the following season [10]. This result was confirmed based on the different frequency responses of mature fruit and immature fruitlets to mechanical harvesting [11].

Canopy shaker systems allow a continuous vibration of the tree row; the rods penetrate the canopy and achieve a high value of fruit detached in areas where there is direct contact of rods with branches. The use of a canopy shaker can generate a greater fall of leaves, shoots, and branches than manual harvesting. The fall of these organs is considered as tree damage, which could have negative implications in the yield, the productive life, and the cost of transport from the orchard to the industry [12]. For this reason, the improvement of these machines has been based on a dual objective; to increase the efficiency of mature fruit removal and to reduce the damage caused to trees. In order to improve the mechanical harvesting process, Savary et al. [13] developed a canopy shaker simulation based on finite element methods to predict and evaluate the interaction between the tree and the machine. Then, Savary et al. [14] evaluated the effect of vibration on the tree canopy according to the distribution of forces and accelerations in branches and fruits. Subsequently, machine improvement proposals were based on mathematical models and prototype tests with a combination of machine operating parameters such as frequency and amplitude of vibration [15,16], and the configuration and material of the rods [17]. The rods have been shown to play an essential role in the process of tree shaking, both in terms of fruit detachment and the possible generation of tree and fruit damage. Liu et al. [17] indicated that rod material affected the vibratory response of trees with respect to the acceleration peaks in branches. They recommended that the rods have high stiffness values, but that their surface should be smooth to reduce damage. The shape of the rods revealed that arc-shaped flexible rods had better performance with respect to fruit detachment efficiency and a lower rate of damage to trees than the free end rods [18]. In a further attempt to adapt the vibration process to tree requirements, Pu et al. [18] designed and tested a two-section independent shaker system in order to minimise tree damage and maximise harvesting efficiency. The adaptation of the machine to the variability of the tree canopy made it possible to reduce tree damage compared to other canopy shaker systems.

During the harvesting process, the machine rods penetrate the tree but do not usually reach all parts of the canopy, so vibration must be transmitted via the branches in order to detach the inner fruit. Whitney et al. [19] indicated that a greater fruit detachment efficiency value was obtained by harvesting smaller rather than larger canopy trees. Canopy shaker systems generally have a limited capacity to detach internal fruits from the tree and yet can remove almost all external fruits. The harvesting process of fruit inside the canopy may cause damage to the outermost parts due to a reduction in machine ground speed or an increment in vibration frequency value and may also require additional manual harvesting.

This study works on the hypothesis that mechanised harvesting processes can be carried out with a high value of harvesting efficiency, where subsequent hand-harvesting is not necessary, and that the possible, but moderate tree damage does not affect the productive life of the orchard. The objective of this work is to analyse the interaction of the canopy shaker rods with the tree branches from the data collected continuously during the mechanical harvesting process of citrus trees. First, the fruit-bearing branches with and without direct contact with the machine rods were characterised in the position and production. Then, the canopy vibration process during mechanical harvesting was monitored. Finally, the vibration parameters necessary to detach the fruit were determined, and recommendations were given on the harvest parameters necessary to increase the fruit removal efficiency.

2. Materials and Methods

Mechanical harvest tests were carried out in Cordoba (Spain) in 2017 during the sweet orange (*Citrus sinensis* L. Osbeck cv. Valencia) harvest season for juice production, during four weeks from the end of flowering to before the natural fall of immature fruitlets in June. Table 1 shows the main characteristics of the four, mechanically harvested citrus orchards. Trees had been planted in wide hedges over 0.4 m ridges and had wide row distances to allow machine manoeuvrability and the use of canopy shaker harvesters (Figure 1).

Table 1. Characteristics of citrus orchards mechanically harvested with the canopy shaker system.

	Plot 1	Plot 2	Plot 3	Plot 4
Date planted	2006	2005	2007	2005
Plot area (ha)	54.7	38.0	33.1	57.3
Trees per ha	440	330	440	330
Tree distance (m)	7 × 3	7 × 4	7 × 3	7 × 4
Hedge height (m)	4.0	4.0	4.3	4.4
Hedge width (m)	3.9	4.1	4.5	4.6

Figure 1. Example of a tree row trained in width hedge for mechanised harvesting.

Mechanical harvesting was carried out with a lateral tractor-drawn continuous canopy shaker system (Oxbo 3210, Byron, New York, NY, USA), working under regular conditions, with a ground speed range between 1 and 1.5 km/h (0.28–0.42 m/s), and a vibration frequency close to 4.5 Hz, which caused the fruit to fall to the ground. Harvesting tests were carried out to ensure close contact of the shaker system with the tree canopy (Figure 2). The machine harvested both sides of the hedge in independent passes, with an approximate working capacity of 0.4–0.5 ha/h. Subsequently, the fruit remaining in the canopy was hand-picked and collected together with the fruit from the ground and loaded into a container.

Figure 2. Lateral tractor-drawn continuous canopy shaker system (Oxbo, 3210) used in citrus harvesting tests. (**a**) front view before the harvesting process; (**b**) rear view after the harvesting process.

Interaction of the harvesting system with the tree canopy was studied for each side of the tree hedges (Figure 3). Because the tree hedges differed between orchards, a representative cross-section of the tree canopy was selected. The cross-section ranged from 4–5 m^2, and was composed of a distance from the line of the trunks to the outside of the canopy of up to 2.0–2.5 m, and of a height ranging from 0.5–2.5 m. This cross-section was selected because it differentiated an external area of the canopy with direct contact of the machine rods with the branches, for a rod length of 1.4 m, and an internal area of the canopy without direct contact with the rods. Furthermore, the cross-section was representative of the tree canopy, with high yield, avoiding the effect of the lower pendulous branches and allowing analysis of the vibration process in a homogeneous canopy area between the tested plots. The cross-section was divided into 16–20 sectors according to the width of the hedge, at intervals of 0.5 m both horizontally and vertically. In each sector, the values of vibration, the properties of the branches, and the number of fruits were recorded. The canopy area that had direct contact with the rods was sampled with 10 sectors, while the canopy area without direct contact was sampled with 6 or 10 sectors.

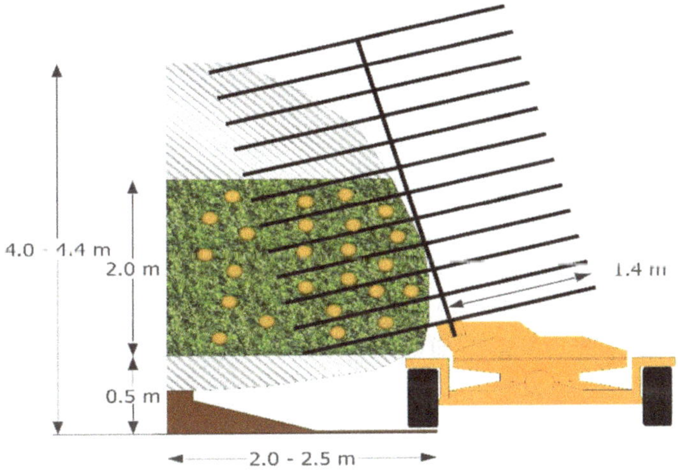

Figure 3. Cross-section of the tree hedge and canopy areas with and without direct contact with the canopy shaker system rods.

Before the mechanical harvesting process, a total of 90 fruit-bearing branches were selected that had mature fruit and were distributed in different sectors. The statistical design established a stratified random sampling, each cluster was a tested plot, and in each plot, 18–24 fruit-bearing branches were randomly selected. The sample guaranteed at least three measurements in each sector. Each branch was assigned a position value in the cross-section at a point close to the fruit that was able to support an acceleration sensor, but which had a diameter less than 10 mm. The fruit detachment ratio was determined by the number of fruits removed from each branch before and after the harvesting process.

Branch vibration measurements were recorded with a triaxial MEMS accelerometer sensor (Gulf Coast Data Concepts LLC X200-4, Waveland, MS, USA) with a measurement range of ±2000 m/s^2, a 16-bit resolution, a sensitivity of 0.06 m/s^2, and a sampling frequency of 400 Hz. Figure 4 shows the placement of the sensor on the fruit-bearing branch.

Figure 4. Location of the acceleration sensor on a fruit-bearing branch on the outermost part of the canopy.

Analysis of the acceleration signals in the time domain and statistical analysis was performed using the R open software (R Core Team, 2016) and in the frequency domain using the NV Gate v8.0 software, with a fast Fourier transformation with 401 lines in a frequency range of 0–156.2 Hz. In the time domain, the resultant acceleration value (A_r) was determined as the module of the vector sum of the three measurement axes in each sensor. Figure 5 shows a sample of A_r in a fruit-bearing branch.

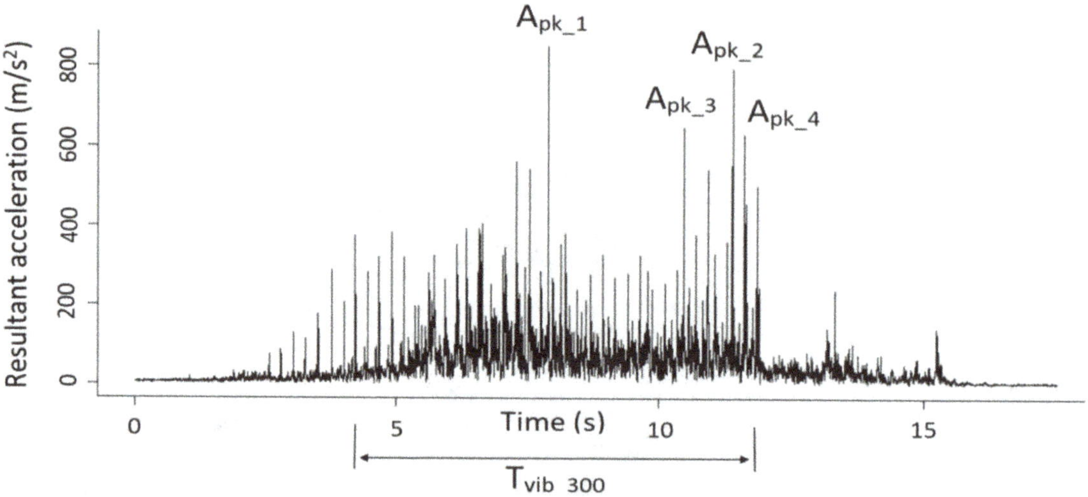

Figure 5. Example of the resultant acceleration (A_r) in the time domain measured in a fruit-bearing branch. T_{vib_300}—time elapsing between the first and the last event with an A_r value of 300 m/s^2; A_{pk_n}— n maximum peak value of A_r.

In the time domain, the vibration variables studied were:

- Vibration time (T_{vib}): time (s) elapsing between the first and last value of A_r measured on the branch, ranging from A_r values of 20 m/s^2 up to 600 m/s^2.
- Mean peak acceleration (A_{pk}): the mean value of the 10 maximum peak values of A_r (m/s^2) for T_{vib_20}.

In the frequency domain, the vibration variables studied were:

- Frequency: number of cycles per second (Hz) of rod movement in the canopy.
- RMS acceleration (A_{RMS}): vector sum of the Root Mean Square values (RMS) of each accelerometer axis at the vibration frequency.

The statistical analysis focused on predicting the fruit detachment ratio according to the vibration variables and the branch measurements as the predicted parameters. Logistic regression was used with a K = 2-fold cross-validation method.

3. Results and Discussion

Most of the fruit (72.7%) was located in the canopy area that had direct contact with the rods, in the height range of 1–2 m from the ground, and in a range of 0.5–2 m from the trunk. Gupta et al. (2015) stated that the area with the highest fructification is in the primary branches of the intermediate zone of the canopy at a height of 1.14–2.29 m, and at a distance of 0.78–0.83 m from the canopy exterior. However, the distribution of fruit in the canopy also depends on planting distances and tree height. A reduced distance between trees can generate a greater percentage of fruit in the upper canopy parts but can reduce the number of fruits inside [19].

The diameter of the branch at the vibration measurement point was 7.9 ± 2.4 mm (mean ± sd), with a variation that ranged from 10.2 mm for the branches closest to the trunk and the ground to 5.53 mm for the outermost and tallest branches of the canopy. Each tested branch carried a mean value of 3.7 ± 1.8 fruits. The results showed a high variability in the distribution of fruit in the canopy and in the morphology of the branches. This variability is important for the outcome of mechanical harvesting systems and was considered by Gupta et al. [15,16] for modelling the tree and simulating the harvesting process in order to improve the canopy shaker system. However, the current harvesting systems based on canopy shakers do not contemplate the variability of branches and fruit within the tree canopy. In an attempt to improve the machine adaptation to the tree, Pu et al. [18] designed and tested a canopy shaker system capable of applying different vibration parameters to the upper and lower parts of the tree. These authors showed the need to use different harvesting parameters and were able to achieve a high fruit detachment ratio (82.6%) with low tree damage.

Canopy shaker systems continuously harvest fruit as they move along the row of trees. Before the machine comes into contact with a branch, the branch may vibrate due to contact with other branches or due to the transmission of vibration from the trunk. Then, the branch vibrates due to contact with the machine rods, and finally, the branch vibrates freely when the machine has passed. Table 2 shows the results of this vibration process measured in branches, both in the canopy area with and without direct contact with the machine rods. In order to define the beginning and end of the vibration process, the acceleration values produced only by natural sources, mainly by wind and gravity, and without machine interaction were recorded. The vibration time where the branch was excited by the machine was defined as the time elapsing between resultant acceleration values greater than 18 m/s^2 (T_{vib_18}). The average vibration time (T_{vib_18}) of the branches was 14.3 ± 2.8 s. No significant differences were found (Student's t, $p > 0.05$) between the vibration time of the branches located in the canopy area with or without direct contact with rods. This indicated that all branches vibrated at the same time, but not all at the same level of acceleration. The length of the machine's shaking system, i.e., the length of the rods and drums, together with the machine ground speed, determined the vibration time during which the rods had direct contact with the branches, whereas tree-training and canopy density could define the vibration time during which the branch vibrated without direct contact with the rods. For the field test conditions, i.e., a rod length of 1.4 m and ground speed ranging from 1–1.5 km/h (0.28–0.42 m/s), the canopy shaker systems were able to maintain direct contact with the branches within the range of 7.7–11.5 s. This indicated that between 20% and 46% of the vibration time may correspond to the transmission of vibration before and after the passage of the machine.

Table 2. Vibration parameters measured on branches with and without direct contact with the rods during mechanical harvesting with the canopy shaker systems.

	Branches with Direct Contact with Rods	Branches without Direct Contact with Rods	Mean Value
Vibration time (s)	14.8 ± 2.8^{a}	13.8 ± 2.9^{a}	14.3 ± 2.8
Frequency (Hz)	4.1 ± 0.2^{a}	4.0 ± 0.3^{a}	4.1 ± 0.5
A_{RMS} (m/s^2)	29.6 ± 10.2^{a}	12.8 ± 6.4^{b}	26.5 ± 13.6
Acceleration peak (m/s^2)	616.7 ± 283.3^{a}	268.1 ± 164.6^{b}	495.1 ± 270.9
Fruit detachment ratio (%)	84.7 ± 30.5^{a}	25.1 ± 22.2^{b}	69.1 ± 40.7

Values shown are mean ± standard deviation, n = 90. The same superscript letters in the same row are not significantly different (Student's t, $p < 0.05$; Wilcoxon–Mann–Whitney test, $p < 0.05$).

During the harvesting process, the branches in contact with the rods experienced a process of forced vibration. The rods, powered with an alternating and rotating movement, penetrated the canopy, producing an impulse excitation in the branches of the fruit of the outermost part of the canopy [20]. The branches showed a mean vibration frequency value of 4.1 ± 0.5 Hz. The value of the vibration frequency did not correlate with the position of the branch in the canopy (Pearson = 0.135, $p > 0.05$). The frequency value used was within the range recommended for citrus detachment with canopy shakers. Liu et al. [17] established that the value of 5 Hz was appropriate for the detachment of fruit without increasing the damage caused to the tree. Similarly, it was demonstrated that a frequency value of 4.8 Hz was adequate to produce a high acceleration in the branches when using rigid rods [18]. The frequency of the vibration in combination with the design of the rods used and the state of the immature fruits [10] play an important role in the damage caused to the trees. Although tree damage is an important element in the process of improving mechanised harvesting, it has not been considered in this study because the damage that occurred was minor and similar to that of previous years, mechanically harvested plantations had not shown any problems in tree development or yield compared to previous seasons.

The values of A_{RMS} measured in the branches was positively related to the fruit detachment ratio. The branch vibration was characterised by an A_{RMS} value of 26.5 ± 13.6 m/s^2 for the machine vibration frequency. However, there was a significant variation in the mean A_{RMS} values in branches depending on their position in the canopy (Figure 6). Branches with direct contact with rods showed a significantly higher mean A_{RMS} value (29.6 ± 10.2 m/s^2) (Student t, $p < 0.05$; Wilcoxon–Mann–Whitney, $p < 0.05$) than branches without direct contact (12.8 ± 6.4 m/s^2). Pu et al. (2018) showed that the highest values of acceleration in the branches (31.4 m/s^2) were provided by contact with the machine rods and these branches reached the highest values of fruit removal efficiency.

During the harvesting process, the acceleration produced in the branch was transmitted to the fruits for their detachment. The internal branches, without direct contact with the rods, showed 43% of the A_{RMS} vibration level of the external branches. A similar result was achieved by Liu et al. [17], whose results showed a reduction of the acceleration in the inner branches of the canopy of 42% with respect to the outer branches. In field trials with measurements inside the fruit before its detachment with canopy shaker systems, Castro–Garcia et al. [20] recorded mean A_{RMS} values between 38.8 and 51.3 m/s^2. These values indicated that there was an amplification of the acceleration values from the branch to the fruit. Savary et al. [13] evaluated the acceleration produced in the branches and indicated that the resultant acceleration values were higher at the ends of the branches, especially on the thinnest and outermost branches. These same authors pointed out that the trunk region of unbranched trees had very low acceleration values, whereas the acceleration values began to be noticeably higher from the first branch of the trunk region.

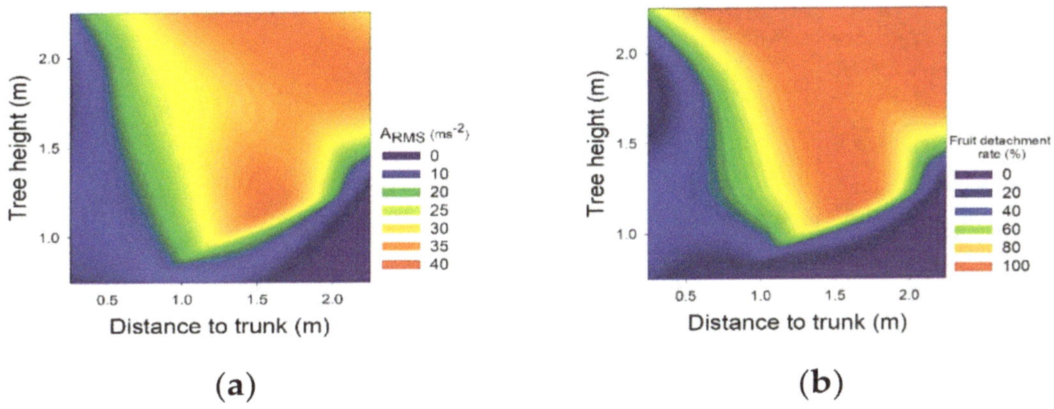

Figure 6. (**a**) Distribution of the A_{RMS} values (m/s^2); and (**b**) fruit detachment rate (%) produced with a canopy shaker system and measured in fruit-bearing branches in the cross-section of the tree canopy.

The interaction of the rods with the branches was characterizsed by a succession of impacts with a high acceleration value in accordance with the vibration frequency of the machine. These impacts presented a mean A_{pk} value of 495.1 ± 270.9 m/s^2. Similarly to Figure 6, the A_{pk} values in branches with direct contact with rods (616.7 ± 283.3 m/s^2) were higher (Student's t, $p < 0.05$) than branches without direct contact (268.1 ± 164.6 m/s^2). The A_{pk} and A_{RMS} values showed a positive linear correlation (Pearson = 0.70, $p < 0.05$) in the tree canopy. In both cases, the direct contact of the rod represented an increment of 2.3 times the acceleration values reached in the branch.

The canopy shaking system achieved a mean fruit detachment value of 69.1 ± 40.7%. However, this variable presented high variability within the cross-section of the tree hedge (Figure 6). As expected, this variable reached its highest values in the branches with direct contact with rods. The fruit detachment ratio was reduced from an average value of 84.7 ± 30.5% in branches with direct contact with rods to 25.1 ± 22.2% for branches without direct contact. The fruit detachment value of 100% was reached in all branches located between 2 and 2.5 m in height. A similar result was reported by Whitney et al. [6] who found that working with small canopies that were accessible to rods achieved a fruit detachment of 96%, while wider canopies achieved reduced values of 55%. Savary et al. [14] reached a fruit detachment value of 88% on the outside of the canopy, while on the inside this figure was reduced to 57%. However, in order to improve the harvesting efficiency, it is not only necessary for the rod to penetrate the canopy, but also for it to interact with the branch. Liu et al. [17], analysing fruit detachment according to the point of contact of the rod with the branch, determined that the operation was more effective when the rod impacted at 30% of the distance to the free end of the branch.

The results obtained from the cross-section of the tree canopy have shown a high variability, both in vibration and in fruit values. Reducing this variability and improving the mechanised harvesting process requires knowledge of the requirements to detach fruit from the tree. The measurements in the canopy showed a high linear correlation between variables, which indicated that fruit detachment prediction could not have a single solution. Analysis of the data focused on the discretisation of quantitative variables that could discriminate whether there was a fruit detachment with a value of 100% and reasonable success. Due to its simplicity and efficiency, logistic regression was used. A_{RMS} and time elapsed between an acceleration greater than 300 m/s^2 (T_{vib_300}) were significant variables to discriminate the events of 100% fruit detachment. The result was defined as a straight line that separates the conditions at which a fruit detachment of 100% was obtained with a precision measured as the area under the ROC curve of 0.95 in the validation set. Equation (1) shows the values obtained and Figure 7 is the graphical representation.

$$ln\left(\frac{\text{Prob of complete fruit detachment}}{1-\text{Prob of complete fruit detachment}}\right) = 7.13417 - 0.52754\ T_{vib_{300}} - 0.17206\ A_{RMS} \quad (1)$$

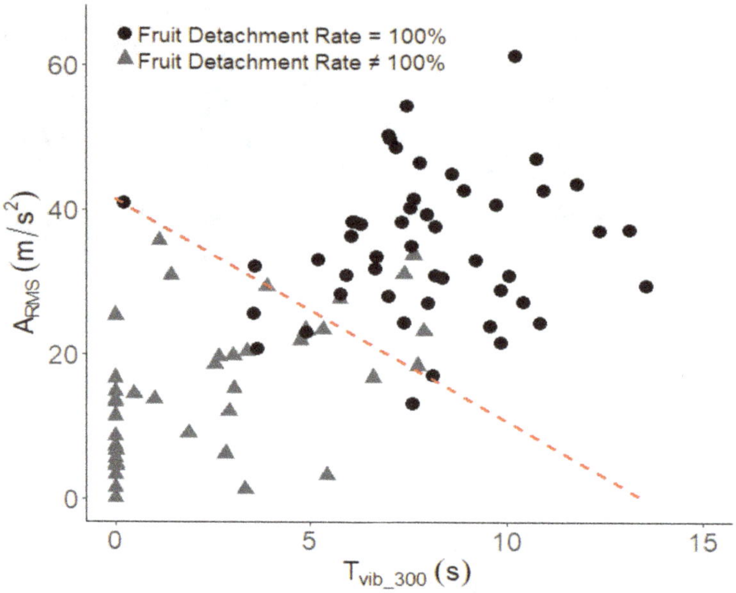

Figure 7. Distribution of the fruit detachment rate values according to A_{RMS} (m/s^2) and T_{vib_300} (s).

Currently, canopy shaker systems for citrus harvesting can employ various types of vibration systems, with variations in the frequency or amplitude of movement, different machine ground speeds or rods with different designs, or mechanical properties. In all cases, the machine produces a forced vibration of the branches with the aim of detaching fruit and avoiding major damage to the tree or fruit. Monitoring of the forced vibration process of the tree canopy showed that it was possible to achieve a 100% fruit detachment ratio based on a combination of acceleration levels and vibration times in branches (Figure 7). Under the conditions of the field tests performed, we propose a combination of a vibration time of at least 5.8 s, after the interaction of the rod with the branch (T_{vib_300}), together with an A_{RMS} value of 23.9 m/s^2. With these harvest parameters, a complete process of fruit detachment was achieved in 88.9% of the branches tested. Although these values could be modified if another type of tree formation, citrus variety, or harvesting machine was considered, the fruit detachment process can be estimated by vibration time and branch acceleration. Both parameters are of great importance, not only for training trees to facilitate the rod penetration in the canopy but also for the design of new canopy shaker systems.

4. Conclusions

The monitoring of the vibration process in the tree during mechanised harvesting with the canopy shaker showed a great variability in results depending on different parts of the canopy. Branches that had direct contact with the machine rods showed a higher mean value of fruit detachment ratio (84.7%) than non-contact branches (25.1%). Vibration transmission from the external branches to the internal branches in the canopy was not effective to remove internal canopy fruit. During the harvesting process, values of 100% fruit detachment ratio could be achieved with a combination of harvest parameters in the branch. Achieving a complete process of fruit detachment is possible if both the tree canopy and machinery are mutually adjusted to facilitate the contact of the shaking system and the necessary vibration time.

Author Contributions: All the authors made significant contributions to the manuscript. S.C.G., F.A.-R., and R.R.S.-G. designed the experiment and wrote the manuscript; S.C.G. and F.A.-R. performed the data acquisition; F.A.-R. and R.R.S.-G. performed the field component; E.S.-O. and A.J.S. performed the statistical analysis; J.A.G.-R. and A.J.S. scientifically supported and reviewed the paper; and S.C.G and E.S.-O. managed the project.

References

1. FAOSTAT. Food and Agriculture Data. Available online: http://www.fao.org/faostat/en/#data/QC (accessed on 25 March 2019).
2. Junta de Andalucía. Average Production Costs. 2016–17 Season. Available online: http://www.juntadeandalucia.es/agriculturaypesca/observatorio (accessed on 25 March 2019).
3. Roka, F.M.; Hyman, B.R. Mechanical harvesting of sweet oranges for juice processing. *Acta Hortic.* **2012**, *965*, 241–244. [CrossRef]
4. Brotons-Martínez, J.M.; Martin-Gorriz, B.; Torregrosa, A.; Porras, I. Economic evaluation of mechanical harvesting of lemons. *Outlook Agric.* **2018**, *47*, 44–50. [CrossRef]
5. Sanders, K.F. Orange harvesting systems review. *Biosyst. Eng.* **2005**, *90*, 115–125. [CrossRef]
6. Whitney, J.D. Field test results with mechanical harvesting equipment in Florida oranges. *Appl. Eng. Agric.* **1999**, *15*, 205–210. [CrossRef]
7. Roka, F.M.; Burns, J.K.; Buker, R.S. Mechanical harvesting without abscission agents-Yield impacts on late season 'Valencia' oranges. *Proc. Fla. State Hortic. Soc.* **2005**, *118*, 25–27.
8. Burns, J.K.; Buker, R.S.; Roka, F.M. Mechanical harvesting capacity in sweet orange is increased with an abscission agent. *HortTechnology* **2005**, *15*, 758–765. [CrossRef]
9. Hartmond, U.; Whitney, J.D.; Burns, J.K.; Kender, W.J. Seasonal variation in the response of 'Valencia' orange to two abscission compounds. *HortScience* **2000**, *35*, 226–229. [CrossRef]
10. Burns, J.K.; Roka, F.M.; Li, K.T.; Pozo, L.; Buker, R.S. Late-season 'Valencia' orange mechanical harvesting with an abscission agent and low-frequency harvesting. *HortScience* **2006**, *41*, 660–663. [CrossRef]
11. Castro-Garcia, S.; Blanco-Roldán, G.L.; Ferguson, L.; González-Sánchez, E.J.; Gil-Ribes, J.A. Frequency response of late-season 'Valencia' orange to selective harvesting by vibration for juice industry. *Biosyst. Eng.* **2017**, *155*, 77–83. [CrossRef]
12. Spann, T.M.; Danyluk, M.D. Mechanical harvesting increases leaf and stem debris in loads of mechanically harvested citrus fruit. *HortScience* **2010**, *45*, 1297–1300. [CrossRef]
13. Savary, S.K.J.U.; Ehsani, R.; Schueller, J.K.; Rajaraman, B.P. Simulation study of citrus tree canopy motion during harvesting using a canopy shaker. *Trans. ASABE* **2010**, *53*, 1373–1381. [CrossRef]
14. Savary, S.K.J.U.; Ehsani, R.; Salyani, M.; Hebel, M.A.; Bora, G.C. Study of force distribution in the citrus tree canopy during harvest using a continuous canopy shaker. *Comput. Electron. Agric.* **2011**, *76*, 51–58. [CrossRef]
15. Gupta, S.K.; Ehsani, R.; Kim, N.H. Optimization of a citrus canopy shaker harvesting system: Properties and modeling of tree limbs. *Trans. ASABE* **2015**, *58*, 971–985. [CrossRef]
16. Gupta, S.K.; Ehsani, R.; Kim, N.H. Optimization of a citrus canopy shaker harvesting system: Mechanistic tree damage and fruit detachment models. *Trans. ASABE* **2016**, *59*, 761–776. [CrossRef]
17. Liu, T.H.; Luo, G.; Ehsani, R.; Toudeshki, A.; Zou, X.J.; Wang, H.J. Simulation study on the effects of tine-shaking frequency and penetrating depth on fruit detachment for citrus canopy-shaker harvesting. *Comput. Electron. Agric.* **2018**, *148*, 54–62. [CrossRef]
18. Pu, Y.; Toudeshki, A.; Ehsani, R.; Yang, F.; Abdulridha, J. Selection and experimental evaluation of shaking rods of canopy shaker to reduce tree damage for citrus mechanical harvesting. *Int. J. Agric. Biol. Eng.* **2018**, *11*, 48–54. [CrossRef]
19. Whitney, J.D.; Wheaton, T.A. Tree spacing affects citrus fruit distribution and yield. *Proc. Fla. State Hortic. Soc.* **1984**, *97*, 44–47.
20. Castro-Garcia, S.; Sola-Guirado, R.R.; Gil-Ribes, J.A. Vibration analysis of the fruit detachment process in late-season 'Valencia' orange with canopy shaker technology. *Biosyst. Eng.* **2018**, *170*, 130–137. [CrossRef]

9

Food System Digitalization as a Means to Promote Food and Nutrition Security in the Barents Region

Dele Raheem [1,*], Maxim Shishaev [2,3] and Vladimir Dikovitsky [3]

1 Northern Institute of Environmental and Minority Law, Arctic Centre, University of Lapland, 96101 Rovaniemi, Finland
2 Department of Information Systems and Technologies, Murmansk Arctic State University, 183038 Murmansk, Russia
3 Institute for Informatics and Mathematical Modelling, Kola Science Center of RAS, 184209 Apatity, Russia
* Correspondence: braheem@ulapland.fi

Abstract: The consumption of food and its safety are important for human security. In this paper, we reviewed the literature on future possibilities for transforming the food system through digital solutions in the Barents region. Such digital solutions will make food business operators more efficient, sustainable, and transparent. Developing cross-border infrastructures for digitalization in the region will break the isolation of the local food system, thus simplifying the availability of processed, novel and safe traditional food products. It is necessary for food growers and processors to respond to the trends driven by consumers' demand while ensuring their safety. Our review highlights the opportunities provided by digital technology to ensure safety and help food business operators predict consumer trends in the future. In addition, digitalization can create conditions that are necessary for the diversification of organizational schemes and the effective monitoring of food processing operations that will help to promote food and nutrition security in the Barents region.

Keywords: food security; food system; digitalization; human security; Internet of things; sustainable; Barents region

---

1. Introduction

The food system encompasses all different stages of food production, including harvesting, processing, distribution and storage. Pre-production and post-harvest processes within the food and agriculture sector have witnessed tremendous changes with the application of science and technology. The first industrial revolution of the early nineteenth century had a remarkable impact on agriculture and food processing. It has been labor intensive, and multinational companies have gained profits through value addition to local foods, which are often packaged, stored, and distributed over long distances [1]. The second industrial revolution witnessed mass production from manufacturing plants [2]. Science and technology have also brought about innovation with food products that respond to consumers' demands, leading to the third industrial revolution with machine intelligence, computers, and digitalization [3]. Currently, we live in an increasingly connected society that revolves around the internet, signaling the fourth industrial revolution [4]. The current fourth industrial revolution combines digital, physical and biological systems, and it will encourage local processing with less need for long distances, which can reduce the miles covered by food. Such a revolution can shift the balance towards local and distributed manufacturing in rural communities and sparsely populated regions, such as the Barents region.

The Barents region comprises the northern parts of Finland, Sweden, Norway, and Northwest Russia. The Barents region within the Russian Federation extends geographically into the European

Arctic, but its socio-economic characteristics differ from those of the Nordic countries [5]. The region is not as homogeneous as the European High North (EHN), since it includes Northwest Russia.

The traditional foods that are available in the Barents region include fish, potatoes, meat, and berries. Meat mainly consists of reindeer meat but it can also include cow and moose. In many communities of the Barents region, fish is deemed the second most important local food, for the women in some villages it can also be the most important [6]. Wild berries e.g., cloudberry (*Rubus chamaemorus*), bilberry (*Vaccinium myrtillus*), lingonberry (*Vaccinium vitis-idaea*), raspberry (*rubusidaeus*), cranberry (*Oxycoccus*) are widely collected and eaten, and could be available fresh or conserved all year during good seasons [7] (p. 40). In the family gardens of many communities in the region they grow potatoes, strawberries (*Fragari avesca*) and blackcurrants (*Ribes nigrum*), as well as some onions and root crops. Mushrooms and herbs are often picked from the forest and incorporated into meals. Animals such as moose (*Alces alces*), rabbit (*Oryctolagus cuniculus*), fowl (*Galloanserae*) and waterfowl (*Aix galericulata*) are not as important because of the low number of wild animals available [8] (p. 347). Dairy farming also exists to produce a wide range of dairy products including milk, cheese, yoghurt, butter and other fermented dairy products [6]. The processing of these foods, in terms of adding value to them, is mainly done by small and medium enterprises in the Barents region. There are some processed products from these traditional foods and herbs in the region. In Finnish Lapland, these include fresh and cold pressed juices made from crowberries, lingonberries and bilberries. Extracts from herbs such as angelica, golden root (*rhodiola rosea*) and nettle are available in the Lapland market. There is also small-scale processing of snack bars from berries and chocolate, spruce sprout syrup and sauces, as well as smoked reindeer chips. When these foods are processed, they are better preserved and can be stored or distributed for longer periods of time. Currently, the value addition to traditional foods in the region can improve with digitalization that targets increased production, better harvesting methods, less waste, as well as better storage and distribution. The introduction of sensors that generates data from food system digitalization will help make accurate predictions in the processes within the food value chains of the countries in the Barents region.

Access to food and its utilization in different parts of the Barents region are largely influenced by food markets and the choice of food that is available to the inhabitants of this region. Due to the increasing connection in our society, the last few decades have witnessed new imported foods making their way into the Barents region. Some of these new imported foods, such as tortillas, kebabs, pizza, and sushi, are increasingly popular in the region. This has resulted in changes within the food system of the Barents region, with consequential effects on traditional foods that people have relied upon in the past [6]. The popularity of these foods, at the expense of traditional foods has been shown to increase the risks of cardiovascular diseases, obesity and diabetes in the Barents region [6,9]. It will be important to address the challenges faced by small and medium enterprises in the region for them to expand their businesses and create a niche for their high-quality nutritional food products. This can be achieved by encouraging a better means of harvesting berries and herbs, and highlighting the nutritional benefits of these bioactive ingredients through mobile applications that can appeal to potential consumers, especially the younger generation.

Food security and human security are prevention oriented; food security helps to develop the overall human well-being, thereby strengthening human security [6]. The concept of human security should endorse a people-centered approach, built on the capacity of individuals, and provide key tools for building resilience around food security and nutrition [10]. Food and nutrition are integral to human security, and the link between diet and health is well established [11,12]. The food we consume plays an important role in our health and well-being. Food availability, accessibility, utilization and stability within the food system are essential pillars of food security. In the Barents region, ensuring sustainable food and nutrition security will help to build resilience, and promote culture and ecology that will offer greater human and societal security [13]. Healthy living is associated with clean water, food and air, but also includes the safety and security of life for the individuals, groups and communities. Changes that occur in the climate, environment and land use, together with socioeconomic factors, have been reported to impact on food and water security in the Arctic region, including the Barents region [14]. The trends associated with food consumption are constantly evolving with technological inputs. How food processing methods and the traditional culture related to food have changed over the years with technology are important topics for consideration. The issues of food security, climate change, and population growth are major concerns discussed in many international agreements as in [15–17]. There are only a few studies on food security in the Barents region, and to date, there is no work on food system digitalization in the region. We therefore emphasize the need for small and medium-based enterprises (SMEs) in the food sector to embrace digital solutions in their operations. The purpose of this article is to identify gaps and challenges related to food and nutrition security in the Barents region, improve the sustainability of the food system through digitalization and promote the development of cross-border infrastructures to support digitalization. The authors have drawn largely on their participation in related project workshops and social networking services in this region.

This review article addresses the relevance of digitalization as a disruptive innovation within the food system that will help to promote food and nutrition security. We highlight in Section 2, the impacts of human activities and climate change in the Barents region; Section 3 is on food system digitalization in the Barents region and how other enablers for digitalization, such as internet penetration, will be crucial in achieving the desired objectives of improving food and nutrition security in the region. Section 4 discusses the role of different digital technologies that can transform the food system in the region and the overlapping relationship that exists amongst the consumers, food business, and authorities; Section 5 discusses how information and communications technology can contribute to a sustainable food system; Section 6 looks at future implications for the Barents region. Finally, Section 7 concludes the article.

2. Impacts of Human Activities and Climate Change in the Barents Region

The Barents region is characterized by its remoteness, high latitude, harsh climate, and varied ecosystems with the Scandinavian mountain chains in the west, the Arctic tundra in the Kola Peninsula, and the Nenets Area and Novaja Zemlja in the east [18]. The midnight sun may persist up to 24 h a day from May to July. Despite the harsh landscape, it is characterized by boreal forest, thousands of lakes, mountains and fells; this provides the region with an abundance of natural food, made available through hunting, gathering, harvesting and fishing.

As a sparsely populated region with pristine environmental characteristics but rich in natural resources such as oil and gas, it attracts human activities that are related to marine areas or inland mining and mineral activities [19,20]. Human activities have significant impacts on food security in the Barents region; they were reported to be detrimental to the environment of the region [20]. Many towns in Northwest Russia often depend on a single industry that causes pollution that can be a major source of contaminant in their food supply. At the Norwegian–Russian border in Nickel and Zapolyamy, nickel smelters release harmful emissions of sulfur dioxide into the environment [21]. Such emissions can end up in the local food system if not accurately monitored. Dudarev et al. [22] reported on the toxic metal levels in local foods such as fish, berries, mushrooms, and game in the Pechenga district. In Finnish and Swedish Lapland, mining activities are also a cause for concern to the environment as potential risks to food safety [21]. With growing international tourism in the region, there will be more waste generated and it will be important that such waste does not result in environmental pollution. Encouraging industrial symbiosis (where small and medium enterprises find ways to use the waste from one enterprise as raw materials for another at local levels) in the region will help to avoid unnecessary waste and encourage a circular economy. The Finnish Innovation Fund (SITRA) in its current roadmap to a circular economy highlighted industrial symbiosis as a means to tackle the challenges of climate change, depletion of natural resources and to minimize the waste of food resources [23]. Shifting to a circular model for food can have economic, health and environmental benefits in the Barents region.

The Barents region is an area where the impact of climate change is most pronounced, with an average rise in temperature twice that of the global average temperature [24,25]. Digitalization of processes within the food system will help to mitigate the impacts of climate, with generated data. As a result of the warming climate, there will be changes in ocean and atmospheric contaminant transport, which may increase the movement of organohalogens and mercury from lower latitudes to the higher latitudes of the Arctic Barents region [26]. As the water in freshwater lakes gets warmer, tundra ponds and streams are likely to be affected by greater bacterial methylation of mercury and mercury released from thawing permafrost [27]. This will present some challenges for the terrestrial and aquatic food resources in the region. The environmental impacts of climate warming, anthropogenic contaminants and zoonotic disease in subsistence food species and rural drinking water can lead to infections and toxic effects in rural communities [28].

Plant, insect and animal species will be able to expand further north as the climate gets warmer, thus bringing new zoonotic diseases with them. "Higher winter temperatures in the Arctic was noted to likely increase the level of microbes, and consequently, the winter survival of infected animals, raising the risk of hunter and consumer exposure" [29] (p. 3). The warming climate may drive changes in the forage resources of subsistence species, their health, and abundance. Similarly, the expansion of new species northwards has already brought new water-borne diseases such as tularaemia [30], and warmer waters in tundra ponds, estuaries, and nearshore ocean waters now support toxin-producing cyanobacteria and algal blooms [31]. In addition, longer Arctic summers and warmer winters with less sea ice cover will increase the use of northern sea routes by commercial shipping from northern Europe and western Russian ports [25], raising the possibility of new rat-borne infections, such as tick-borne encephalitis [32]. The use of sensors and other digital solutions will lead to a greener means of transport that can prevent such disease outcomes.

Some of the gaps in knowledge for adaptation strategies to better cope with food and water security in the Arctic-Barents region include the availability of better metrics on contaminants and pathogens in the region [29]. It will be important to share data on the microbial and chemical pollutants in the agri-food sector across the region. Therefore, there is a need to ensure food safety and sustainability by utilizing the benefits of digitalization within the food system to accurately measure and predict outcomes from these activities. Efficient and smart manufacturing in the future will need to accurately gather data, with which to predict and inform processing operations. Developing and transforming the food system through digitalization will help in the response to the consequences of these impacts in the Barents region.

3. Developing the Infrastructure for Food System Digitalization in the Barents Region

Food system digitalization refers to the application of innovative technology to enhance harvesting, processing, distribution and storage operations along the agri-food value chain. The power of mechanization in the early nineteenth century, automation in the 1970s, and the growth of information and the internet in recent decades have had significant impacts on our day-to-day activities, including the food system. Since the current adoption of connected intelligence into the business and social fabrics is happening at an advanced speed, it will change the way we conduct business. The food business operations in the Barents region and its system will not be an exception as it develops novel food products from traditional foods.

The food system has been distorted globally, mainly due to capitalism's drive for increasing profit. Some major worries for the global food system highlighted by United States climate researchers include absurd transnational lengths for supply chains, genetically modified organisms inserted into plant and animal deoxyribonucleic acid (DNA), overuse of pesticides on mono-culture crops, common resources such as water privatized through legislation and patent restrictions on seeds [33]. All components of the food system are challenged to quickly adapt and keep pace with the evolution of technology, trying to evade the disruptive effects of digital Darwinism (a phenomenon where technology and society are evolving faster than businesses can naturally adapt). In doing so, it was anticipated by the World Food

System Centre that "the dynamics of the whole food-value chain change, and unforeseen consequences enter the economic landscape through digitalisation" [34]. In responding to these challenges, local food systems need to be revisited to ensure food safety and sustainability in the Barents region. Along the food and beverage supply chain, there are many involved processes, such as workers and touchpoints, which can make it difficult, not only to keep track of food but also to monitor its quality. Quality is of incredible importance in the food industry, as faulty or contaminated foods entering the market can be detrimental to human health [35]. Therefore, food must always remain traceable and safe, and this can be complex to guarantee within the food system with its many working cogs. Digitalization with accurate data on the process inputs and outputs will help to guarantee quality, safety and traceability.

Food system digitalization will consider all the aspects of growing, harvesting, processing, storage and distribution of foods from the region in order to explore their potential in promoting human security. In the Barents region context, food system digitalization can help to transform the local economy by improving the supply of traditional values, increase the yield of natural products and create novel food products with health giving properties. With an improved local economy, the purchasing power of entrepreneurs will increase, leading to an improvement in food and human security in the region. Ultimately, this will result in better food sovereignty across the region as SMEs in the food sector make economic gains.

In an anticipated future scenario, technological advances that are emerging from the Fourth Industrial Revolution (Industry 4.0) will benefit the current food system in becoming greener and sustainable. Industry 4.0 harnesses uncertainties and, in their place, removes assumptions and risky forecasts, enabling a relevant level of actual knowledge and a newfound level of insight. The use of digital technologies, such as precision planting and irrigation techniques, will improve yields organically. "The speed, efficiency, and sustainability of transport will improve radically. Mobile information technologies will improve farmers' understanding of the land they are farming and the markets they are selling to, while also allowing them to communicate with and learn from each other" [36].

The importance of advances in communication, education, and finance to support farmers and their communities cannot be underestimated in the present age. Telecommunication is an important platform in which digitalization can be built and further developed. In order to maximize the benefits of digitalization in the Barents region's food sector, it is important that infrastructure, such as telecommunication networks be developed across the entire Barents region. For instance, in Finland, Norway, and Sweden, the networks are highly developed while in Northwest Russia it needs to improve when digitalization harmonization across the region amongst stakeholders is considered. There is a discrepancy in technological and infrastructural developments in the Barents region. For instance, internet penetration in Finland, Norway, Russia, and Sweden was 92.5%, 98%, 76.4%, and 93.1% respectively [37].

The capacity of the telecommunication network in Northwest Russia remains limited, and the cost of using telecommunications within low-populated northern regions of Russia, as well as that of cross-border services between the Nordic countries and Northwest Russia, remains extremely high. This is one of the initiatives addressed in the Northern dimension [38]. The Northern dimension is a joint policy initiated in 1999 between four equal partners—the European Union, Russia, Norway, and Iceland—regarding the cross-border and external policies geographically covering Northwest Russia, the Baltic Sea, and the Arctic regions, including the Barents region.

The issues of telecommunication development in the north of Russia are given great attention with the possibility of adopting smart technology with space communication [39]. The task of developing communications appears as one of the strategic documents of the Russian Federation, both at the federal and regional levels [40–42]. The development of telecommunications infrastructure in the Russian Arctic has been an important topic of discussion, both at the state level and in the business environment. Annually, there are many conferences and forums devoted to this issue [43].

In an effort to develop the communication infrastructure in northern Russia, one of the largest projects in the field of telecommunications development is the laying of a fiber-optic backbone along the Northern Sea Route. The project, in different forms, has been discussed since 1999. Currently, the active stakeholders of the project are Russia, Finland, and China [44]. In 2016, the Russian Federation implemented a similar project for the construction of an underwater optical highway linking the largest settlements in the northeast of the Russian Federation (Sakhalin–Magadan–Kamchatka) [45]. Along with the terrestrial telecommunications infrastructure, the satellite communication system is also developing [46,47]. Mobile communications are actively developing as well. Coverage of mobile networks of the third and fourth generations is growing, and already covers the territories inhabited by indigenous peoples [48]. The Nordic industries are also developing a strategy towards the fifth-generation (5G) network. Finland and Sweden are both in the European Union, while Norway is not but aligns strongly with Nordic cooperation for mobile networks. For example, a declaration of intent [Press release, May 2018. https://www.government.se/press-releases/2018/05/new-nordic-cooperation-on-5g/] signed recently states that the Nordic region will be the first interconnected 5G region in the world, and identifies areas in which Nordic cooperation needs to be intensified. The development of communication infrastructures in the Russian North naturally will lead to an increase in internet penetration. In order to reap maximum digital dividends, the economies of a country need to implement the reforms necessary to improve internet access. In addition, they need to focus on the analogue foundations that promote the digital economy, such as skills, institutions, and regulations [49]. It was also suggested that the development of a strategy that will reap digital dividends requires addressing the digital divide by removing the barriers to an internet that is universal, affordable, open, and safe for firms, citizens, and governments [50].

According to Fälström and Jörgensen [51], food system digitalization will need to consider important issues on the Internet of food (IOF). IOF refers to the discussion regarding how digital aspects, technical innovation and new data layers around food will change the global food system. It is related to the complexity of combined food products, their mobility, best practices on quantified food handling, Big Data, hygiene factors in the food sector, environmental control (open reporting sensors), self-configuring sensor nets, edible sensors, and multicast instructions. The authors also suggested the deployment of Internet Protocol Version 6 (IPv6) as a factor for improved food competitiveness, security, better access, identity, and new business models from breaking up structures. IPv6 as a communication tool is able to handle extended space in comparison to IPv4, and will be able to prevent exhaustion that may result from the complexity. In addition, IPv6 provides improved service quality. Automation and modern analytics tools can be deployed to track products and goods from inception to fulfilment. Since the systems in question are designed to track and monitor on their own with little to no input, the user can tap in anywhere along the chain to seek the information he or she needs. The success of the system will largely depend on the exchange of information and knowledge across the region.

One way to encourage knowledge sharing and new partnerships is through an initiative that harmonizes information across the Barents region through data visualization. Data visualization is the modern way of disseminating information that can be used more frequently in the Barents region. The data visualization website contains a broad variety of parameters, such as data on population, unemployment, health, and education that can be visualized on a map (available at https://barents.no/en/focus-areas/european-border-dialogues).

4. Transforming the Food System through Digitalization to Promote Food and Nutrition Security

Agriculture, rural livelihoods, sustainable management of natural resources and food security are connected within the development and climate change challenges of the twenty-first century [24]. The impact of food processing on the climate is well established and it will be necessary that value addition to traditional foods in the region be well managed throughout the value chain. Food processing contributes to climate change, eutrophication (the process by which there is a gradual increase in the concentration of phosphorus, nitrogen, and other plant nutrients in an aging aquatic ecosystem such as a lake), acid rain, and the depletion of biodiversity. Each of these parameters can be monitored with digital accuracy by sensors. The efficient processing of traditional and local foods incorporating digitalization in the value chain will be important for keeping the global temperature at 1.5 to 2 °C above the pre-industrial level. Several researchers have shown that food processing, which often involves water and energy expenditures, needs to be climate smart [52–54]. Since the Barents region and other parts of the world at higher latitudes are witnessing the effects of climate change to a greater degree than elsewhere, climate-smart agriculture and digitalization of the food system can also be part of the solution.

Industry 4.0 is transforming products, the operations of companies, and how their production is managed. This is a giant leap for manufacturing innovation that employs smart devices that take control of machines on the shop floor, communicating autonomously "device-to-device" to manage manufacturing operations and their distribution. "The entire manufacturing value chain is transformed by industry 4.0 and they can be monitored from concept to completion and beyond" [55]. Small and medium-based food enterprises can benefit from Industry 4.0 especially when industrial symbiosis is adopted as a way of promoting sustainability in the region. The management of a business enterprise is now able to access important information in real-time as they are able to monitor the productivity and efficiency of both employees and machinery. Another critical benefit this new technology delivers is a consistent feedback flow between companies and their customers, in which products could be improved or highly influenced by the end-user, transforming how products are designed and produced [56].

"The fusion of 'Big data', the 'Internet of Things', and advanced analytics is providing manufacturers with unprecedented insights into manufacturing performance, customer behaviour, and new product development" [57]. Other enabling tools, such as cloud computing and cyber physical systems, have also been introduced as digital enablers in manufacturing. For example, cyber physical systems research can increase food consumption efficiency and the overall food production capability through precision agriculture, intelligent water management, and more efficient food distribution [58]. Automation, intelligence, and collaborations are also relevant with particular reference to smart manufacturing, smart products/services, and smart cities.

Blockchain technology creates a shared, distributed ledger of transactions over a decentralized peer-to-peer network. As an emerging technology, its suitability needs to be examined against the use cases requirements [59,60]. In the food sector, it makes it possible to track the source of various food items starting from the farm to supermarket shelves. By creating a blockchain network, contaminated products can be traced to their sources faster. Some of the uses of blockchain that will improve business include reducing carbon footprints, the security of the Internet of Things, contract fraud reduction, secure real-time payment, supply chain efficiency, clarity in business agreements and transactions, increased traceability, and improved customer experience. The use of blockchain in food processing will ensure safety, since food ingredients can be easily tracked, monitored and reported on, thereby building trust among consumers [61].

Blockchain, as a distributed, shared ledger for recording transactions, will revolutionize how the food industry can optimize the trading and shipping of food, trace contaminated items, and reduce fraud and waste. Blockchain holds incredible promise in delivering the transparency needed to help promote food safety across the whole supply chain. Within the food supply chain, the different actors, i.e., growers, suppliers, processors, distributors, retailers, regulators, and consumers, can access information about the origin and state of food [62]. This will help to resolve issues of authenticity and food fraud that may result from geographic origin, production method, processing technology, ingredients composition, etc., [63].

In practical terms, a framework can be constructed that can be used for structuring the discussion around the Internet and food, thereby furthering the discussion on how to facilitate openness and innovation in the field of food, with the goal to feed the planet in a healthy and sustainable way. The increasing activity of info-communications in virtual space leads to an avalanche-like growth for information stored and circulating on the Internet. This opens up new opportunities for increasing the security of the population of the territories in all aspects, including food security. Possible ways of realizing this potential based on modern information technologies are diverse. They include monitoring the actual diet of the inhabitants of the territories, monitoring and predicting consumer demand, the formation of new chains of production and food supplies.

The overlapping relationship that exists between the consumers, the food business, and the authorities is shown in Figure 1. The figure shows how the main actors contribute to the digitalization of the food system in different ways, and the benefits of digitalization in the food system. General feedback shown in the picture may reflect in different layers, processes and effects.

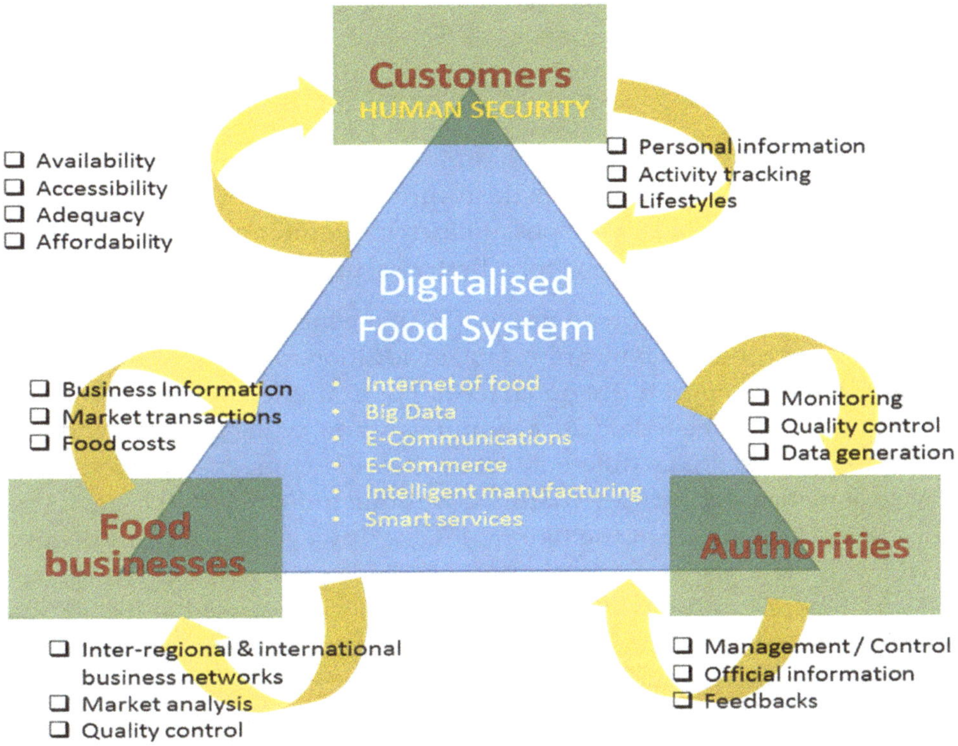

Figure 1. An overlapping relationship within a digitalized food system.

The benefits of digitalization seem to appear in two general ways. The first, "extensive" one consists of an improvement of almost all traditional processes within the food system by using information and communications technology (ICT). The second, "intensive" way is connected to a restructuring of processes within the system. Along with the revolutionary changes of Industry 4.0, one particular restructuring is a shift from hierarchical organizations to horizontal ones, based on peer-to-peer (P2P) interactions. In this article, the authors use P2P as a common name of architectures implementing horizontal instead of hierarchical structures of interactions, i.e., 'peer-to-peer. This kind of interaction appears in the food system in different layers, from IOF to informal customer and producer networks within social networking services (SNS) and to the network of food businesses. The most valuable advantages of any P2P system are its scalability, flexibility, and reliability [64]. Transformations from hierarchical to horizontal structures will potentially improve the sustainability of the regional food system, making it more adequate to the actual demands of the actors and expanding it to trans-regional and cross-border contexts.

Due to the harsh climate and poor development of transport infrastructure in parts of the Barents region, the development of horizontal relations will be especially important for the region. Regional food markets in the region are specific, characterized by small volume, low density and unevenness of the population as the main consumer of products [65]. The consequence of these features results in vulnerability of the food system in the Barents region, due to dependence on unreliable transportation, especially for the northern delivery dependent territories [66]; the higher cost of products, due to transportation costs, result in an imbalance in the structure of food consumption by the population [67].

Strengthening horizontal interaction within the region will reduce these problems due to more effective use of local opportunities for food supply amongst the population. Digitalization, in a broad sense, covers the sphere of interpersonal communications, which will provide a more efficient exchange of information at all levels—from state to personal. This will create the right conditions and lead to the emergence of alternative channels for providing the population of the region with better access to food, including the crowd principle basis, which uses the resources of the population itself. An example of the realization of this potential at the interpersonal level is of a number of groups in the "VKontakte" social network, which provide the ordering and delivery of products from Scandinavian countries to the cities of northwest Russia [68]. Such means are especially important within the human security perspective, since they provide a more efficient personal satisfaction at individual levels, including specific requirements for food.

The opportunities provided by food system digitalization can be considered in terms of what is currently available and what will be available in the future, such as the Internet of food (IOF) mentioned earlier. These opportunities will create new possibilities to monitor a regional food system to inform more contemporary and adequate decisions for a better secure regional food system.

Digitalization results in the widespread use of ICT that generates and accumulates huge amounts of data that opens the way to obtain distinct raw data, in addition to official statistics, to characterize the peculiar aspects of the food system at the regional level. For example, in Russia, official statistics mostly focus on supporting the tasks of the federal level administration, while the tasks and instruments of the regional administration are essentially different (customs tariffs, compensation fees, excises, sales taxes, quotas, etc.). These statistical records are mainly economic indicators to characterize food security amongst the population as a whole at the federal level. However, specific indicators that focus on local and regional levels are missing. Such an approach makes the Russian food security monitoring system state security focused rather than being human security focused at local and regional levels. For instance, the peculiar requirements for the diet of indigenous people from the northern regions based on cultural preferences are not accounted for in the monitoring system. Therefore, additional data is necessary for effective monitoring of the food system of the Barents region taking into account its peculiarities, since "food security is not only about reliable access to nutritious food. It is also linked to climate change, wildlife management, pollution, economic vulnerability and cultural security" [69]. Digitalization provides access to data that concerns several actors in the food system, including persons and households. Beneficial information can be gathered from this data through text analysis in general and opinion-, knowledge- or data-mining technologies, as well as demand analysis and personalization techniques.

An important task that provides a technological basis for creating information technologies to support food security is the formation and maintenance of a knowledge base (KB) that characterizes the overlapping structure of a digitalized food system. Ideally, such a base should describe all components of the food system and their relations to provide the possibility of automated reasoning aimed at supporting decision making in food security and risk management. The creation of a KB is a key task for the potential use of blockchains, the IOF, and other technologies. The KB provides the necessary common conceptual basis for any intelligent distributed system. An essential feature of the knowledge base, which determines the requirements of the technologies for its implementation, is the high dynamism of its content. In our modern world, new products, ingredients, trademarks, supply chains appear and disappear very quickly over time. In such conditions it is extremely difficult to create and maintain the relevance of the knowledge base directly, relying only on the knowledge of experts. This increases the demand for intelligent information systems with feedback that implement the principle of "user as an expert" [70].

Another opportunity for digitalization in transforming the food system is that widely available e-communications will provide a basis for peer-to-peer networking within food systems that will include customer-to-customer, business-to-business, and business-to-customer (C2C, B2B, and B2C, networking respectively). One modern example of such a P2P network that partly cover the Barents region is the REKO system, which implements a new model of a sustainable marketing channel based on SNS interactions. The system unites small food producers (farmers) and consumers and "offer consumers a way of ordering products directly from the producer, without the need for middlemen" [71]. In addition, the system provides the production of food products in volumes and in assortment, taking into account the actual needs of consumers. Such a model of food producing and marketing makes the food system more diversified and resistant to disruptions from individual components including transportation. The example of REKO demonstrates the potential of digital modern ICT in the creation of sustainable distributed food systems for providing the population of the Barents region with quality nutritious food. Technically, there are no obstacles in using such technologies to create similar cross-border networks, including for larger businesses. However, this potential is not used due to organizational and political problems, such as sanction restrictions and differences within food systems regulations in different countries.

5. Sustainable Food System and ICT

The food system strongly relates to many sustainability challenges such as climate change, biodiversity loss, water scarcity, and food insecurity [72,73]. There is a need to deliver sustainable and healthy diets to global populations by responding to the challenges of climate change, rising populations, and decreasing crop yields [15]. In respect of this global challenge, the Barents region specifically can offer a good example by which digitalization in the food system can help to increase yield, adding value to promoting food and human security. This necessitates digital technology to push beyond traditional agricultural practices and be deeply embedded into the food system [74]. For example, in the Barents region, food business operators in the region can add value to carrots, turnips, cloudberries, bilberries, lingonberries, cranberries and other traditional foods as described under the introductory section. The berries are made into juices, jams, sauces, and liqueurs; natural herbs extracts are also utilized. Finnish Lapland angelica herb (*Angelica archangelica*) is used as a flavouring for pies and ice cream [6]. These food business operators within their local communities are small and medium sized, they are often faced with challenges on novel food requirements and trade restrictions by the European Union [75]. In the Norwegian aquaculture, the biggest challenges are sea lice, diseases, access to sufficient areas and adequate feed resources, pollution in the form of feed spills and faeces from the fish [76]. The challenges faced by the Norwegian aquaculture industry can benefit from digitalization by providing sensors to monitor biomarkers that are associated with these pollutants.

In the Barents region, many communities have low population density; data can be readily collected for future cross-border collaboration within a digitalized food system. The value addition to traditional foods can be promoted with innovative digital technology that promotes environmentally-friendly food products. Community-based initiatives in the region support food security and help to promote sustainability within the communities. For instance, the Norwegian Institute of Bioeconomy Research (NIBIO), in collaboration with community stakeholders in Troms county, aims to market Arctic foods with the goal of highlighting their unique arctic quality, natural production, and lack of pesticides in the food products [77]. Swede and carrots grown at low temperature result in a fresher and better, sweeter taste. Similarly, the trial marketing of midnight sun-grown, oval-shaped almond potatoes was a success in the Finnish Lapland market. The Finnish Lapland potato has been designated as a brand under the EU name protection "Protected Designation of Origin" [78]. The unique quality of these products can serve as a marketing tool that will benefit from e-marketing and digital technology that can prove their authenticity.

The existing digital and communication infrastructure in the region will be necessary for stimulating the drive towards food system digitalization in the future. It will require cross-border collaborative initiatives to share best practices and access the latest trends to boost innovative entrepreneurship in the food sector. This will require that the digital infrastructure be up to speed in ensuring that information gathered can be easily shared across borders in rural communities. In a study related to innovation and entrepreneurship conducted at the Imperial College, London, the authors emphasized that information and communications technology (ICT) is required to successfully combine secure and sustainable food systems [74]. Digital ICT was used to help coordinate the activities of multiple suppliers in dynamic strategic networks to coordinate the local delivery of food through the application of innovative digital technologies that take into account the interactions between food and agriculture systems, with broader industrial systems, as illustrated in Figure 2. This was achieved by exploiting the timely delivery of information and expertise using mobile devices in rural areas.

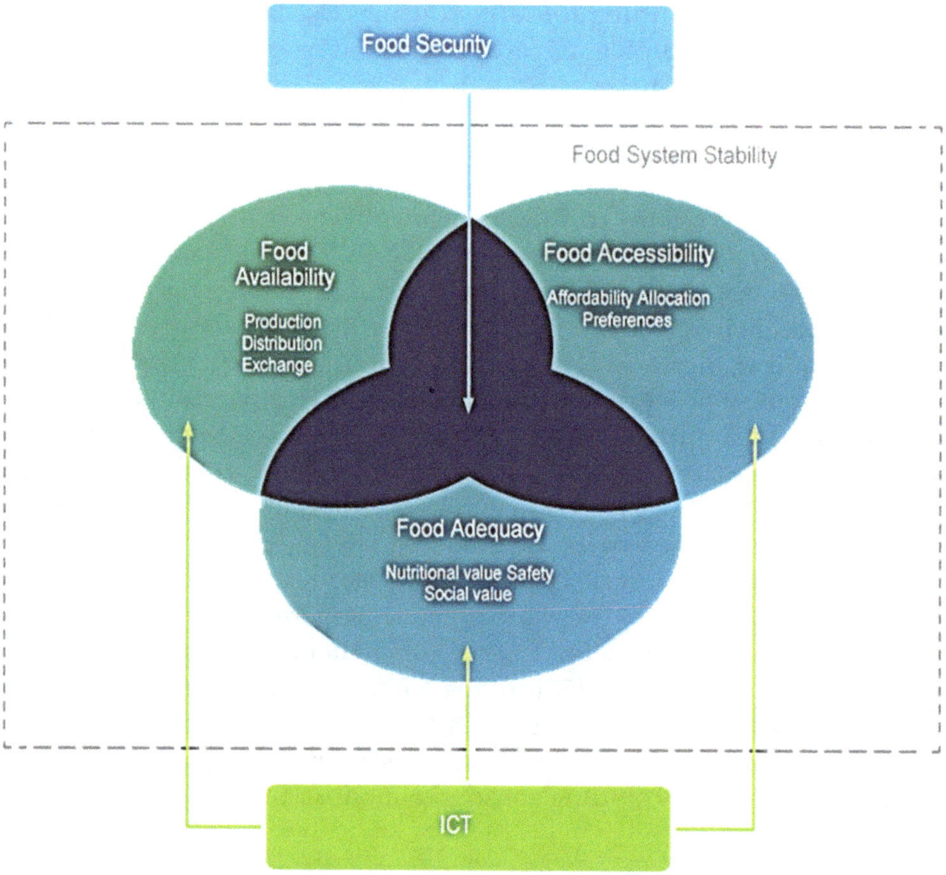

Figure 2. Sustainable food system and information and communications technology (ICT) (Adapted from Berti and Mulligan, 2015).

The food system can be transformed based on the inputs of digitalization on how we produce, process, and integrate information, which will be critical to innovation. Digitalization in food and agriculture will offer new opportunities through the ubiquitous availability of highly interconnected and data-intensive computational technologies as part of the so-called Fourth Industrial Revolution [4]. An interesting aspect of Industry 4.0 is the notion of individualized 'small batch' product offerings. There are interests in data and data analytics as a means to improve production. Small and medium-sized entrepreneurs, as important drivers of economic activities in the Barents region, will benefit from Industry 4.0. It can be applied to all aspects of dairy production, processing of berries, their storage, packaging and distribution in the value chain systems. It also reflects a shift from generalized management of farm resources to highly optimized, individualized, real time, hyper-connected, and data-driven management. The desired outcomes of digital agriculture are more productive, profitable, and sustainable systems. Digital agriculture will leverage the smart use of data and communication to achieve system optimization. The tools that enable digital agriculture are multiple and varied, and include cross-cutting technologies such as computational decision and analytics tools, the cloud, sensors, robots, and digital communication tools. Factories and plants will connect machines and production systems via the internet, allowing information to be exchanged and triggering actions that can enable each entity to control the other independently. Easy data exchange and access to that information can provide 3D printing of foods with more customization possibilities for manufacturers and their customers. As production becomes digitalized, the socio-economic drawbacks can be minimized through new trainings for jobs that are linked to digitalization.

The Finnish Arctic research has a long-term strategy that aims to strengthen Arctic cooperation, as well as new kinds of partnerships, including public and private, in particular by strengthening the business environment and networking of actors in the Finnish Barents region, i.e., Lapland, Northern Ostrobothnia, and Kainuu [79]. The strategy strives to ensure sustainability in line with the Sustainable Development Goals (in particular goals 2, 9 and 12) by 2030. One of the themes at the Natural Resources Institute Finland i.e., *Luonnonvarakeskus* in Finnish (LUKE) is the 'Innovative Food System' that supports a sustainable, profitable, and innovative food chain at all stages. Digitalization will help as a tool to support innovation across the food value chain. The overall aim of the innovative food system is to process wholesome, sustainable food products and support the rotating economy in the food system by utilizing digital and intelligent technology in the Barents region of Finland [76]. The main areas of this theme on innovative primary production systems, circular economy, smart agriculture, healthy and sustainable food and feed, consumers and markets will benefit from digitalization that can be shared through cross-border collaboration in the region.

6. Future Implications for the Barents Region

The previous three sections (Section 3. Developing the infrastructure for food system digitalization, Section 4. Transforming the food system through digitalization, and Section 5. Sustainable food system and ICT) will be relevant for a systemic change in the food system at a regional level; they will play important roles in shaping the future. As shown in Figure 1, a digitalized food system will help to promote both food security and food sovereignty by involving the customers or consumers, food businesses, and the authorities in a digitalized food system. Food sovereignty empowers local and rural communities in value addition to their food crops. It is defined by the World Forum on Food Sovereignty (WFFS) as the ability and the right of people to define their own policies and strategies for the sustainable production, distribution and consumption of food that guarantee the right to food for the entire population [80]. Currently, in Finnish Lapland, there are initiatives such as the "Arctic Bioeconomy, Smart specialisation" that encourage industrial symbiosis with digital solutions. In Loue near Tervola in Finnish Lapland, there are processing facilities for meat, ice-cream, biogas and a cattle barn with robots for milking. These facilities are shared within the same premises with a vocational college with active collaboration with the Natural Resources Institute Finland (LUKE).

There are also 'Facebook', and 'Whatsapp' groups that encourage virtual meetings between local farmers and consumers. Similar practices exist in other communities of the Barents region. In the near future, as the interest in where food comes from grows, and how it has produced becomes more relevant, it will encourage the development of artisan foods and food sovereignty in the Barents region. Food sovereignty puts emphasis on the promotion of small and medium-sized production where respect for the inhabitants' own cultures as well as the diversity of peasant, fishing and indigenous forms of agricultural production, marketing and management of rural areas play a fundamental role [80]. The unique food crops, herbs, natural products and other indigenous foods that are part of the food culture in the Barents region will also contribute to biodiversity when there is value addition with digital solutions. An integrated farming system that employs sensors to increase yield, fertility and help in the management of pests and diseases will be important for future collaboration in the region. As food products move across border in the Barents region, the role of digitalized packages in e-commerce and e-communication will be very important. Digitalized packages with labels with quick response (QR) codes can provide important information on the nutritional quality of the ingredients, ethical standards, and the possibilities to interact with customers. In addition, they can be useful in product tracking, item identification, time tracking, and general marketing. We developed a framework for knowledge base creation as shown in Figure 3 below.

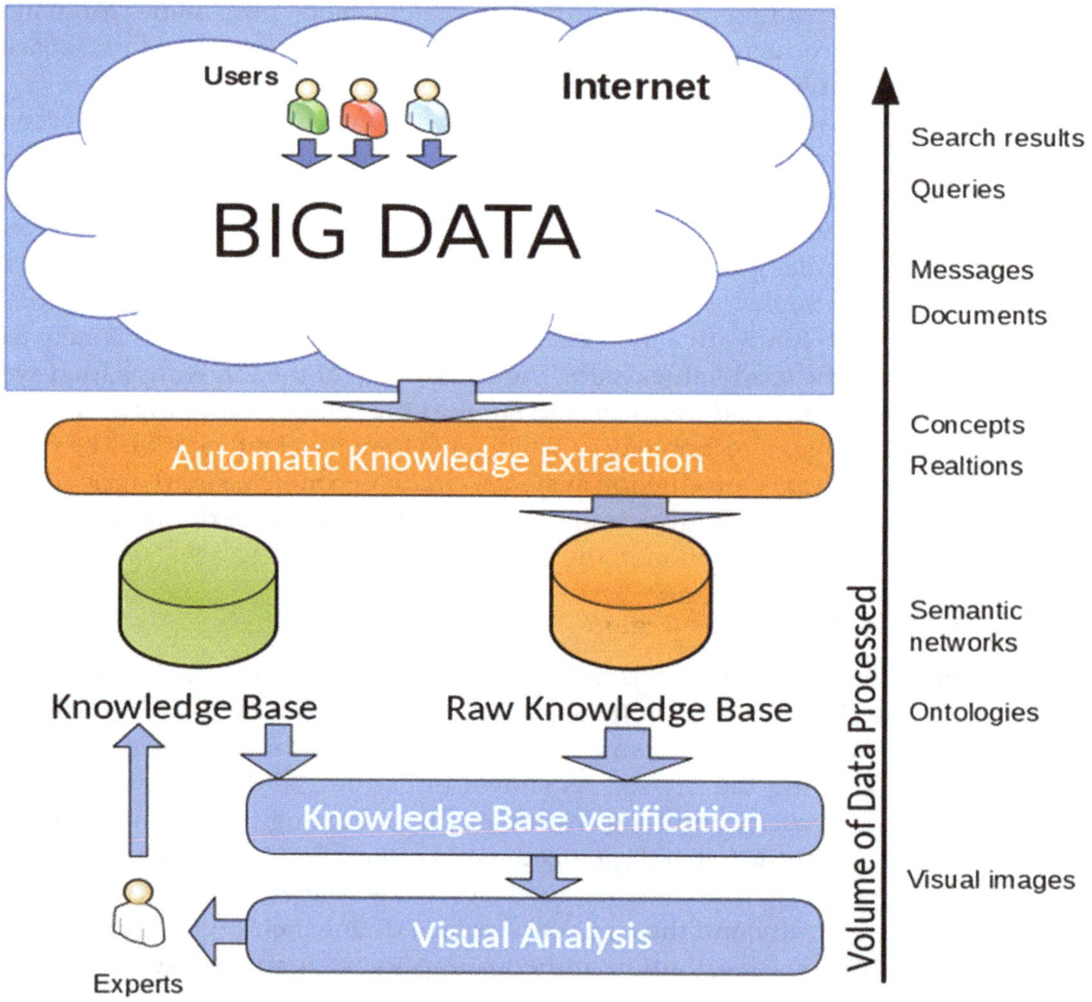

Figure 3. Information technology for knowledge base creation.

We developed a framework for knowledge base creation as shown in Figure 3. This an example of how customers as users will be able to tap into available and documented information of foods and food ingredients. It will collate all information gathered by experts to create the knowledge base within a food system digitalization that can be applied in the Barents region. As there are different languages in the region, there is a need to have a common understanding across the region. In addition, increasing amounts of accumulated information require the use of automated methods for extracting knowledge from raw and inconsistent data, represented in particular by heterogeneous textual sources. Generally, in this case, the task of automated processing should be to summarize the information: a gradual decrease for data to process while simultaneously increasing their information impressiveness. Visual images provide the greatest information impressiveness and at the same time offers the highest speed of information processing by humans. However, the tradeoff for efficiency becomes a reduction in processing accuracy due to the ambiguity of interpreting visual images. A compromise variant of a simultaneously compact, rich information and unambiguously interpreted representation of knowledge is the formalized concept systems presented in the form of an ontology or in the form of another logical system that determines the semantics of a domain. Interactive visual analysis methods can be used in the region which is multi-lingual. This will combine the formal rigor of logical systems with the efficiency of processing visual images.

Digital solutions are important aspects of the Smart specialization strategy in the European Union (EU). The EU recognizes the importance of specific regional knowledge, technical assets and critical mass towards the diversification of regional economies alongside smart specialization [81]. However, there are also challenges in the exchange of data and communication when there is a need to link different systems together in a unified system to cover the different aspects of the food sector in a region. For future collaborations in digitalization of the food system in the Barents region, there should be emphasis on the harmonization of standards through the national and regional governments. The ICT solutions, when developed in isolation from realities and practices of producers and consumers, run the risk of hampering rather than advancing possibilities for sustainability transitions in the food system [82]. Furthermore, there is a need to adopt a holistic approach that takes into consideration links in the food chain; such that ICT solutions consider production practices, communication in food chain, and consumer behavior [83]. This is important in order to avoid sustainability gains in one part of the food chain offsetting changes in another part.

7. Concluding Remarks

There is an urgent need to have a systemic change in our food system that encourages food security and sovereignty. The basis of the technology that underpins digitalization as described in this paper, when clearly aligned to the food industry, can revolutionize the industry. Digitalization across the food system will positively affect food security and safety, thereby ensuring human security in the Barents region. Such a disruption, coupled with the development of other infrastructures particularly in Northwest Russia, will help to promote cross-border collaboration that will be of benefit to the Barents region. Cross-border collaboration at a regional level that foster sustainability, transparency, and an efficient management of resources will improve with digitalization. This review will be useful in conceptualizing a framework for food system digitalization in the region. The knowledge base can be enhanced through the gathering of metrics that can lead to a more integrated approach for the region in future collaboration on food security and safety. The authors have reviewed the opportunities of food system digitalization, challenges of telecommunication, and the impact of human activities in the Barents region. It is envisaged that policy makers and stakeholders in the region will be better informed to support digitalization in the food sector across the Barents region. The socio-economic challenges associated with digitalization, as with any technological development such as digital inequality and threats caused by the complexity and vulnerability of ICT-based infrastructures, the governance of generated data, cybersecurity, were not fully discussed in this paper. We hope that the food system in the Barents region will embrace digitalization in the future, support and targeted trainings to rural communities will improve the adoption rate of digitalization in the region.

Author Contributions: D.R., M.S. and V.D. jointly carried out the conceptual framework, analysis, as well as writing of the manuscript.

Acknowledgments: The authors would like to acknowledge the anonymous reviewers who provided useful comments before the manuscript was submitted to Agriculture journal.

References

1. Allen, R.C. *The British Industrial Revolution in Global Perspective: How Commerce Rather than Science Caused the Industrial Revolution and Modern Economic Growth*; Mimeo: Oxford, UK, 2006.

2. Mokyr, J. *The Second Industrial Revolution, 1870–1914*; Castronovo, V., Ed.; Laterza Publishing: Italy, Rome, 1999; pp. 219–245.
3. Rifkin, J. The Third Industrial Revolution: How the Internet, Green Electricity, and 3-d Printing are Ushering in a Sustainable Era of Distributed Capitalism. *World Financ. Rev.* **2012**, *1*, 4052–4057.
4. Schwab, K. *The Fourth Industrial Revolution*; World Economic Forum: Geneva, Switzerland, 2016; 184p.
5. AMAP. Adaptation Actions for a Changing Arctic. Barents Area: Overview Report. Arctic Monitoring and Assessment Programme (AMAP). Available online: https://www.amap.no/documents/download/2885/inline (accessed on 25 July 2019).
6. Hossain, K.; Raheem, D.; Cormier, S. *Food Security Governance in the Arctic-Barents Region*; Springer: Cham, Switzerland, 2018.
7. Nilsson, L.M. *Sami Lifestyle and Health: Epidemiological Studies from Northern Sweden*; Umeå University: Umeå, Sweden, 2012.
8. Muller-Wille, L.; Granberg, L.; Helander, M.; Heikkila, L.; Lansman, A.S.; Tuiku, T.; Berrouard, D. Community viability and well-being in northernmost Europe: Social change and cultural encounters, sustainable development and food security in Finland's North. *Int. J. Bus. Glob.* **2008**, *2*, 331–353. [CrossRef]
9. Frenk, J.; Gómez-Dantés, O.; Moon, S. From sovereignty to solidarity: A renewed concept of global health for an era of complex interdependence. *Lancet* **2014**, *383*, 94–97. [CrossRef]
10. FAO. *Food and Agriculture Organization and United Nations Human Security Unit, Human Security & Food Security: Hunger, Food Insecurity, and Malnutrition*; FAO: Rome, Italy, 2016.
11. De Ridder, D.; Kroese, F.; Evers, C.; Adriaanse, M.; Gillebaart, M. Healthy diet: Health impact, prevalence, correlates, and interventions. *Psychol. Health* **2017**, *32*, 907–941. [CrossRef] [PubMed]
12. Kaiser, M.L.; Dionne, J.; Carr, J.K. Predictors of Diet-Related Health Outcomes in Food-Secure and Food-Insecure Communities. *Soc. Work Public Health* **2019**, *34*, 1–16. [CrossRef] [PubMed]
13. Raheem, D. Food and Nutrition Security as a Measure of Resilience in the Barents Region. *Urban Sci.* **2018**, *2*, 72. [CrossRef]
14. Nilsson, L.M.; Destonuni, G.; Berner, J.; Dudarev, A.; Mulvad, G.; Odland, J.O.; Rautio, A.; Tikhonov, C.; Evengård, B. A call for urgent monitoring of food and water security based on relevant indicators for the Arctic. *AMBIO J. Hum. Environ.* **2013**, *42*, 816–822. [CrossRef] [PubMed]
15. Godfray, H.C.J.; Beddington, J.R.; Crute, I.R.; Haddad, L.; Lawrence, D.; Muir, J.F.; Toulmin, C. Food security: the challenge of feeding 9 billion people. *Science* **2010**, *327*, 812–818. [CrossRef]
16. Vermeulen, S.J.; Campbell, B.M.; Ingram, J.S. Climate change and food systems. *Annu. Rev. Environ. Resour.* **2012**, *37*, 195–222. [CrossRef]
17. Wheeler, T.; Von Braun, J. Climate change impacts on global food security. *Science* **2013**, *341*, 508–513. [CrossRef]
18. BEAC, Barents Euro-Arctic Council. History. 2016. Available online: http://www.beac.st/en/About/Barents-region/history (accessed on 10 February 2019).
19. Salminen, M.; Hossain, K. Digitalisation and human security dimensions in cybersecurity: an appraisal for the European High North. *Polar Rec.* **2018**, *54*, 108–118. [CrossRef]
20. Cormier, S.; Raheem, D. Food security in the Barents region. In *Society, Environment and Human Security in the Arctic Barents Region, Routledge Explorations in Environmental Studies*; Hossain, K., Cambou, D., Eds.; Routledge: Abingdon, UK, 2018.
21. Ojala, T. Challenges in sustainable use of natural resources of the North. In *The Finnish Committee for European Security (STETE) Nordic Forum for Security Policy Final Report*; 2014; pp. 45–47. Available online: http://www.widersecurity.fi/uploads/1/3/3/8/13383775/nordic_forum_2014.pdf (accessed on 18 August 2018).
22. Dudarev, A.A.; Dushkina, E.V.; Sladkova, Y.N.; Chupahin, V.S.; Lukichova, L.A. Evaluating health risk caused by exposure to metals in local foods and drinkable water in Pechenga district of Murmansk region. *Meditsina Truda I Promyshlennaia Ekologiia* **2015**, *11*, 25–33.
23. SITRA. Finnish Innovation Fund. Finnish Road Map to a Circular Economy 2.0 (2016–2025). 2019. Available online: https://www.sitra.fi/en/articles/improving-resource-efficiency-industrial-symbiosis-opportunities-smes/ (accessed on 2 May 2019).
24. Tubiello, F. Climate Change Adaptation and Mitigation: Challenges and Opportunities in the Food Sector. In Proceedings of the High-level conference on world food security: the challenges of climate change and bioenergy, Rome, Italy, 3–5 June 2008.

25. IPCC. *Climate Change 2013: The Physical Science Basis. Contribution of Working Group I to the Fifth Assessment Report of the Intergovernmental Panel on Climate Change*; Stocker, T.F.D., Qin, G.K., Plattner, M., Tignor, S.K., Allen, J., Boschung, A., Nauels, Y., Xia, V.B., Midgley, P.M., Eds.; Cambridge University Press: Cambridge, UK; New York, NY, USA, 2013; 1535p. [CrossRef]
26. Kallenborn, R.; Halsall, C.; Dellong, M.; Carlsson, P. The influence of global distribution on fate processes and distribution of anthropogenic persistent organic pollutants. *J. Environ. Monit.* **2012**, *14*, 2854–2869. [CrossRef]
27. Stern, G.A.; Robie, W.M.; Outridge, P.M.; Wilson, S.; Chetelat, J.; Cole, A.; Hintelmann, H.; Loseto, L.L.; Steffen, A.; Wang, F.; et al. How does climate change influence arctic mercury? *Sci. Total Environ.* **2012**, *414*, 22–42. [CrossRef]
28. Fisk, A.; DeWit, C.; Wayland, Z.; Kuzyk, N.; Burgess, R.; Letcher, R.; Braune, B.; Norstrom, R.; Blum, S.P.; Sandau, C.; et al. An assessment of the toxicological significance of anthropogenic contaminants in Canadian Arctic wildlife. *Sci. Total Environ.* **2005**, *351*, 53–93. [CrossRef]
29. Berner, J.; Brubaker, M.; Revitch, B.; Kreummel, E.; Tcheripanoff, M.; Bell, J. Adaptation in Arctic circumpolar communities: Food and water security in a changing climate. *Int. J. Circumpolar Health* **2016**, *75*, 33820. [CrossRef]
30. Rydén, P.; Sjöstedt, A.; Johansson, A. Effects of climate change on tularemia activity in Sweden. *Glob. Health Action* **2009**, *2*, 2063. [CrossRef]
31. Carey, C.C.; Ibelings, B.W.; Hoffmann, E.P.; Hamilton, D.P.; Brookes, J.D. Eco-physiological adaptations that favour freshwater cyanobacteria in a changing climate. *Water Res.* **2012**, *46*, 1394–1407. [CrossRef]
32. Revich, B.; Tokarevich, N.; Parkinson, A. Climate change and zoonotic disease in the Russian Arctic. *Int. J. Circumpolar Health* **2012**, *71*, 18792. [CrossRef]
33. Holt-Giméénez, E.; Wang, Y. Reform or Transformation? The Pivotal Role of Food Justice in the U.S. Food Movement. *Race Ethn. Multidiscip. Glob. Contexts* **2011**, *1*, 83–102. [CrossRef]
34. WFS. World Food System Centre ETH Studio AgroFood, Zurich, Switzerland. 2018. Available online: http://www.worldfoodsystem.ethz.ch/research/flagship-projects/eth-studio-agrofood.html (accessed on 12 December 2018).
35. Di Renzo, L.; Colica, C.; Carraro, A.; Goga, B.C.; Marsella, L.T.; Botta, R.; Sarlo, F. Food safety and nutritional quality for the prevention of non-communicable diseases: the Nutrient, hazard Analysis and Critical Control Point process (NACCP). *J. Transl. Med.* **2015**, *13*, 128. [CrossRef]
36. Struebi, P. How the Fourth Industrial Revolution Can Radically Improve Our Food Supply Chain. Huffington Post. New York, NY, USA, 2016. Available online: https://www.huffingtonpost.com/patrick-struebi/how-the-fourth-industrial-revolution-can-radically-improve-our-food-supply-chain_b_9011556.html?guccounter=1 (accessed on 12 November 2018).
37. Internet Live Stats. Number of Internet Users Worldwide. 2016. Available online: http://www.Internetlivestats.com/Internet-users/ (accessed on 28 February 2018).
38. Muilu, T. A Northern dimension for the European Union: Background and proposals. Revival. In *Globalization and Marginality in Geographical Space (2001): Political, Economic and Social Issues of Development at the Dawn of New Millennium*; Taylor & Francis: Abingdon, UK, 2017; pp. 79–89.
39. Lagunov, A.; Terekhin, V. Modelling the Barents territory coverage area of satellite KA-SAT. In *Smart Technologies, Proceedings of the IEEE EUROCON 2017-17th International Conference on Smart Technologies, Ohrid, Macedonia, 6–8 July 2017*; Curran Associates, Inc.: New York, NY, USA, 2017; pp. 588–593.
40. Russian Arctic Policy. Basis of State Policy of the Russian Federation in the Arctic for the Period up to 2020 and the Further Perspective. 2018. Available online: http://static.government.ru/media/files/A4qP6brLNJ175I40U0K46$\times$4SsKRHGfUO.pdf (accessed on 27 February 2018).
41. Russian Arctic Strategy. Strategy for development of the Arctic Zone of the Russian Federation and Ensuring National Security for the Period up to 2020. 2018. Available online: fromhttp://static.government.ru/media/files/2RpSA3sctElhAGn4RN9dHrtzk0A3wZm8.pdf (accessed on 20 March 2018).
42. Murmansk Strategy. Strategy of Social and Economic Development of the Murmansk Region until 2020 and for the Period up to 2025. 2018. Available online: http://minec.gov-murman.ru/activities/strat_plan/sub02/ (accessed on 17 February 2018).

43. Communications in the North. The 6th Conference "Communication in the Russian North". Official Conference Website. 2018. Available online: http://www.xn--80adblbaj2c5ace3kob.xn--p1ai/ (accessed on 27 February 2018).
44. Commission for Arctic Development. The Russian Federation and Finland Discussed the Project of Laying a Fibre-Optic Communication Line along the Northern Sea Route. 18 October 2017. Available online: https://arctic.gov.ru/News/1959c8dc-31b4-e711-80d7-00155d006312?nodeId=4f828d76-8f58-e511-8259-e82aea5c46bb&page=1&pageSize=10. (accessed on 27 February 2018).
45. Rostelecom. Underwater Fiber-Optic Line of Kamchatka-Sakhalin–Magadan. 2018. Available online: https://www.rostelecom.ru/projects/FarEast_FOCL/ (accessed on 27 February 2018).
46. D-Russia. RTComm Built a Satellite Station in One of the Northernmost Cities in Russia. 2016. Available online: http://d-russia.ru/rtkomm-postroil-sputnikovuyu-stanciyu-v-odnom-iz-samyx-severnyx-gorodov-rossii.html (accessed on 7 March 2018).
47. Electrosvyaz. Communication in Extreme Conditions. Interview with Evgeny Buydinov, Deputy Director General for Innovative Development of the Federal State Unitary Enterprise "Space Communication" (GP KS) Electrosvyaz. 2016. Available online: http://www.elsv.ru/svyaz-v-ekstremalnyh-usloviyah/ (accessed on 27 February 2018).
48. Rostelecom. Coverage Maps for MTS, Megafon, Yota, Tele2, Rostelecom, SkyLink LTE. 3G, 4G, 2G and Cellular. 2018. Available online: https://4g-faq.ru/karty-pokrytiya/ (accessed on 23 February 2018).
49. World Bank. *World Development Report 2016: Digital Dividends*; World Bank Group: Washington, DC, USA, 2016. [CrossRef]
50. Kelly, T.; Liaplina, A.; Tan, S.W.; Winkler, H. *Reaping Digital Dividends: Leveraging the Internet for Development in Europe and Central Asia*; World Bank Publications: Washington, DC, USA, 2017.
51. Fälström, P.; Jörgensen, J. Internet of Food. When Food Gets Networked, Life Changes. 2015. Available online: https://stupid.domain.name/stuff/IoF_Framework-201505.pdf (accessed on 16 March 2018).
52. Dinesh, D.; Frid-Nielsen, S.; Norman, J.; Mutamba, M.; Loboguerrero Rodriguez, A.M.; Campbell, B. Is Climate-Smart Agriculture effective? A Review of Selected Cases. CCAFS Working Paper No. 129. Copenhagen. CGIAR Research Program on Climate Change, Agriculture and Food Security (CCAFS): Copenhagen, Denmark, 2015. Available online: www.ccafs.cgiar.org (accessed on 14 April 2019).
53. Neate, P.J. Climate-Smart Agriculture Success Stories from Farming Communities around the World. CGIAR Research Program on Climate Change, Agriculture and Food Security (CCAFS) and the Technical Centre for Agricultural and Rural Cooperation (CTA). 2013. Available online: http://hdl.handle.net/10568/34042 (accessed on 20 May 2018).
54. Whitfield, S.; Challinor, A.J.; Rees, R.M. Frontiers in Climate Smart Food Systems: Outlining the Research Space. Front. Sustain. *Food Syst.* **2018**, *2*, 2. [CrossRef]
55. Engineering News. 2017. Available online: http://www.engineeringnews.co.za/print-version/industry-40-and-iot-transforming-the-manufacturing-industry-2017-06-01 (accessed on 10 October 2018).
56. Design and Manufacturing News. Industry 4.0 and IoT Central to Manufacturing Transformation. 2017. Available online: http://www.bizcommunity.com/Article/196/399/162851.html (accessed on 12 October 2018).
57. GAN. Global Africa Network. Industry 4.0 and IoT: Transforming the Manufacturing Industry. 2017. Available online: https://www.globalafricanetwork.com/2017/06/01/company-news/industry-4-0-and-iot-transforming-the-manufacturing-industry/ (accessed on 14 November 2018).
58. Chen, H. Applications of cyber-physical system: A literature review. *J. Ind. Integr. Manag.* **2017**, *2*, 1750012. [CrossRef]
59. Lo, S.K.; Xu, X.; Chiam, Y.K.; Lu, Q. Evaluating suitability of applying blockchain. In Proceedings of the IEEE 24th International Conference on Engineering of Complex Computer systems, Fukuoka, Japan, 6–8 November 2017; pp. 158–161.
60. Casino, F.; Dasaklis, T.K.; Patsakis, C. A systemic literature review of blockchain based applications: Current status, classification and open issues. *Telemat. Inform.* **2019**, *36*, 55–81. [CrossRef]
61. Lu, S. How Blockchain will Restore Consumer Confidence in Food Safety. 2017. Available online: https://www.fooddive.com/news/how-blockchain-will-restore-consumer-confidence-in-food-safety/503846/ (accessed on 1 August 2018).

62. Manski, S. Building the blockchain: The co-construction of a global commonwealth to move beyond the crises of global capitalism. In Proceedings of the 12th Annual California Graduate Student Conference, University of California, Irvine, CA, USA, 7 May 2016.
63. Galvez, J.F.; Mejuto, J.C.; Simal-Gandara, J. Future challenges on the use of blockchain for food traceability analysis. *TrAC Trends Anal. Chem.* **2018**, *107*, 222–232. [CrossRef]
64. Steinmetz, R.; Wehrle, K. What Is This "Peer-to-Peer" About? In *Peer-to-Peer Systems and Applications*; Springer: Berlin/Heidelberg, Germany, 2005; pp. 9–16.
65. Arcticstat-Statistics on Population\Density in All Regions for All Years. 2018. Available online: http://www.arcticstat.org/default.aspx/Indicator/[Population]Density/P/3/default.aspx (accessed on 17 December 2018).
66. Odland, J.Ø.; Donaldson, S.; Dudarev, A.; Carlsen, A. AMAP assessment 2015: human health in the Arctic. *Int. J. Circumpolar Health* **2016**, *75*. [CrossRef]
67. Rosstat. Publications Catalog: Federal State Statistics Service. 2018. Available online: http://www.gks.ru/wps/wcm/connect/rosstat_main/rosstat/ru/statistics/publications/catalog/doc_1286360627828 (accessed on 17 December 2018).
68. V-Kontakte. Russian Version of Facebook for E-Commerce. 2019. Available online: https://www.ecommerce-nation.com/still-dont-know-vkontakte-the-russian-facebook-with-70-million-active-users/ (accessed on 15 January 2019).
69. O'Keeffe, A. Food Security in the Arctic. 2018. Available online: https://griffithreview.com/articles/food-security-in-the-arctic/ (accessed on 17 December 2018).
70. Shishaev, M.; Lomov, P. High Automated Integration of Ontologies on the Basis of Extendable Thesaurus//Information Modelling and Knowledge Bases XXIII. In *Frontiers in Artificial Intelligence and Applications*; IOS Press: Amsterdam, The Netherlands, 2009; pp. 321–330.
71. Szymoniuk, B.; Valtari, H. The REKO System in Finland: a New Model of a Sustainable Marketing Channel. *Probl. Sustain. Dev.* **2018**, *2*, 103–111.
72. FAO, Food. *The State of Food and Agriculture: Climate Change, Agriculture and Food Security*; FAO: Rome, Italy, 2016.
73. El Bilali, H.; Mohammad, S.A. Transition towards sustainability in agriculture and food systems: Role of information and communication technologies. *Inf. Process. Agric.* **2018**, *5*, 456–464. [CrossRef]
74. Berti, G.; Mulligan, C. ICT & the Future of Food and Agriculture. Industry Transformation – Horizon Scan: ICT & the Future of Food. Telefonaktiebolaget LM Ericsson: Stockholm, Sweden, 2015. Available online: http://gow.epsrc.ac.uk/NGBOViewGrant.aspx?GrantRef=EP/J000604/2 (accessed on 12 February 2018).
75. EU. Novel Food Regulation. European Union (EU) 2015/2283. 2018. Available online: https://eur-lex.europa.eu/legal-content/EN/TXT/?uri=CELEX:32015R2283 (accessed on 15 December 2018).
76. Bellona Report. The Environmental Status of Norwegian Aquaculture. 2003. Available online: https://network.bellona.org/content/uploads/sites/3/The-Environmental-Status-of-Norwegian-Aquaculture.pdf (accessed on 10 May 2019).
77. Johansen, T.J.; Hykkerud, A.L.; Uleberg, E.; Molmann, J. Arctic quality–The effect of high latitude growth conditions on quality of food products. In Proceedings of the Oral Presentation at the 10th Circumpolar Agriculture Conference, Arctic Centre, University of Lapland, Rovaniemi, Finland, 13–15 March 2019.
78. Töyli, P. Introducing Twelve Protected Finnish Products. Aitomakuja Newsletter. 2016. Available online: http://www.aitojamakuja.fi/blogi/?author=3&lang=en (accessed on 10 December 2018).
79. Kurppa, S.; Reinikainen, A. *Tilannekatsaus Luonnonvarakeskuksen (Luke) Arktiseen Biotalouteen Liittyvistä Hankkeista ja Toiminnasta Arktisella Alueella.* Natural Resources Institute Finland Report on Arctic Bioeconomy Research and Development. 2017. Available online: http://jukuri.luke.fi/handle/10024/541454 (accessed on 22 July 2018).
80. Windfuhr, M. *Food Sovereignty and the Right to Adequate Food*; Discussion Paper; FIAN: Heidelburg, Germany, 2003.
81. Stančová, K.C.; Cavicchi, A. Smart Specialisation and the Agri-food System. In *Smart Specialisation and the Agri-Food System*; Palgrave Pivot: London, UK, 2019; pp. 43–57.
82. Davies, A.R. Co-creating sustainable eating futures: technology, ICT and citizen–consumer ambivalence. *Futures* **2014**, *62*, 181–193. [CrossRef]

10

Investigating 2-D and 3-D Proximal Remote Sensing Techniques for Vineyard Yield Estimation

Chris Hacking [1], Nitesh Poona [1], Nicola Manzan [2] and Carlos Poblete-Echeverría [3,*]

1 Department of Geography and Environmental Studies, Stellenbosch University, Private Bag X1, Matieland 7602, South Africa
2 Dipartimento di Scienze AgroAlimentari, Ambientali e Animali, University of Udine, Via delle Scienze 208, 33100 Udine, Italy
3 Department of Viticulture and Oenology, Faculty of AgriSciences, Stellenbosch University, Private Bag X1, Matieland 7602, South Africa
* Correspondence: cpe@sun.ac.za.

Abstract: Vineyard yield estimation provides the winegrower with insightful information regarding the expected yield, facilitating managerial decisions to achieve maximum quantity and quality and assisting the winery with logistics. The use of proximal remote sensing technology and techniques for yield estimation has produced limited success within viticulture. In this study, 2-D RGB and 3-D RGB-D (Kinect sensor) imagery were investigated for yield estimation in a vertical shoot positioned (VSP) vineyard. Three experiments were implemented, including two measurement levels and two canopy treatments. The RGB imagery (bunch- and plant-level) underwent image segmentation before the fruit area was estimated using a calibrated pixel area. RGB-D imagery captured at bunch-level (mesh) and plant-level (point cloud) was reconstructed for fruit volume estimation. The RGB and RGB-D measurements utilised cross-validation to determine fruit mass, which was subsequently used for yield estimation. Experiment one's (laboratory conditions) bunch-level results achieved a high yield estimation agreement with RGB-D imagery (r^2 = 0.950), which outperformed RGB imagery (r^2 = 0.889). Both RGB and RGB-D performed similarly in experiment two (bunch-level), while RGB outperformed RGB-D in experiment three (plant-level). The RGB-D sensor (Kinect) is suited to ideal laboratory conditions, while the robust RGB methodology is suitable for both laboratory and in-situ yield estimation.

Keywords: Kinect sensor; RGB; RGB-D; image segmentation; colour thresholding; bunch area; bunch volume; point cloud; mesh; surface reconstruction

---

1. Introduction

Modern-day viticulture has seen an increase in the use of robust scientific methods combined with new technologies to improve overall production [1]. Precision farming, a direct result of the modernisation in agriculture, can be discipline-specific—i.e., specific to horticulture or viticulture. Precision viticulture aims to effectively manage production inputs to improve yield and grape quality while reducing the environmental impact of farming [2]. The use of remote sensing technology and techniques in precision viticulture allows variability to be monitored at vineyard level, per individual block or on a vine basis. Aspects such as vine shape, size and vigour can be observed, providing more accurate yield and fruit quality information [3].

Yield estimation provides information to the winegrower that can be used to manage the vineyard, optimising quality and yield [4]. Awareness of the estimated yield allows the vineyard manager to manipulate the vines to obtain the desired grape characteristics, and provides an effective plan for use during the winemaking process [5]. Accurate yield forecasting assists with logistical planning, both during and after the harvest; for example, what volume will be harvested, where the grapes will be stored, and an expected market price [5].

Wolpert and Vilas [6] outlined a two-step method for vineyard yield estimation. The start of the process determines the number of bunches situated on individual vines early in the season. Subsequent determination of bunch weights occurs at vèraison. Unfortunately, the two-step method is labour-intensive, error-prone and destructive in the estimation process. Additionally, De la Fuente et al. [7] presented yield prediction models using destructive, manually collected data between fruit-set and vèraison, aligning with the more 'classical' two-step method. To overcome the limitations of the manual methods, modern techniques have employed sensors attached to automatic harvesters to monitor yield during the harvesting process [3]. Yield estimation before harvest is becoming possible, with increasing accuracies when making use of non-invasive proximal remote sensing (PRS) technology and techniques [8–12].

A common PRS approach employs 2-D (2-dimensional) RGB (Red, Green, Blue) imagery, captured with a digital camera for yield estimation; for example [8,9,13]. The 2-D approach can be categorised into two steps: (i) image segmentation (to differentiate the bunch from the background); and (ii) yield estimation, using a suitable bunch metric (i.e., the pixel count of the bunch area in an image).

Diago et al. [13] used image classification (for segmentation) and a bunch metric to classify 'background noise' and 'grape' classes. The authors achieved a testing r^2 of 0.73. An alternative segmentation approach to image classification uses colour thresholding to differentiate grapes from the background. Dunn and Martin [8] presented an RGB colour thresholding approach that applied specific thresholds to the colour properties of an RGB image, generating a binary image—background and grapes. The authors achieved an r^2 of 0.72. A similar thresholding approach was adopted by Liu, Marden and Whitty [9] and Font et al. [14]. Liu, Marden and Whitty [9] introduced a bunch-level experiment under laboratory conditions with manual colour thresholding for image segmentation, which resulted in a yield estimation r^2 of 0.77. Additionally, these authors [9] presented a more complex automatic process for image segmentation on the same dataset presented by Dunn and Martin [8], resulting in an improved r^2 of 0.86. Automatic segmentation stemming from colour thresholding removes the human factor of manual thresholding, resulting in a more robust methodology [15,16]. However, these techniques are more sophisticated than manual thresholding, and therefore require adequate knowledge of computer vision techniques.

The second step, yield estimation, depends on the selected bunch metric. A simple pixel count of the segmented bunches is a favoured metric [8,13], with adaptions to the pixel count presented by Liu, Marden and Whitty [9] and Font et al. [14]. Liu, Marden and Whitty [9] tested five metrics: (i) volume, (ii) pixel count, (iii) perimeter, (iv) berry number and (v) berry size. Yield estimation using the pixel count produced superior results over the remaining metrics [9]. Unlike Liu, Marden and Whitty [9], Nuske et al. [11] avoided image segmentation and used a berry detection algorithm to determine a berry count. The use of the berry count as a yield estimation metric provided an r^2 of 0.74. Subsequent work on multiple multi-temporal datasets produced r^2 values between 0.6–0.73 [4].

A less common PRS approach utilises an RGB-D (RGB-Depth) camera to capture a 3-D (3-dimensional) model in either mesh or point-cloud format, representing the bunch or vine in a 3-D coordinate system [17]. The use of the Microsoft Kinect™ constitutes an ideal low-cost RGB-D sensor for in-situ imaging of vines [17]. The resulting 3-D models can be used to extract volumetric measurements for yield estimation. A limited number of studies have investigated the utility of the Kinect sensor for volume estimation. For example, Wang and Li [18] and Andújar et al. [19] employed the Kinect sensor under laboratory conditions for the volume estimation of sweet onions and cauliflowers, respectively. To date, the only use of an RGB-D Kinect sensor for yield estimation within viticulture was presented by Marinello et al. [10]. The authors assessed the sensor position for volume estimation of table grape bunches by testing two viewing angles, side-on and bottom-up. Multiple sensor-target distances were tested for the side-on viewing angle. Marinello et al. [10] concluded that a side-on viewing angle with a sensor-target distance of 0.8–1.0 m generated the best results.

An alternative PRS technique combines 2-D imagery with computer vision techniques, whereby 3-D models are created from the 2-D RGB images. Volume estimations are extracted from the subsequent 3-D models, enabling yield estimation calculations. Advanced computer vision techniques allow substantial automation in the process [20–22].

RGB and RGB-D technology and techniques incorporated into a suitable methodology present a viable solution for vineyard yield estimation. The established use of RGB imagery is evident, while the novel use of RGB-D imagery shows promise for future yield estimation. However, to date no study has investigated 2-D and 3-D PRS techniques side-by-side for vineyard yield estimation. Examining these two techniques in a commercial vineyard could provide insight into their capabilities and operational potential.

A key aspect for consideration when implementing these methodologies within a vineyard is the canopy coverage. The combination of essential canopy management practices and the vineyard's trellis system—particularly a vertical shoot positioned (VSP) system—directly influences canopy coverage, and inevitably the success of PRS techniques for yield estimation. High canopy coverage in the bunch zone results in bunch occlusion from the sensor. The incorporation of a canopy treatment was therefore proposed for this study.

This study aimed to investigate 2-D RGB and 3-D RGB-D PRS techniques for yield estimation in a VSP vineyard, using bunch area/volume estimation. The study was undertaken as three experiments, occurring under laboratory and field conditions. Field conditions were conducted at both bunch- and plant-level. Two canopy treatments were implemented from direct canopy manipulation and were defined as full canopy (FC) and leaf removal (LR). Hypothetically, the LR treatment will produce better estimation results. Furthermore, to achieve the aim of the study, two objectives were determined: to develop independent 2-D and 3-D yield estimation methodologies and to analyse and compare the success of the two PRS techniques for yield estimation.

2. Materials and Methods

2.1. Study Site

The study was carried out at the end of the 2016/17 growing season in a drip-irrigated Shiraz vineyard at the Welgevallen Experimental Farm located in Stellenbosch, South Africa (33°56′26″ S; 18°51′56″ E). The vineyard was planted in the year 2000, with a grapevine spacing of 2.7 × 1.5 m in a North-South orientation, and lies approximately 157 m above sea level. A seven-wire hedge VSP trellis system with three sets of moveable canopy wires is used in the vineyard. The Stellenbosch area falls within the coastal wine grape region of the Western Cape, which is characterised by a Mediterranean climate with long, dry summers [23]. Thirty-one individual vines were selected across three rows and sampled for this study (Figure 1).

2.2. Data Acquisition

Data was acquired between 28 February and 3 March 2017 (harvest), where data collected in situ was implemented with the two canopy treatments. The purpose behind the canopy treatments was to gain a direct line of sight to the bunches, which were generally hidden by the vine's canopy. No manipulation of the canopy, essentially the normal canopy condition, was classified as the FC treatment. The alternative LR treatment occurs after manual manipulation of the canopy, resulting in complete leaf removal in the bunch zone, effectively displaying the bunches.

Figure 2 illustrates the data-acquisition process that resulted in a total of ten datasets, as outlined by each step:

1. RGB and RGB-D imagery acquired for the FC treatment taken at bunch-level (n = 21; randomly selected and labelled bunches from the 31 vines) and plant-level (n = 31; individual vines). The resulting four datasets included: (i) RGB: bunch; (ii) RGB-D: bunch; (iii) RGB: vine; and (iv) RGB-D: vine.

2. Manual manipulation of the canopy resulted in the environment for the four LR treatment datasets. These datasets were identical to the FC datasets and concluded the in-situ datasets (datasets = 8). At this point, the 31 vines were harvested.
3. RGB and RGB-D imagery of the harvested bunches (n = 21) captured under laboratory conditions resulted in the final two datasets. At this point, all datasets (datasets = 10) have been captured.
4. Reference measurements, captured under laboratory conditions, included mass (g) and displacement (mL) for the individual bunches (n = 21) and individual vines (n = 31).

Figure 1. Location of the Shiraz vineyard in Stellenbosch, South Africa. Inset map (red rectangle) shows the three rows used for data collection.

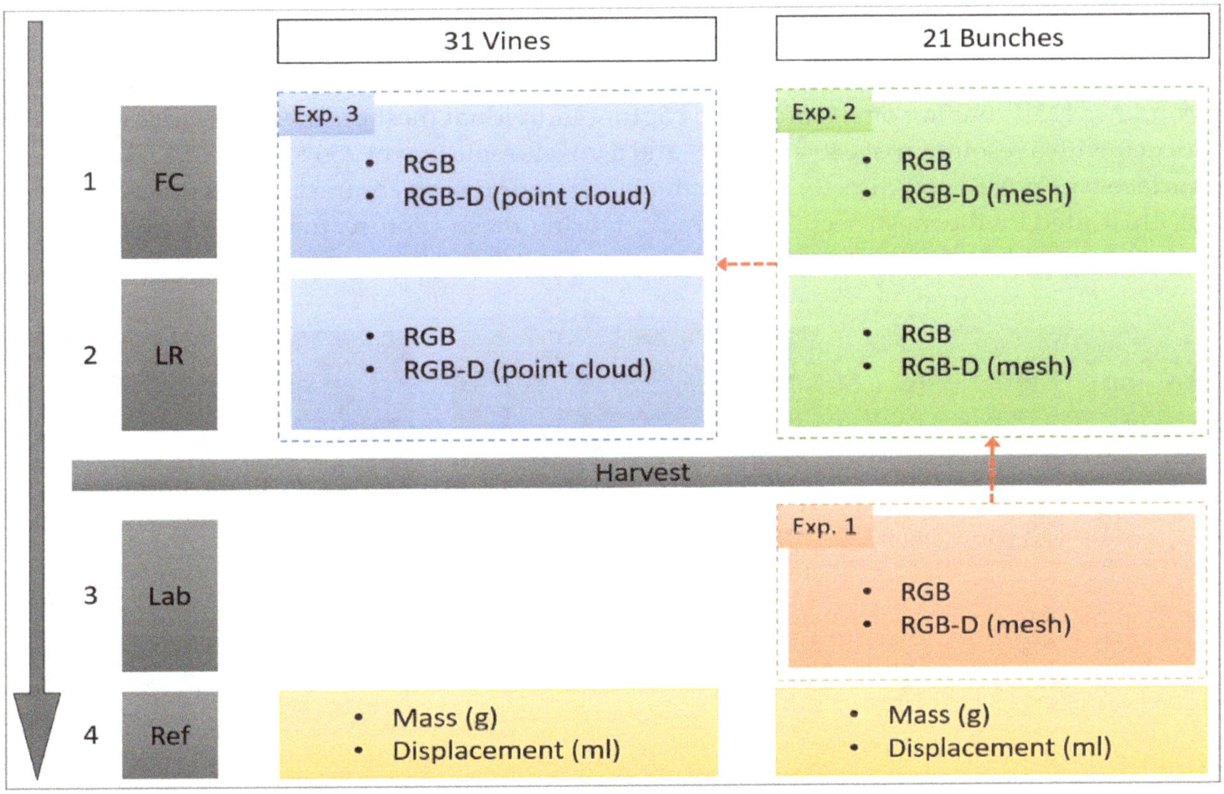

Figure 2. Data-acquisition protocol used in this study. Order of acquisition indicated by the grey arrow. {Key: FC = full canopy; LR = leaf removal; Lab = laboratory; Ref = reference measurements; Exp = experiment}.

RGB and RGB-D images were captured by two PRS sensors. A D3200 digital single-lens reflex camera (Nikon, Tokyo, Japan) was used for capturing 24.2-megapixel RGB images. The camera captured images in *auto* mode, with the flash disabled. The second sensor, a Kinect™ V1 (Microsoft, Redmond, WA, USA), was used to capture RGB-D imagery as either a mesh (bunch-level) or a point cloud (plant-level). The following data-acquisition subsections provide experiment-specific details.

2.2.1. Reference Measurements

Reference measurements were collected under laboratory conditions for the 21 individual bunches and the 31 individual vines. Individual bunch mass (g) was recorded with a Mentor scale (Ohaus, Parsippany, NJ, USA), and individual vine mass (g) was recorded with a Viper SW scale (Mettler Toledo, Columbus, OH, USA). Bunch/vine volume measurements were recorded as the displacement (mL) of water when bunches were submerged in a container of water [24]. The mass and volume measurements were used as reference measurements for the estimated measurements derived from the two PRS techniques.

2.2.2. Experiment One: Individual Bunches under Laboratory Conditions

The 21 individual bunches were imaged using the RGB camera and Kinect sensor under laboratory conditions; i.e., the laboratory was illuminated with white fluorescent lights and natural light entered through the windows. Each bunch was suspended from a tripod against a green background to maximise image contrast (Figure 3a):

(a) *RGB imagery:* The camera was placed parallel to the suspended bunch at a distance of 60 cm. A single image per bunch was captured for image processing. A ruler was included in each image for length reference (Figure 3b).

(b) *RGB-D (Kinect) imagery:* The Kinect sensor was placed 60 cm proximal to the target and captured individual meshes per bunch, resulting in a solid 3-D model of the bunch. The Kinect sensor, coupled with the Kinect Fusion software (part of Microsoft's Software Development Kit 1.8 for Windows [25]) running on a laptop, can capture individual meshes. Meshes were captured in *.stl* format with a resolution of 640 voxels/m, and a voxel resolution of 256 × 256 × 256 voxels. A white surface was positioned directly behind the bunch to improve contrast and depth determination, as illustrated by the mesh seen in Figure 3c. During mesh capture, the entire bunch system was rotated, providing different angles of view.

(a) (b) (c)

Figure 3. Data acquisition under laboratory conditions. (**a**) Experimental setup for image capture; (**b**) RGB image of an individual bunch with a ruler for reference length; and (**c**) RGB-D (Kinect mesh) of an individual bunch.

2.2.3. Experiment Two: Individual Bunches in Field Conditions

Images of the same 21 individual bunches were captured in in-situ conditions under both canopy treatments with the same RGB and RGB-D proximal sensors that were used in experiment one. Here, individual bunches were still attached to the respective vines:

(a) *RGB imagery:* The camera captured images of the individual bunches for both FC (Figure 4a) and LR (Figure 4b) treatments. The camera was positioned approximately 40 cm from the bunch being imaged, maintaining the reference length (ruler) in each image. Image acquisition occurred between 12H00 and 13H00, under natural solar illumination.
(b) *RGB-D (Kinect) imagery:* The same software setup from experiment one was used, which allowed the Kinect to capture an individual mesh per bunch for FC (Figure 4c) and LR (Figure 4d) treatments. The imagery was captured after sunset (approximately 20H00) with artificial illumination. The Kinect was held approximately 60 cm from the bunch and was moved around the bunch axis by hand. For the LR treatment, a board was placed behind the bunches, thereby creating an artificial background to improve volume extraction (Figure 4d).

(a) (b) (c) (d)

Figure 4. Data acquisition of individual bunches in the field. (**a**) RGB image with full canopy (FC); (**b**) RGB image with leaf removal (LR); (**c**) RGB-D (Kinect mesh) with FC; and (**d**) RGB-D (Kinect mesh) with LR.

2.2.4. Experiment Three: Individual Vines in Field Conditions

The same RGB and RGB-D sensors were used to capture in-situ images at plant-level (31 individual vines imaged for both FC and LR treatments):

(a) *RGB imagery:* Imagery was captured at 12H00, under natural solar illumination. The camera was positioned 2 m from the vine, capturing an image for each side of the canopy. This protocol was repeated for both canopy treatments (FC: Figure 5a; LR: Figure 5b).
(b) *RGB-D (Kinect) imagery:* The Kinect sensor captured point clouds—instead of meshes—of the 31 vines, due to the scale difference. Point clouds consisted of thousands of individual points to create 3-D models; for both the FC and LR treatments (Figure 5c,d). The Real-Time Appearance-Based Mapping (RTAB-Map) software [26] was used for point-cloud modelling and regeneration. In RTAB-Map [26], the default filtering parameter and a 3-D cloud decimation ('thinning' of the point cloud) value of '2' were used during exportation of the point clouds in *.ply* format. The point cloud was captured per individual row, repeated for both sides of the canopy. The Kinect sensor was hand-held approximately 2 m from the vines at a perpendicular angle and was moved along the row in a north to south direction. The imagery was collected at 19H00, immediately before sunset, using the last natural illumination of the day.

Figure 5. Experiment three data examples at plant-level. RGB imagery of FC (**a**) and LR (**b**) treatments. RGB-D (Kinect point cloud) of FC (**c**) and LR (**d**) treatments.

2.3. Data Analysis

Data pre-processing and analysis occurred sequentially from experiment one (indicated via red arrows in Figure 2). The canopy treatments existing in experiment two and three had no effect on how the datasets were analysed for yield estimation. The proposed RGB and RGB-D methodologies were created on the LR datasets, before being directly applied to the FC datasets.

2.3.1. RGB Imagery

RGB images were processed using a custom script in MATLAB® [27], as follows:

1. The reference length (obtained from the ruler) in each image was used to scale the image pixels, creating a calibration value in cm^2.
2. Manual selection of the region of interest (ROI) containing the relevant bunch/bunches was undertaken. ROIs were strategically digitised so as to capture minimal background.
3. The masked RGB images were then converted to the HSV (hue, saturation, value) colour space and segmented using MATLAB's [27] *Colour Thresholder* app, part of the *Image Processing Toolbox*™. It was visually evident when selecting the threshold values that the lighting conditions influenced the values. Separate threshold values were therefore determined for the respective experiments. Threshold values were computed using a random training sample from the specific dataset; equivalent to 25% of the experiment's dataset.
4. After the image segmentation process, the number of segmented pixels was determined (adaption of the pixel count metric [9]) and converted into a pixel area (cm^2) using the calibration value.

Figure 6 illustrates the image segmentation process for experiment one at bunch-level (Figure 6a,b), and experiment three at plant-level (Figure 6c,d).

For experiment three, segmentation produced a segmented bunch area image (Figure 6d) for a single side of the vine, with a similar image for the reverse side of the vine. To obtain a single area per vine, the *Total Bunch Area of Vine (TBAV)* was calculated, as follows:

$$TBAV = (Ae + Aw)/2 \tag{1}$$

where *Ae* was the area of the east-facing side of the vine, and *Aw* represented the vine's west-facing area. The *TBAV* was in cm^2. Segmentation success was assessed on the testing subset (25% of the total dataset). A confusion matrix was computed for each experiment, using the predicted values from the segmented binary image versus the actual values of the original RGB image. F1-score and accuracy metrics calculated from the confusion matrix evaluated the segmentation accuracy.

(a) (b) (c) (d)

Figure 6. (**a**) Represents the original RGB image, with (**b**) illustrating the segmented binary image at bunch-level. (**c**) An RGB image of an east-facing vine, with (**d**) the segmented binary image at plant-level.

2.3.2. RGB-D (Kinect) Imagery

Due to the two image data types captured—mesh vs. point cloud—RGB-D imagery was processed differently for experiment one and two (mesh: bunch-level), and experiment three (point cloud: plant-level). Data processing for experiments one and two progressed as follows:

1. Each mesh was manually cleaned and sectioned to remove any background, using the *Cross-Section* tool in CloudCompare [28].
2. The cleaned mesh was reconstructed in MeshLab [29], using the *Screened Poisson Surface Reconstruction* [30] method—see Figure 7. Additional cleaning of the mesh was completed ad hoc in MeshLab [29].
3. Volume (cm^3) of the mesh was calculated in CloudCompare [28].

Screened Poisson Surface Reconstruction [30] allows the back of the mesh, which was 'open' (Figure 7a), to be reconstructed. It becomes 'watertight', as seen in Figure 7b.

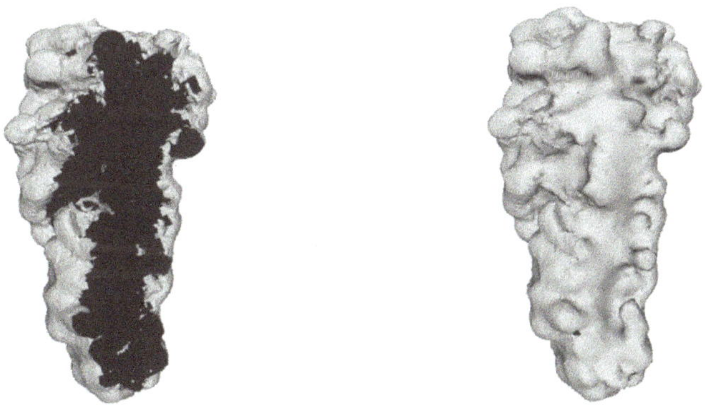

(a) (b)

Figure 7. (**a**) Example of mesh prior to reconstruction, and (**b**) the same mesh after Poisson reconstruction.

The nature of the point-cloud data requires different processing steps to that of the mesh data. This effect was attributed mainly to the necessity of closing the points of the point cloud, thereby producing a 'watertight mesh' for volume extraction.

The point cloud datasets from experiment three were processed as follows:

1. Point clouds were imported into CloudCompare [28], and subsequently cleaned and sectioned to individual vines, focusing on the bunch zone.
2. Bunches were segmented from the point cloud using their colour properties. CloudCompare's [28] *Filter Points by Value* tool incorporates user-defined thresholds, manipulating the RGB colour space of the point cloud. Threshold values were determined on a random sample (25%) of the LR dataset.
3. A custom-built script in R statistical software [31] was used for calculating the segmented point cloud's volume, representing the vine's bunches.

Figure 8 illustrates the raw point cloud (Figure 8a) captured by the Kinect sensor, with the segmented point cloud (Figure 8b) displaying the bunches.

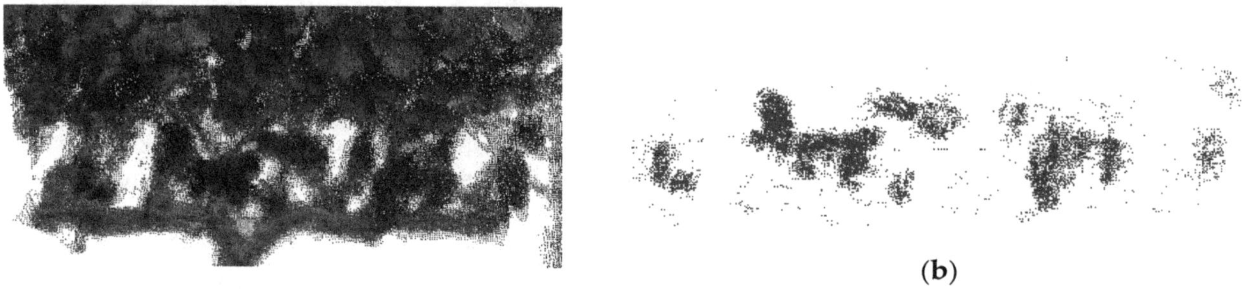

(a) (b)

Figure 8. (**a**) The Kinect point cloud for an LR-treated vine (1E—east side) and (**b**) the segmented point cloud of the same vine.

The custom script in R statistical software [31] incorporated the *alphashape3d* package v1.3 [32] to compute the 3-D shape for volume calculation purposes. The *alphashape3d* [32] package includes the *α-shape* algorithm [33] to recover the geometric structure of the 3-D point cloud for volume calculation. The *α-shape* algorithm [33] requires a specific alpha value for computation; hence, the alpha value directly influences the total volume calculated. To adjust an alpha value for our experimental conditions, various levels of alpha were tested on 25% of the dataset and linearly regressed against the reference values. Similar investigative experiments were conducted by Rueda-Ayala et al. [34] and Ribeiro et al. [35] to determine experiment-specific alpha values.

The process described above for experiment three was repeated for both sides of the vine, resulting in two volume measurements per vine. A single volume value per vine was obtained via the *Total Bunch Volume per Vine* (*TBVV*) calculation:

$$TBVV = (Ve + Vw)/2 \tag{2}$$

where Ve was the volume of the vine's east-facing side, and Vw was the volume of the west-facing side. The resultant $TBVV$ was in cm^3 per vine.

2.3.3. Cross-Validation

Five-fold cross-validation was used to develop the yield estimation model for each dataset. Cross-validation was implemented using the *Caret* package [36] in R statistical software [31], repeated ten times for model robustness. The model produced 'fitted values', which represented the estimated yield (in grams). Following this, the estimated yield (g) values were linearly regressed against the actual mass (g) to produce a final r^2 value (coefficient of determination), indicating the potential for yield estimation. The Root Mean Square Error (RMSE) was computed from the linear regression, indicating the yield estimation error (in grams).

3. Results

3.1. *Reference Measurements*

Results of the reference measurements indicated a strong relationship between mass and volume at bunch-level ($r^2 = 0.971$) and plant-level ($r^2 = 0.996$). The established relationships between mass and volume served as the basis for the subsequent experiments, which were used to evaluate the 2-D and 3-D techniques.

3.2. *Pre-Processing*

The complexity of the RGB-D (Kinect) datasets required two additional pre-processing steps. The first step was the determination of an alpha value required for volume calculation, using the *alphashape3d* package [32]. The second step was volume correction for all Kinect datasets, due to volume estimation errors in the datasets. The segmentation accuracy for the RGB pre-processing has also been included in this section.

3.2.1. RGB Segmentation Accuracy

Experiment one's segmentation results yielded an F1-score of 0.976, with an accuracy of 0.971. Experiment two resulted in a lower F1-score of 0.842, with an accuracy of 0.781. Lastly, experiment three's F1-score and accuracy were 0.833 and 0.932, respectively. Solar illumination could contribute to the lowered results presented in the in-situ measurements.

3.2.2. alphashape3d's Adjusted Alpha Value

Figure 9 represents the r^2, and RMSE curves for the various alpha values tested. Values tested ranged from 0.001 to 0.050, in increments of 0.001. It is evident from this figure that the alpha value selected (alpha = 0.010) satisfies a high coefficient of determination ($r^2 = 0.605$) combined with the lowest RMSE (703.301 cm^3). Figure 10 provides a visual interpretation of the results from Figure 9. A low alpha value, such as 0.005 (Figure 10b), produced an r^2 of 0.520 and an RMSE of 2052.751 cm^3 (sample dataset mean volume = 1961.875 cm^3). Conversely, a high alpha value of 0.050 (Figure 10d) produced an r^2 of 0.506 and an RMSE of 13 936.31 cm^3. The selected alpha value of 0.010 (Figure 10c) was subsequently used for all further analyses.

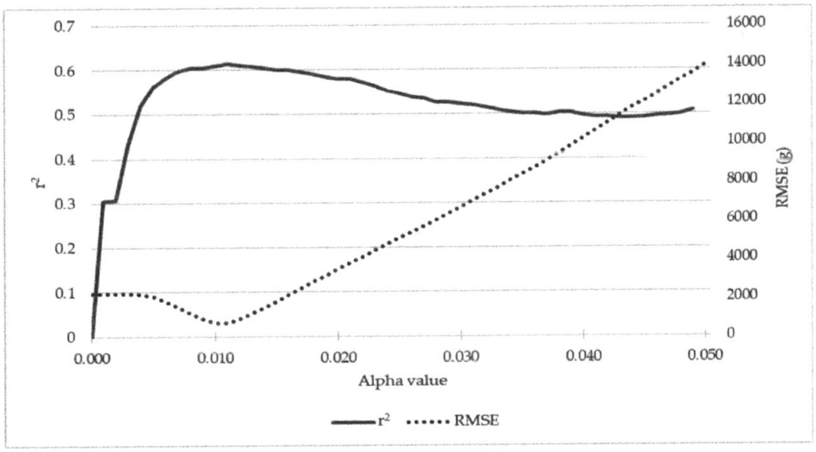

Figure 9. The relevant r^2 and RMSE values for each alpha value tested for the *alphashape3d* package in the custom R script. Alpha values incremented in 0.001, ranging from 0.001–0.050.

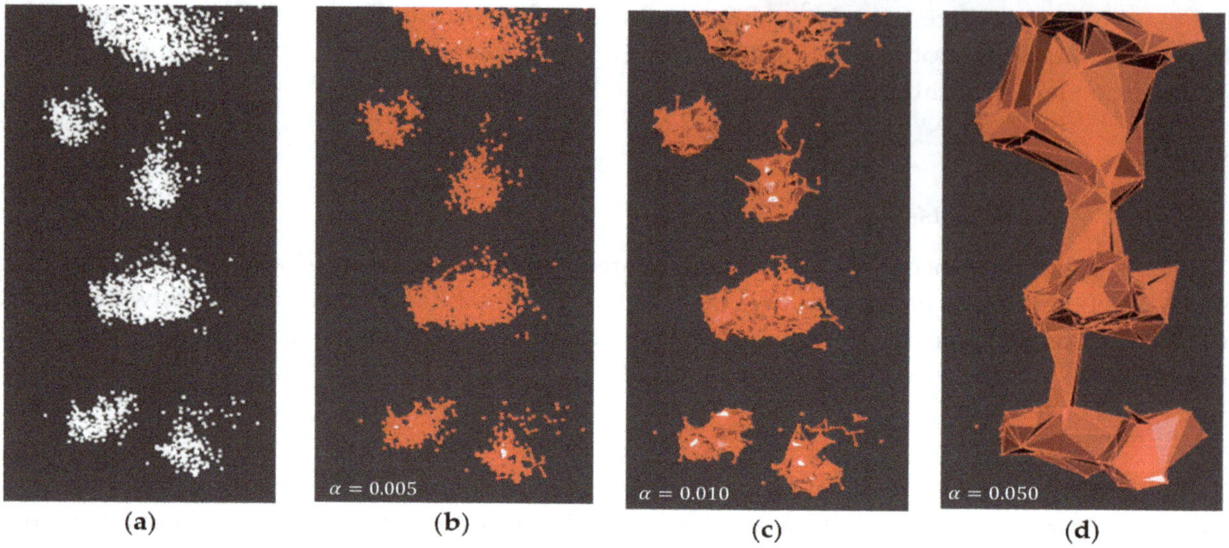

Figure 10. Point-cloud reconstruction testing the alpha value for the *alphashade3d* package. The original point cloud before reconstruction—represented as white points for a visual purpose (**a**), and after reconstruction (**b**–**d**).

3.2.3. Kinect Volume Correction

Figure 11 shows a volume estimation error present in experiment one's data, which aligns with a subsequent review of the literature [10,18,19]. The 21 bunches have a mean actual volume (the reference volume) of 144.952 cm^3 and a mean estimated Kinect volume of 175.672 cm^3. Overestimation by the Kinect V1 sensor was evident from the results. Volume correction via cross-validation was therefore subsequently incorporated into the methodology, where a mean corrected volume of 144.952 cm^3 was achieved. Thereafter, the correction was applied to the remaining Kinect datasets.

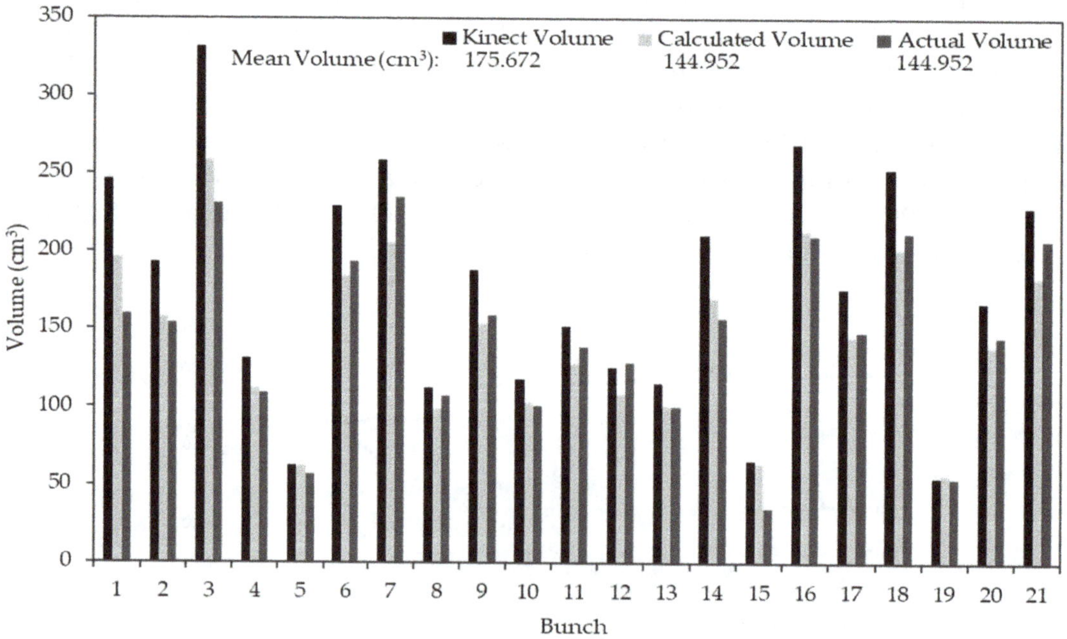

Figure 11. Experiment one's results, for the 21 individual bunches, illustrating the volume estimation error by the Kinect sensor.

3.3. RGB Results

Figure 12 shows the results for the three experiments that used 2-D RGB digital imagery. The best results were obtained in experiment one (Figure 12a), which produced an r^2 of 0.889 and an RMSE of

17.978 g. The level of accuracy achieved in experiment one can be attributed to the controlled laboratory conditions and supports the proposed methodology for yield estimation. Applying this methodology to in-situ bunches produced less accurate results, as seen in experiment two. Experiment two's FC treatment (Figure 12b) produced the lowest yield estimation results for 2-D RGB imagery, with an r^2 of 0.625 and an RMSE of 27.738 g. The LR treatment's (Figure 12c) r^2 and RMSE values were 0.742 and 25.066 g, respectively.

The lesser FC values were directly attributed to the canopy coverage present, as the bunches were partially occluded from the sensor's view. The effect of the canopy treatment was evident when comparing the results. At plant-level (experiment three), the same pattern was present between the canopy treatments. The FC (Figure 12d) treatment of experiment three produced an r^2 of 0.779, while the LR (Figure 12e) treatment produced an even higher r^2 of 0.877. The respective RMSE values were 559.357 g and 443.235 g. The success of yield estimation in experiment three, specifically the LR treatment, supports the methodology's capability for 2-D RGB yield estimation. The in-situ yield estimation was preferable at plant-level, which may be attributed to the lighting conditions and the success of the colour thresholding for bunch segmentation at plant-level.

3.4. *RGB-D Results*

Figure 13 illustrates the RGB-D results obtained for the three experiments. The unrivalled results obtained in experiment one (Figure 13a) produced an r^2 of 0.950 and an RMSE of 12.458 g—the best-performing results presented in this study. The Kinect sensor favoured the controlled conditions of the laboratory, specifically the artificial illumination as a source of light. Applying the same methodology to experiment two (in-situ bunches) resulted in a lower yield estimation performance for both canopy treatments. The FC treatment (Figure 13b) produced an abnormally low r^2 of 0.020, with an RMSE of 8.081 g. A statistical outlier in the data was evident when the results were analysed. With the removal of this outlier, the modified FC results (n = 20) improved drastically, with a new r^2 of 0.609 and an RMSE of 26.790 g (Figure 13c). In contrast, the LR treatment (Figure 13d) resulted in an r^2 of 0.756 and an RMSE of 24.601 g, which aligned with the LR results for bunch-level obtained with RGB imagery (Figure 12c). At plant-level, the unfavourable results of the FC treatment (Figure 13e) generated an r^2 of 0.487 and an RMSE of 673.535 g.

However, the LR treatment (Figure 13f) provided some promise for the Kinect sensor at plant-level, achieving an r^2 of 0.594 and an RMSE of 661.739 g. The same effect of the canopy treatment was evident in the RGB-D results as in the RGB results, with the LR treatment producing a better yield estimation agreement. The results of experiment three indicated a limitation within the proposed methodology for RGB-D yield estimation at plant-level. Such limitation could be from the data-acquisition process, or the image segmentation within the data analysis.

The poor results obtained in experiment two's FC dataset can be attributed to the mesh reconstruction step in the data-analysis process. Exaggerated reconstruction of bunches presents a potential limitation of the *Screened Poisson Surface Reconstruction* algorithm [30], as depicted in Figure 14. The statistically-outlying bunch circled in red (Figure 13b) produced a Kinect volume of 856 cm^3, as illustrated in Figure 14a, when the actual volume was only 35 cm^3. An example of an accurately reconstructed mesh (Figure 14b) was included for visual comparison. This anomaly was unavoidable during data processing.

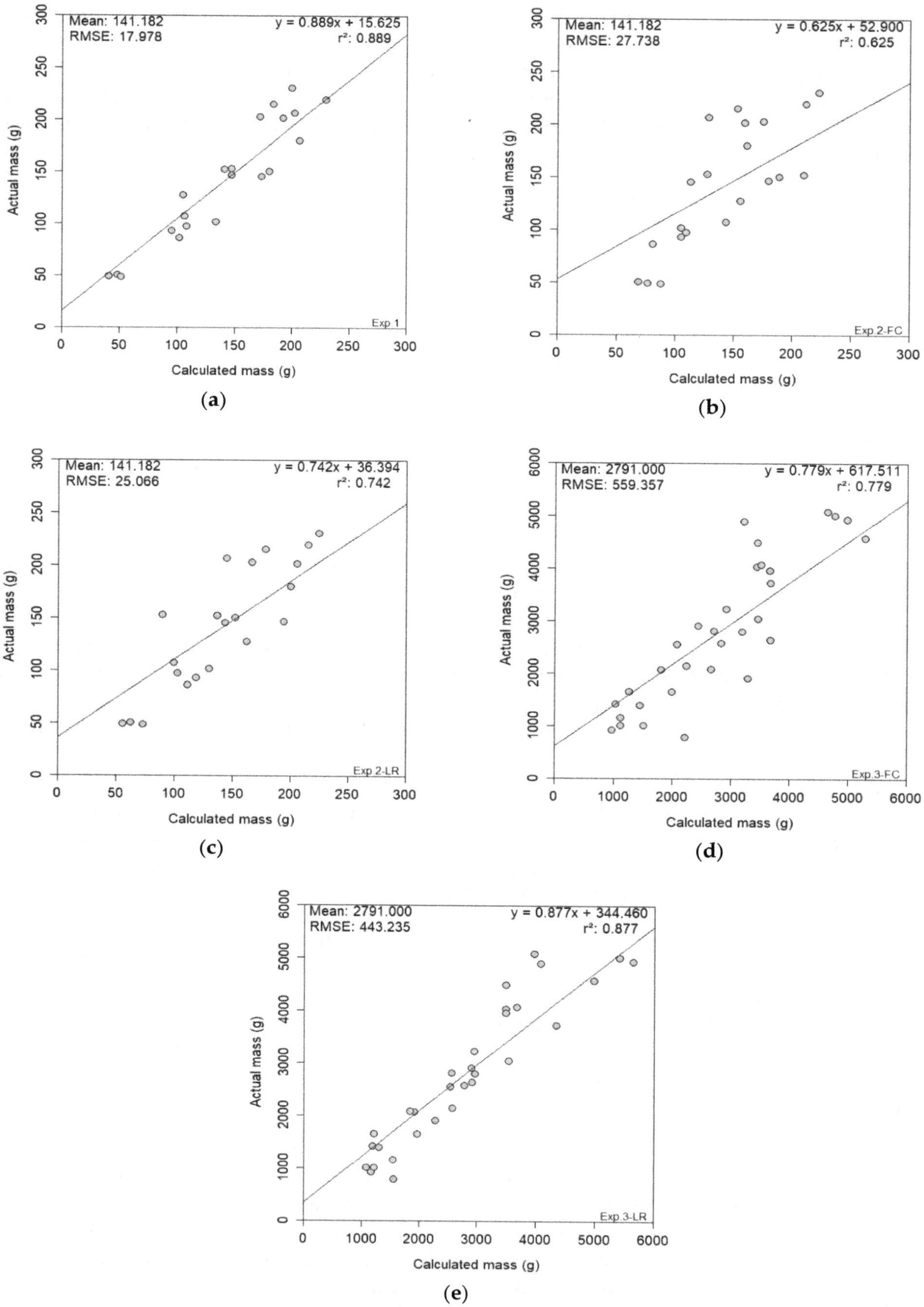

Figure 12. RGB results presented for the three experiments; experiment one (**a**), experiment two FC (**b**) & LR (**c**) and experiment three FC (**d**) & LR (**e**). *{Key: Exp. = experiment; FC = full canopy; LR = leaf removal}.*

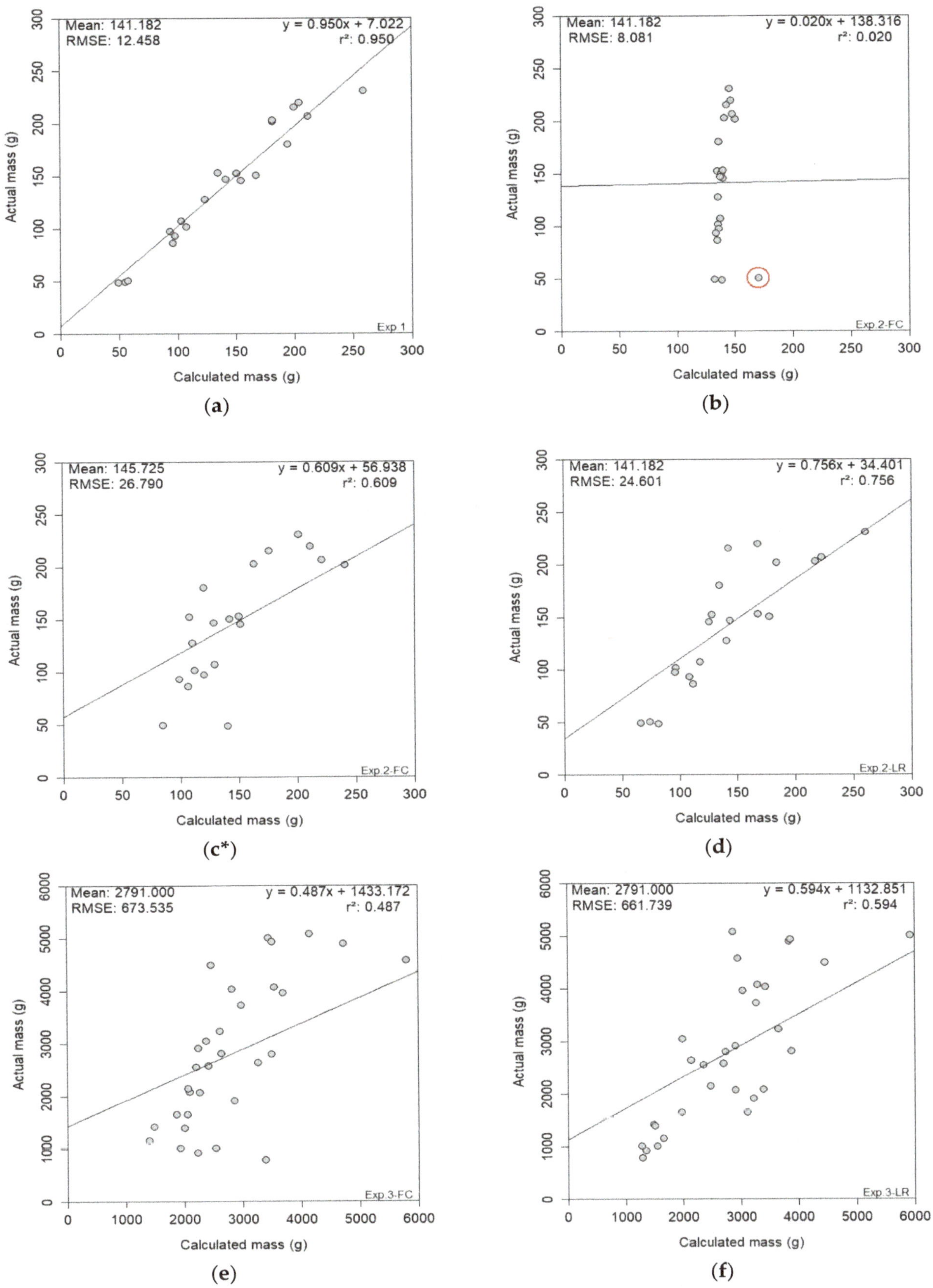

Figure 13. Presented RGB-D results of the three experiments; experiment one (**a**), experiment two FC (n = 21) (**b**), experiment two FC* (n = 20) (**c**), experiment two LR (**d**) and experiment three FC (**e**) & LR (**f**). {Key: Exp. = experiment; FC = full canopy; LR = leaf removal; *statistical outlier removed, resulting in 20 bunches}.

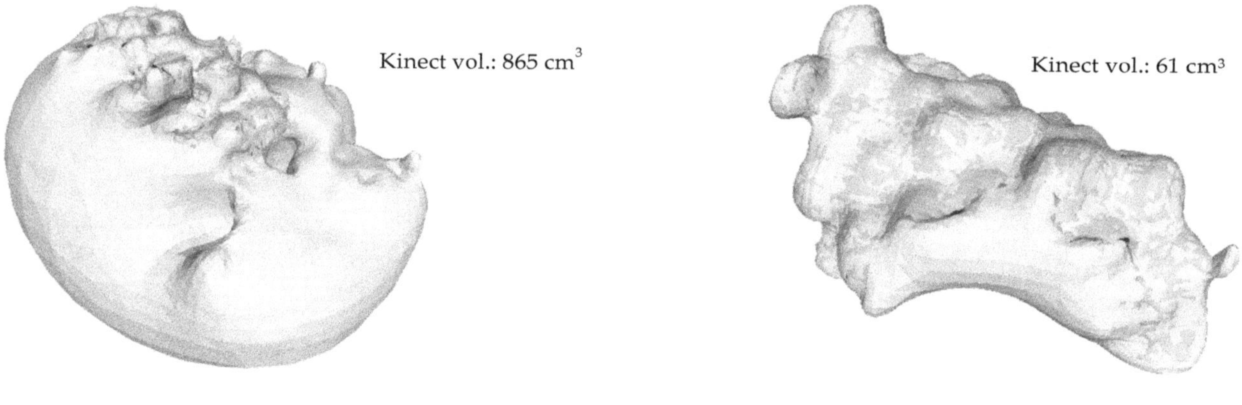

(**a**) (**b**)

Figure 14. Illustration of *Screened Poisson Surface Reconstruction*. The reconstructed bunch (circled in red—Figure 13b) with the incorrect volume (**a**), and an example of a reconstructed bunch of the correct volume (**b**).

4. Discussion

To date, most studies have employed 2-D RGB imagery for vineyard yield estimation at both bunch- and plant-level. However, only Marinello et al. [9] have investigated the use of 3-D RGB-D (specifically making use of a Kinect sensor) imaging for vineyard yield estimation. The research we have presented investigated 2-D and 3-D PRS techniques for vineyard yield estimation. The study was undertaken as three experiments, consisting of bunch-level and plant-level datasets, with in-situ measurements captured for the two canopy treatments (FC and LR). The following subsections discuss the results of the two PRS techniques in further detail.

4.1. Using 2-D RGB Imagery for Yield Estimation

The presented results using RGB imagery for yield estimation are robust and support the 2-D PRS technique. Experiment one ($r^2 = 0.889$; RMSE = 17.978 g) illustrated the success of RGB imagery in a controlled environment for yield estimation at bunch-level. At plant-level, similar results were presented for the LR treatment in experiment three ($r^2 = 0.877$; RMSE = 443.235 g). The success of the methodology under both laboratory and field conditions supports the use of colour thresholding for image segmentation, and the adapted pixel area metric for yield estimation.

Colour thresholding for image segmentation was favoured by several studies [8,9,14]. In our study, the use of the HSV colour space for thresholding has proven fruitful. The HSV colour space is supported by Font et al. [14], who achieved favourable results (estimation error of 13.55%) when working with the H layer for segmentation purposes. At bunch-level, experiment one's result ($r^2 = 0.889$) shows an improvement to the result ($r^2 = 0.77$) presented by Liu, Marden and Whitty [9]; similarly conducted under laboratory conditions. This aspect was a noteworthy improvement, especially considering the manual nature of the colour thresholding. At plant-level, the manually produced results of experiment three (FC: $r^2 = 0.779$; LR: $r^2 = 0.877$) align with the automated classification results ($r^2 = 0.865$) of Liu, Marden and Whitty [9]. Here, Liu, Marden and Whitty [9] use the same 1×1 m image dataset as Dunn and Martin [8], as opposed to the plant-level imagery used in our study. Additionally, the colour thresholding approach of our study outperformed the image classification approach (test $r^2 = 0.73$) [13] to segmentation.

We presented an adaption of the pixel count metric for yield estimation, which expands on current literature [8,9,13,14]. Of the five metrics tested by Liu, Marden and Whitty [9], the pixel count produced the best results. We found the incorporation of a calibration length (the ruler) for pixel count resulted in an improved quantitative pixel area (cm^2) for yield estimation. Again, our results at bunch-level ($r^2 = 0.889$) improved the bunch-level results ($r^2 = 0.77$) conducted in laboratory conditions—as presented by Liu, Marden and Whitty [9], who were specifically testing the various yield estimation metrics. At plant-level, our LR treatment ($r^2 = 0.877$) outperformed all current literature to date, with our FC treatment ($r^2 = 0.779$) representing a slight improvement for in-situ measurements. The presented pixel area (cm^2) metric also improved on the berry count results [4,11], with the highest berry count r^2 (0.74) presented still being lower than our plant-level pixel area r^2 (0.779) for the FC treatment. However, the potential of the berry count is applicable across all cultivars, as it does not depend on the colour of the berry [11].

The limitation of slight distance variations between the camera and bunches within each image was resolved by the incorporation of the reference length (the ruler). The necessity of determining the calibration length for each image outweighs the added processing requirement of this step. Overall, this allows improved yield estimation success, as represented in our results. Future work could attempt to automate this process. However, a more restrictive limitation was the human involvement in determining the appropriate threshold values; this could explain the lowered in-situ estimation performance at bunch-level. Future work could investigate a more automated methodology, thus alleviating this limitation.

4.2. Using 3-D RGB-D Imagery for Yield Estimation

To date, the work presented by Marinello et al. [10] is the only literature supporting the use of the Kinect V1 sensor for vineyard yield estimation. Marinello et al. [10] used table grape bunches to determine the optimal viewing angle and distance for the sensor. The authors concluded that a side-on view of the bunch, with a distance of between 0.8–1.0 m, produced the least variability in mass estimations. The current findings of our study, which incorporated RGB-D imagery for yield estimation across the three experiments, are therefore the most comprehensive findings to date. The presented study exemplifies the potential of 3-D PRS techniques for yield estimation, specifically a cost-effective sensor like the Kinect V1.

The nature of our study's datasets resulted in separate data analysis between the bunch- and plant-level datasets. Astounding results were obtained in experiment one ($r^2 = 0.950$; RMSE = 12.458 g). It was evident that the Kinect sensor favours ideal laboratory conditions, allowing accurate yield estimation at bunch-level. The ramification of less-favourable conditions, such as in-situ monitoring, became apparent in experiment two (bunch-level) and three (plant-level).

Although no bunch-level studies have used a Kinect sensor for vineyard yield estimation, a similar approach under laboratory conditions, estimating the volume of cauliflowers, was presented by Andujar et al. [19]. The authors were able to achieve an r^2 of 0.868 when regressing the estimated volume against the known fruit mass. Conversely, our methodology enables the relationship between fruit volume and mass to be modelled, allowing the use of the adjusted mass (calculated from the model) for subsequent yield estimation. A fundamental difference between our methodology and that of Andujar et al. [19] is the method in which the 3-D model was captured. Our methodology captured a 3-D mesh of the fruit, while Andujar et al. [19] captured a 3-D point cloud of the vegetable. This required model reconstruction, constructively producing a 'watertight mesh'. Our improved results at this level ($r^2 = 0.950$) can be accredited to this fundamental difference. Interestingly, both methodologies used the same 3-D reconstruction method—*Screened Poisson Surface Reconstruction* [30] found in MeshLab [29].

The primary limitation of the proposed methodology at bunch-level was a combination of dataset quality and the *Screened Poisson Surface Reconstruction* method [30]. The consequence of a poor-quality mesh became apparent when bunch reconstruction resulted in significant defects, as seen in Figure 14a. Such imperfections directly affect the potential for accurate yield estimation, exemplified by the results of experiment two's FC dataset ($r^2 = 0.020$; in Figure 13b). Future research is necessary to gain a better understanding of this shortfall in the methodology.

This study presented the novel use of a Kinect RGB-D sensor for in-situ vineyard yield estimation at plant-level (experiment three). The presented plant-level methodology produced promising results for the LR treatment ($r^2 = 0.594$; RMSE = 661.739 g), while the effect of the canopy coverage was evident in the FC treatment ($r^2 = 0.487$; RMSE = 673.535 g). Future work should improve on the presented methodology, potentially overcoming several limitations. Additionally, the nature of the Kinect V1 sensor results in the fundamental constraint of being suited to indoor use only, as solar illumination produces excessive interference in the captured imagery [19]. Future work should make use of the Kinect V2 sensor, since several improvements to the sensor have been implemented, effectively allowing improved outdoor imagery to be captured [37]. RGB-D sensors, specifically the Kinect V2, are being incorporated in terrestrial vehicles as cheap sensor alternatives for vineyard modelling and yield estimation, which is demonstrated by [38].

4.3. *The Operational Potential of Developed Methodologies*

Both the presented 2-D RGB and 3-D RGB-D methodologies achieved acceptable accuracies across the three experiments. Our results (Figures 12 and 13) support the use of PRS technology and techniques for vineyard yield estimation, especially for VSP-trained Shiraz vineyards. The nature of the presented work was conceptualised to assess 2-D and 3-D PRS sensors side-by-side, a novelty in the vineyard yield estimation domain.

Experiment one illustrated the capability of these two techniques for successful yield estimation of individual bunches, where the Kinect RGB-D sensor ($r^2 = 0.950$) outperformed the digital RGB sensor ($r^2 = 0.889$). The suitability of the lighting under laboratory conditions coupled with the Kinect's ability to capture a 3-D model of the bunch both contribute to the success of the Kinect sensor over the RGB sensor. Nonetheless, robust methodologies were established in a controlled environment.

Experiment two tested the established methodologies in situ, under both FC and LR canopy treatments. The produced results of experiment two—and experiment three—confirmed the hypothesis of a superior yield estimation agreement under the LR canopy treatment. If good canopy management practices are established early in the season, better yield estimation results than those obtained under the FC treatment could be achieved. Both sensors produced similar results for FC [approximate $r^2 = 0.61$; using Kinect's modified results (n = 20)] and LR (approximate $r^2 = 0.75$) treatments in experiment two.

The success of the two PRS methodologies can be differentiated at plant-level by the results of experiment three. Unlike experiment one, the RGB methodology outperformed the RGB-D methodology for yield estimation. The RGB results ($r^2 = 0.877$) significantly outclassed the RGB-D results ($r^2 = 0.487$) under the FC treatment, whereas a smaller margin between the RGB ($r^2 = 0.779$) and RGB-D ($r^2 = 0.594$) sensors occurred under the LR treatment. Inference behind the differing results can lead to the following observations. The different lighting conditions influenced the results: the RGB imagery was collected at midday, while the RGB-D imagery was collected immediately before sunset. Additionally, the continuous movement of the Kinect sensor (RGB imagery captured in a stationary position) could have contributed to lowered RGB-D results. Further research is encouraged, using a standardised experimental setup, where feasible, to create more favourable conditions under which both sensors can operate.

For the RGB and RGB-D datasets, the LR treatment yielded higher estimation agreements compared with the FC treatments. The LR treatment was adopted to create an ideal in-situ environment for yield estimation. The commercial feasibility of complete leaf removal in the bunch zone is not practical in some viticulture regions, since there are adverse effects associated with this practice (such as 'sunburn'). However, this practice can be implemented in specific zones of the vineyards, enabling an approximation of the total yield using a monitoring strategy. The FC results provide a better indication of the commercial readiness of the developed methodologies. Future work should determine an optimal canopy coverage that favours both sensor operationality and commercial farming methods.

Overall, the use of the Kinect sensor as a cost-effective RGB-D sensor for vineyard yield estimation, specifically at bunch-level, is supported by the results obtained for experiment one. However, the robustness of the RGB methodology is evident across all three experiments, with substantial plant-level results obtained in situ. A better understanding of the current limitations would allow further improvements to the methodologies. To this end, the results obtained in this study support the potential for operationalisation of both PRS sensors. A refined methodology could see commercially favourable data-acquisition methods implemented on a larger dataset, with fully automated data processing. See for example [15,16]. The commercialisation of such methodologies could become more feasible, due to the simplicity and robustness of the current methodologies, coupled with improved yield estimation accuracy.

5. Conclusions

A novel approach to a side-by-side investigation of 2-D and 3-D PRS techniques for successful vineyard yield estimation has been presented. This study assessed RGB imagery captured by a digital camera and RGB-D imagery captured by a Kinect V1 sensor across three experiments, with in-situ measurements obtained under two canopy treatments. Our results show that the Kinect RGB-D sensor produced the highest yield estimation agreement under laboratory conditions (bunch-level). At bunch-level, RGB and RGB-D PRS techniques performed equally under both canopy treatments for in-situ yield estimation. At plant-level, the best in-situ results were obtained using the RGB imagery, which significantly outperformed the RGB-D results. Both sensors support the use of PRS technology and techniques for vineyard yield estimation, with improved accuracies presented. The results of this study confirm the operational potential of 2-D RGB imagery for accurate yield estimation, with the recommendation that future work should investigate a more automated RGB methodology suitable for operational environments. Regarding the presented RGB-D methodology, the Kinect demonstrates the potential for vineyard yield estimation using 3-D RGB-D imagery. Future work should investigate the use of the Kinect V2 sensor coupled with suitable lighting conditions for in-situ yield estimation.

Author Contributions: C.H.—data analysis, writing and editing of the manuscript; N.P.—content supervision, and editing of the manuscript; N.M.—data collection and analysis; C.P.-E.—conceptualisation, experimental study design, content supervision and editing of the manuscript.

Acknowledgments: The authors would like to thank Albert Strever, Berno Greyling, Aloïs Houeto and Alessandro Bellotto for their technical and research support.

References

1. Arnó, J.; Martínez-Casasnovas, J.A.; Ribes-Dasi, M.; Rosell, J.R. Review. Precision viticulture. Research topics, challenges and opportunities in site-specific vineyard management. *Span. J. Agric. Res.* **2009**, *7*, 779–790. [CrossRef]
2. Blackmore, S. Developing the Principles of Precision farming. In Proceedings of the International Conference on Precision Agriculture, Minneapolis, MN, USA, 14–17 July 2002; p. 5.
3. Matese, A.; Di Gennaro, S.F. Technology in precision viticulture: A state of the art review. *Int. J. Wine Res.* **2015**, *7*, 69–81. [CrossRef]
4. Nuske, S.; Wilshusen, K.; Achar, S.; Yoder, L.; Narasimhan, S.; Singh, S. Automated visual yield estimation in vineyards. *J. Field Robot.* **2014**, *31*, 837–860. [CrossRef]
5. Cunha, M.; Marçal, A.R.S.; Silva, L. Very early prediction of wine yield based on satellite data from vegetation. *Int. J. Remote Sens.* **2010**, *31*, 3125–3142. [CrossRef]
6. Wolpert, J.A.; Vilas, E.P. Estimating Vineyard Yields: Introduction to a Simple, Two-Step Method. *Am. J. Enol. Vitic.* **1992**, *43*, 384–388.
7. De la Fuente, M.; Linares, R.; Baeza, P.; Miranda, C.; Lissarrague, J. Comparison of Different Methods of Grapevine Yield Prediction in the Time Window. *J. Int. Sci. Vigne Vin* **2015**, *49*, 27–35. [CrossRef]
8. Dunn, G.M.; Martin, S.R. Yield prediction from digital image analysis: A technique with potential for vineyard assessments prior to harvest. *Aust. J. Grape Wine Res.* **2004**, *10*, 196–198. [CrossRef]
9. Liu, S.; Marden, S.; Whitty, M. Towards automated yield estimation in viticulture. In Proceedings of the Australasian Conference on Robotics and Automation, Sydney, Australia, 2–4 December 2013.

10. Marinello, F.; Pezzuolo, A.; Cillis, D.; Sartori, L. Kinect 3D reconstruction for quantification of grape bunches volume and mass. In Proceedings of the Engineering for Rural Development, Jelgava, Latvia, 25–27 May 2016; pp. 876–881.
11. Nuske, S.; Achar, S.; Bates, T.; Narasimhan, S.; Singh, S. Yield estimation in vineyards by visual grape detection. In Proceedings of the IEEE International Conference on Intelligent Robots and Systems, San Francisco, CA, USA, 25–30 September 2011; pp. 2352–2358. [CrossRef]
12. Aquino, A.; Millan, B.; Diago, M.P.; Tardaguila, J. Automated early yield prediction in vineyards from on-the-go image acquisition. *Comput. Electron. Agric.* **2018**, *144*, 26–36. [CrossRef]
13. Diago, M.P.; Correa, C.; Millán, B.; Barreiro, P.; Valero, C.; Tardaguila, J. Grapevine yield and leaf area estimation using supervised classification methodology on RGB images taken under field conditions. *Sensors* **2012**, *12*, 16988–17006. [CrossRef]
14. Font, D.; Tresanchez, M.; Martínez, D.; Moreno, J.; Clotet, E.; Palacín, J. Vineyard Yield Estimation Based on the Analysis of High Resolution Images Obtained with Artificial Illumination at Night. *Sensors* **2015**, *15*, 8284–8301. [CrossRef]
15. Liu, S.; Whitty, M. Automatic grape bunch detection in vineyards with an SVM classifier. *J. Appl. Log.* **2015**, *13*, 643–653. [CrossRef]
16. Millan, B.; Velasco-Forero, S.; Aquino, A.; Tardaguila, J. On-the-go grapevine yield estimation using image analysis and boolean model. *J. Sens.* **2018**. [CrossRef]
17. Bengochea-Guevara, J.M.; Andújar, D.; Sanchez-Sardana, F.L.; Cantuña, K.; Ribeiro, A. A low-cost approach to automatically obtain accurate 3D models of woody crops. *Sensors* **2018**, *18*, 30. [CrossRef] [PubMed]
18. Wang, W.; Li, C. Size estimation of sweet onions using consumer-grade RGB-depth sensor. *J. Food Eng.* **2014**, *142*, 153–162. [CrossRef]
19. Andújar, D.; Ribeiro, A.; Fernández-Quintanilla, C.; Dorado, J. Using depth cameras to extract structural parameters to assess the growth state and yield of cauliflower crops. *Comput. Electron. Agric.* **2016**, *122*, 67–73. [CrossRef]
20. Herrero-Huerta, M.; González-Aguilera, D.; Rodriguez-Gonzalvez, P.; Hernández-López, D. Vineyard yield estimation by automatic 3D bunch modelling in field conditions. *Comput. Electron. Agric.* **2015**, *110*, 17–26. [CrossRef]
21. Ivorra, E.; Sánchez, A.J.; Camarasa, J.G.; Diago, M.P.; Tardaguila, J. Assessment of grape cluster yield components based on 3D descriptors using stereo vision. *Food Control* **2015**, *50*, 273–282. [CrossRef]
22. Rose, J.C.; Kicherer, A.; Wieland, M.; Klingbeil, L.; Töpfer, R.; Kuhlmann, H. Towards automated large-scale 3D phenotyping of vineyards under field conditions. *Sensors* **2016**, *16*, 2136. [CrossRef]
23. Conradie, W.; Carey, V.; Bonnardot, V.; Saayman, D.; van Schoor, L. Effect of Different Environmental Factors on the Performance of Sauvignon blanc Grapevines in the Stellenbosch/Durbanville Districts of South Africa. I. Geology, Soil, Climate, Phenology and Grape Composition. *S. Afr. J. Enol. Vitic.* **2002**, *23*, 78–91.
24. Ferreira, J.H.S.; Marais, P.G. Effect of Rootstock Cultivar, Pruning Method and Crop Load on Botrytis cinerea Rot of Vitis vinifera cv. Chenin blanc grapes. *S. Afr. J. Enol. Vitic.* **2017**, *8*, 41–44. [CrossRef]
25. Microsoft. Kinect for Windows SDK 1.8. Available online: https://www.microsoft.com/en-us/download/details.aspx?id=40276 (accessed on 16 May 2019).
26. Labbe, M. RTAB-Map. Available online: http://introlab.github.io/rtabmap/ (accessed on 1 December 2018).
27. The MathWorks Inc. Matlab R2018b v9.5.0.944444. Available online: https://ww2.mathworks.cn/en/ (accessed on 8 January 2019).
28. CloudCompare. CloudCompare Version 2.10.alpha [GPL Software]. Available online: http://www.cloudcompare.org/ (accessed on 29 November 2018).
29. Cignoni, P.; Callieri, M.; Corsini, M.; Dellepiane, M.; Ganovelli, F.; Ranzuglia, G. MeshLab: An Open-Source Mesh Processing Tool. In Proceedings of the Sixth Eurographics Italian Chapter Conference, the Eurographics Association, Pisa, Italy, 2–4 July 2008; pp. 129–136. [CrossRef]
30. Kazhdan, M.; Hoppe, H. Screened poisson surface reconstruction. *ACM Trans. Graph.* **2013**, *32*, 29. [CrossRef]
31. R Core Team. *R: A Language and Environment for Statistical Computing*; R Foundation for Statistical Computing: Vienna, Austria, 2018.

32. Lafarge, T.; Pateiro-Lopez, B. alphashape3d: Implementation of the 3D Alpha-Shape for the Reconstruction of 3D Sets from a Point Cloud. R Packag version 1.3. 2017. Available online: https://rdrr.io/cran/alphashape3d/ (accessed on 20 August 2019).
33. Lafarge, T.; Pateiro-López, B.; Possolo, A.; Dunkers, J.P. R Implementation of a Polyhedral Approximation to a 3D Set of Points Using the α-Shape. *J. Stat. Softw.* **2014**, *56*, 1–19. [CrossRef]
34. Rueda-Ayala, V.; Peña, J.; Höglind, M.; Bengochea-Guevara, J.; Andújar, D. Comparing UAV-Based Technologies and RGB-D Reconstruction Methods for Plant Height and Biomass Monitoring on Grass Ley. *Sensors* **2019**, *19*, 535. [CrossRef] [PubMed]
35. Ribeiro, A.; Bengochea-Guevara, J.M.; Conesa-Muñoz, J.; Nuñez, N.; Cantuña, K.; Andújar, D. 3D monitoring of woody crops using an unmanned ground vehicle. In Proceedings of the 11th European Conference on Precision Agriculture, Advances in Animal Biosciences, Edinburgh, UK, 16–20 July 2017; Volume 8, pp. 210–215. [CrossRef]
36. Kuhn, M. Building predictive models in R using the caret package. *J. Stat. Softw.* **2008**, *28*, 1–26. [CrossRef]
37. Wasenmüller, O.; Stricker, D. Comparison of Kinect V1 and V2 Depth Images in Terms of Accuracy and Precision. In Proceedings of the ACCV Workshops, Taipei, Taiwan, 20–24 November 2016; pp. 34–45. [CrossRef]
38. Lopes, C.; Torres, A.; Guzman, R.; Graca, J.; Reyes, M.; Vitorino, G.; Braga, R.; Monteiro, A.; Barriguinha, A. Using an unmanned ground vehicle to scout vineyards for non-intrusive estimation of canopy features and grape yield. In Proceedings of the 20th GiESCO International Meeting, Mendoza, Argentina, 5–10 November 2017; pp. 16–21.

11

High-Resolution Multisensor Remote Sensing to Support Date Palm Farm Management

Maggie Mulley [1,*,†], Lammert Kooistra [1] and Laurens Bierens [2]

1 Laboratory of Geo-Information Science and Remote Sensing, Wageningen University and Research, P.O. Box 47, 6700 AA Wageningen, The Netherlands; lammert.kooistra@wur.nl

2 TEC-IB B.V., Oude Veiling 29, 2635 GK Den Hoorn (ZH), The Netherlands; l.bierens@tec-ib.com

* Correspondence: maggie.m.mulley@gmail.com

† Current employment: Aerovision BV, Stadsring 47, 3811 HN Amersfoort, The Netherlands.

Abstract: Date palms are a valuable crop in areas with limited water availability such as the Middle East and sub-Saharan Africa, due to their hardiness in tough conditions. Increasing soil salinity and the spread of pests including the red palm weevil (RPW) are two examples of growing threats to date palm plantations. Separate studies have shown that thermal, multispectral, and hyperspectral remote sensing imagery can provide insight into the health of date palm plantations, but the added value of combining these datasets has not been investigated. The current study used available thermal, hyperspectral, Light Detection and Ranging (LiDAR) and visual Red-Green-Blue (RGB) images to investigate the possibilities of assessing date palm health at two "levels"; block level and individual tree level. Test blocks were defined into assumed healthy and unhealthy classes, and thermal and height data were extracted and compared. Due to distortions in the hyperspectral imagery, this data was only used for individual tree analysis; methods for identifying individual tree points using Normalized Difference Vegetation Index (NDVI) maps proved accurate. A total of 100 random test trees in one block were selected, and comparisons between hyperspectral, thermal and height data were made. For the vegetation index red-edge position (REP), the R-squared value in correlation with temperature was 0.313 and with height was 0.253. The vegetation index—the Vogelmann Red Edge Index (VOGI)—also has a relatively strong correlation value with both temperature (R^2 = 0.227) and height (R^2 = 0.213). Despite limited field data, the results of this study suggest that remote sensing data has added value in analyzing date palm plantations and could provide insight for precision agriculture techniques.

Keywords: remote sensing; date palms; precision agriculture; plantation management; thermal; hyperspectral

---

1. Introduction

Across the globe, people are suffering from a lack of food availability and food security as populations grow and land becomes unsuitable for farming [1–3]. There is a need for crops with high nutritional value that can withstand the arid and semiarid conditions in countries with limited water resources. The date palm is one such plant [2]. In the Middle East and North Africa alone, 100 million palms are cultivated on one million hectares of cropland [4].

There are several biotic and abiotic factors that affect the management of a healthy date palm crop. Global climate change may pose a threat to plantations of date palm; climate models predict that regions suitable for date palm growth will shrink, especially in the Middle East [5]. Water is a limiting factor for growth, although excess water can also reduce yield. Studies investigating the effects of water on date palm growth have found that insufficient water application slows the growth of the plants [6,7].

Increasing salinity also appears to negatively impact the growth of date palms [8–10]. This is a growing problem in arid regions as groundwater reservoirs are depleted due to excess irrigation.

Plantations of date palm and other palm species are also under threat due to infestations by different pests, the most extensive of which is the red palm weevil, or *Rhynchophorus ferrugineus* (Olivier, 1790). *R. ferrugineus* is spread via the transport of cuttings of palm tree leaves [4,11], and has been detected in 50% of date palm growing countries [12]. The effects of *R. ferrugineus* are severe; in the later stages of infestation the trunk can be damaged to the extent that the tree collapses and will most certainly die. Other pests that affect date palms include the Dubas bug, lesser date palm moth, and many different borer species [13]. The Dubas bug feeds on leaf sap which results in yellowing and wilting of leaves [3]. Lesser date palm moths infest and damage the fruit of the date palm [13]. Borers have similar effects to *R. ferrugineus*, although they may infest different parts of the tree, such as the middle rib of the frond and the upper section of the tree near the bunches of fruit [13].

At present, a heavy focus is placed on monitoring plantations for *R. ferrugineus* infestation. Unfortunately, current methods of detection do not allow for effective, reliable identification of infested trees [14]. In fact, the detection of palms in the early stages of infestation has been stated as the main issue in controlling weevil infestation [15]. It is very difficult to recognise whether a tree is infested based on visual inspection as the symptoms vary depending on conditions such as the tree species and initial site of infestation [14,16]. All of the current detection methods, including acoustic detection, x-ray, and sniffer dogs are often ineffective on a large scale because each tree has to be inspected individually [14,16].

Remote sensing has been suggested as a potential detection method, as it allows for large scale monitoring and effective data analysis techniques [16]. Thermal imaging studies have shown it might be possible to detect plants infested with *R. ferrugineus* using aerial thermal photographs [17]. Multi- and hyperspectral images have successfully identified infested trees. WorldView-3 satellite imagery has been used for *R. ferrugineus* detection [18]. Hyperspectral imagery was conducted on oil palms to classify *R. ferrugineus*-infested palms based on vegetation indices [19]. Light Detection and Ranging (LiDAR) has not been applied to the palm assessment case, but it does have potential; for example, LiDAR has been used to classify eucalyptus tree health [20].

Remote sensing has also been applied to assess the effect of other stressors on palm plantations. Cohen et al. (2012) used thermal images to investigate how changes in irrigation affect the health of date palm plantations [6]. The effect of soil salinity on date palm health has been assessed by using Landsat Thematic Mapper (TM) and Enhanced Thematic Mapper Plus (ETM+) images to calculate the Soil Adjusted Vegetation Index (SAVI) for two different years [21].

The studies listed above have all explored how individual stressors affect date palm plantations. However, it is highly likely that a date palm plantation will be subject to multiple stressors at once and identifying the influence of these effects will allow for more comprehensive and efficient plantation management. Remote sensing approaches can provide the tools to conduct this form of management; analysis of the entire plantation is possible using automated techniques in order to differentiate between a combination of stressors.

Therefore, the goal of this study is to explore which remote sensing sensors and derived indicators can detect stressors on palm health at different spatial levels, in order to support a holistic solution for effective date palm management. Plantations are often organised in large rectangular sub-sections or blocks—this will be the first spatial level to be evaluated. The second level of analysis will be conducted on an individual tree level. For this study, several different remote sensing imagery sources have been used, in order to determine whether different techniques are most suitable for highlighting differences in palm status. The available imagery sources are thermal, hyperspectral, and LiDAR, with data collected using airborne platforms. High resolution imagery is achievable with airborne platforms, allowing for analysis of individual tree health.

2. Materials and Methods

2.1. Study Area and Data Description

The farm shown in Figure 1 is in the Al-Kharj region in Saudi Arabia, southeast of the capital Riyadh. Saudi Arabia was one of the earliest countries growing date palms to be affected by *R. ferrugineus* infestation outside of South East Asia, with first infestations occurring since the 1980s [22]. The country has the highest per capita consumption of dates in the world. Flood irrigation is making way for more modern irrigation methods such as drip irrigation in many farms [23].

Of the farm, an area of 168.8 hectares (outlined in blue in Figure 1) has been chosen as the key study area of this analysis. This farm is relatively well managed, with continuous maintenance and targeted irrigation practices in place. As is exemplified in Figure 1b, date palm plantations are commonly divided into regular-sized rectangles of approximately 10 hectares; these sub-sections are hereafter referred to as blocks. Figure 1c illustrates the uniform growth patterns of the date palms; the crowns are distinct and organised in a grid-like structure.

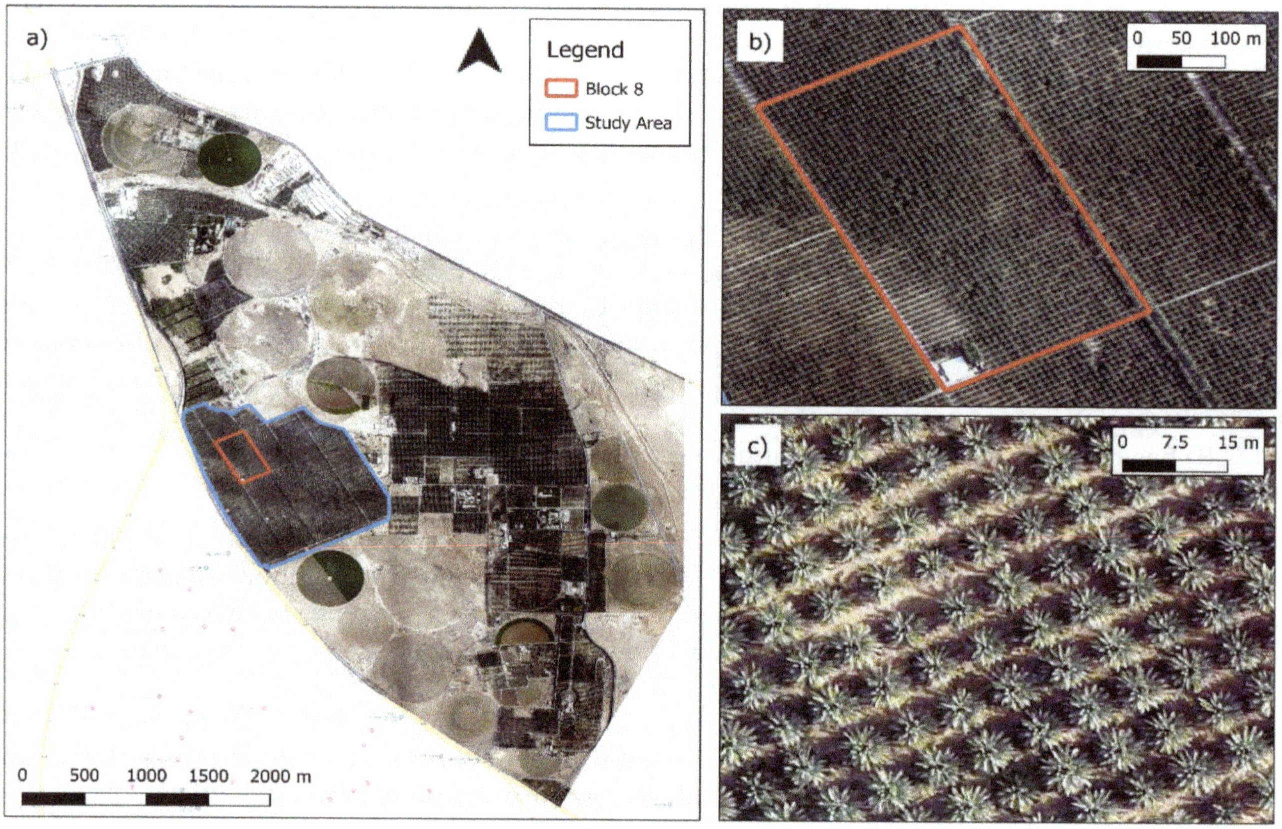

Figure 1. Visual red-green-blue (RGB) images with 15-cm resolution, taken using an aerial platform on 26 May 2016, of (**a**) the date palm plantation in the study area (outlined in blue), located south-east of Riyadh, the capital of Saudi Arabia; (**b**) an example of a block within the plantation (outlined in red) and (**c**) an example of individual palm trees.

For this study, four remote sensing data types were available for analysis: hyperspectral, thermal, LiDAR, and high-resolution visual red-green-blue (RGB) images. Each dataset contains different information that will be useful when assessing date palm health. The data has been collected using a Diamond DA42 MPP GEOSTAR (Diamond Aircraft Industries GmbH, Wiener Neustadt, Austria) aircraft, which is designed to carry many different remote sensing instruments. The flight was conducted on 26 May 2016 between approximately 6:00 and 9:00AM.

Table 1 shows information about the four sensors used for data collection. The spatial resolution refers to the pixel size of the data used in the analysis after preprocessing; the original LiDAR data was provided as a point cloud but has been processed into raster data.

Table 1. Characteristics of the four sensor systems used for data collection.

Data Type	RGB	LiDAR	Thermal	Hyperspectral	
Name of sensor [1]	Phase One iXA-R Camera [2]	RIEGL LMS-Q1560 [3]	VarioCAM®HD head [4]	HySpex VNIR-1800 [5]	Hy-Spex SWIR-384 [5]
Type of sensor	Medium-format camera system	Rotating polygon mirror	Uncooled microbolometer focal-plane array	Pushbroom camera actively cooled and stabilized scientific CMOS detector	Pushbroom camera Mercury cadmium telluride sensor
Spectral range	Visible	Near-infrared	7.5–14 μm	0.4–1 nm, 182 bands	0.93–2.5 nm, 288 bands
Spatial resolution	0.15 m, 0.6 m, 1.8 m	1 m, 2 m	1.8 m	1 m	1 m

[1] Manufacturer information; [2] Phase One Industrial, Frederiksberg, Demark; [3] RIEGL Laser Measurement Systems GmbH, Horn, Austria; [4] InfraTec, Dresden, Germany; [5] Norsk Elektro Optikk AS (NEO), Skedsmokorset, Norway.

2.2. Methodology

2.2.1. Preprocessing

Different preprocessing steps were necessary for each of the multiple remote sensing data sources available. These steps will be described briefly in the next few paragraphs. Steps were carried out in various software: R (version 3.3.1, R Foundation for Statistical Computing, Vienna, Austria), QGIS (version 2.1.4 'Essen', Open Source Geospatial Foundation, Chicago, IL, USA), ArcGIS (version 10.4, ESRI, Redlands, CA, USA), LAS tools (version 2016, rapidlasso GmbH, Gilching, Germany), and Agisoft PhotoScan (version 1.2, Agisoft LLC, St. Petersburg, Russia).

The RGB data was provided in JPEG format and was not geographically referenced. The files were given a geographic coordinate system and then projected to WGS 1984 UTM Zone 38N, and also converted to GeoTIFF. This projection was also used for all the other datasets.

The LiDAR dataset was originally received as a point cloud. To generate a tree height raster, Digital Surface Models (DSM) at different resolutions was generated; additionally, the mean, minimum, maximum and standard deviation values of the points within these pixels were used.

During acquisition, the VarioCAM thermal camera was controlled using the IRBIS®Thermography Software (InfraTec, Dresden, Germany). Within this software, the temperature range for acquisition was set to values between 0 and 50 degrees Celsius [24]. This range was broader than the actual temperature range which could be expected but was set by the flight operators to avoid saturation at low or high values. The raw thermal data was provided as individual snapshot images that were then stitched together using the structure-from-motion method [25] as implemented through Agisoft PhotoScan. In some areas where there are a lot of overlapping images, tie-points do not match perfectly, causing some blurring in the image; where this occurs, individual trees are not identifiable. This effect cannot be reduced without compromising the resolution of the entire image. The image was received as a (8 bit) greyscale image with 256 levels. To convert the values into a relative approximation of degrees Celsius, the acquired greyscale image was converted using the multiplication factor of 0.1953125 to fit the temperature range settings during acquisition (0 and 50 degrees Celsius).

The hyperspectral images were received without the inertial measurement unit (IMU) data due to an error in the acquisition procedure. This means that it was not possible to correct for image distortion due to plane movement. The images were also provided in an unknown multiplication factor, and thus the values are unusual. However, an evaluation of this data source is still considered worthwhile as it was acquired at the same time as the other remote sensing types. This meant that extensive steps were required to attain usable imagery.

By using the recorded times and positioning of the RGB imagery, the order in which hyperspectral images were acquired was determined. This meant that a representative image could be constructed and aligned with the study area. The image was then transformed using digitally assigned geometric control points as identified from the RGB images. The QGIS *Georeferencer* tool was used, with transformation type set as Thin Plate Spline which is useful for heavily distorted data, and Nearest Neighbour as a resampling method to maintain the original data values. The pixel data was multiplied by a scaling factor of 0.002 as this ensured they were in the relative range of standard reflectance values. As this is not the standard way to determine actual reflectance values, it is only possible to carry out relative comparisons within the dataset—this dataset cannot be numerically compared with any hyperspectral data collected in later studies.

2.2.2. Identification of Blocks and Trees

As analysis is carried out on two different spatial levels, namely blocks and trees, it is necessary to identify these regions. Block shapefiles were provided for the farm, and these were used for analysis of the thermal and LiDAR-derived height data. As the hyperspectral blocks were geometrically distorted, these were drawn manually.

Several methods were used in order to identify the trees for tree-level analysis. For the non-distorted data, the canopy height map as determined from LiDAR was used to locate individual tree points. This was done using the watershed method, which involves inverting the height rasters, calculating the sinks and converting them to points, and filtering these points at an appropriate buffer width. The software used was LAS tools and ArcMap.

It was also decided that deriving canopy polygons could be interesting as the area of the tree canopy could prove a useful indicator of tree health. The LiDAR data was analysed using an R package called Individual Tree Crown (ITC) Segment (version 0.6, Michele Dalponte, San Michele all'Adige, Italy) [26]. This tool selects local maxima within a height raster and then spreads outwards to select the surrounding pixels that are within a specific height range representing the canopy. A polygon shapefile is generated, for which the height and area are calculated. A 3-m width limit was selected to constrain the growth of the polygon outside of the general tree size. To investigate sensitivity to spatial resolution, the model was run for pixel sizes of both 1 m and 2 m. In this case, the raster's containing the maximum value of the point cloud within each pixel were used for analysis, as these data were available for both 1-m and 2-m resolution.

For the hyperspectral data, a different method for tree identification was used. This is necessary because the distortions in the hyperspectral dataset mean that it is not possible to use the same tree locations for each dataset. It was decided to use vegetation indices, specifically normalised difference vegetation index (NDVI) (see Equation (1)), to identify the trees. Once the NDVI was calculated, based on the values of Equation [1], the *TreeTopFinderTool* in the R package "Forest Tools" (version 0.1.5, Andrew Plowright, BC, Canada) was used to identify local maxima. It is hypothesised that the points where the NDVI is highest represent the top of the tree canopy. A width threshold of four was set, based on the upper limit of the average canopy size, which is 3 to 4 m. Additionally, a minimum NDVI value of 0.3 was set to ensure that maxima in treeless regions were excluded—this value was based on the bimodal distribution of the NDVI values that distinguishes between soil and tree points.

$$NDVI = \frac{NIR_{750} - R_{660}}{NIR_{750} + R_{660}} \tag{1}$$

2.2.3. Block-Level Analysis

The six blocks shown in Figure 2 have been selected for further analysis. These blocks were chosen based on the attributes provided in Table 2.

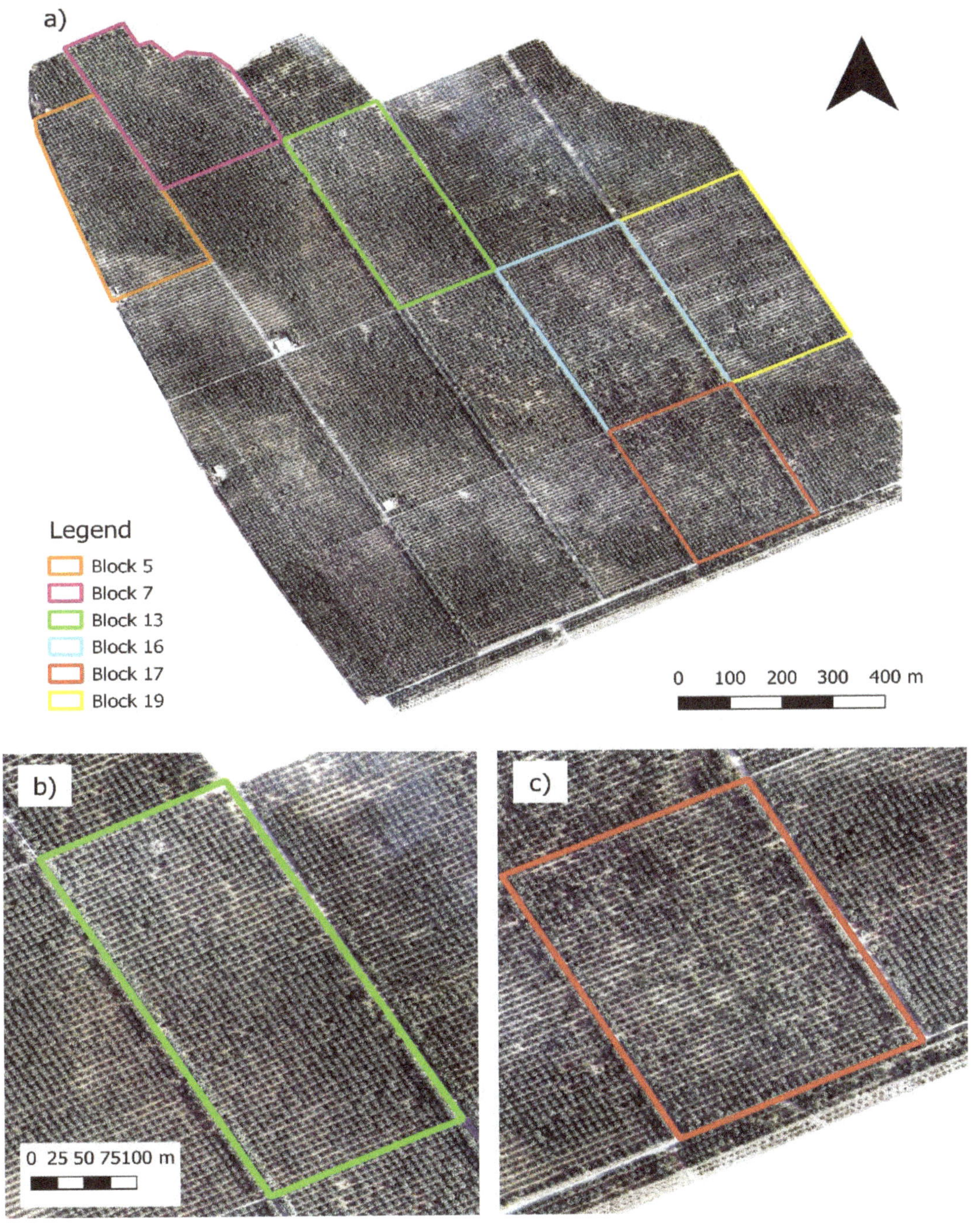

Figure 2. (**a**) Shows the selected test blocks; Blocks 5, 7, and 13 have been classified as healthy and Blocks 16, 17 and 19 have been classified as unhealthy. (**b**) and (**c**) shows an example of blocks classified as healthy (green) and unhealthy (red) respectively.

Two different groups of blocks have been determined based on their characteristics as described in Table 2. The data that was available from the farm manager mainly related to instances of *R. ferrugineus*, so this was a factor that was considered in block division. A visual assessment of the farm was made based on the homogeneity of canopy area and the sparseness of the block. The first group of trees was chosen with the requirements of having no instances of *R. ferrugineus* and of having high canopy homogeneity and low sparseness. The second group of blocks had relatively high instances of *R. ferrugineus*, and the appearance of the blocks revealed many gaps in trees and a wide variation of tree canopy widths. These two indicators were used to differentiate between potentially healthy and unhealthy groups. An additional factor that was considered was the amount of blurring in the thermal image, as discussed in Section 2.2.1; if the blurring in a block was visually estimated as being more than 70%, it was not considered.

The assumed healthy test blocks are 5, 7, and 13, and the unhealthy blocks are 16, 17, and 19. Their respective values are shown in Table 2.

Table 2. Criteria used to determine the selection for the six test blocks.

	May 2016	June 2016	RGB Check [1]	Thermal Image Distortion [2]
Block No.	No. RPW infested	No. RPW infested	1–5 rating	Percent distorted
5	0	0	1	20
7	0	0	1	30
13	0	0	2	30
16	8	4	5	30
17	10	4	5	30
19	1	9	5	20

[1] Indicator of block health status based on visual check for sparseness and heterogeneity, where 1 = homogenous and wide canopies, 5 = many patches and smaller canopies; [2] Visual check to determine how much of the image is affected by blurring due to structure-from-motion processing technique.

To summarise the statistical attributes of these blocks, boxplots were generated for temperature and height of the trees. These illustrate the spread of temperature and height across the test blocks, also in relation to all blocks in the farm. These boxplots illustrate the 25th, 50th and 75th percentiles, extreme values, and outlier values for the canopy pixels. For the temperature values, the derived canopy areas were used as a mask layer for the temperature raster. The height values are also based on the polygon shape. A visual check was conducted to explain the findings of the boxplot analysis.

For each block, a total of 10 'healthy' and 10 'unhealthy' trees were randomly selected after filtering based on their ratio between height and canopy area—based on discussions with the farm manager, it was hypothesised that trees with a high height area ratio (HAR), meaning they have a small canopy in relation to height, would be more likely to be unhealthy. Trees were selected based on whether they were above or below a threshold for this ratio and were also filtered based on height to attempt to select trees of similar height.

The mean temperatures of each group of trees (healthy and unhealthy) have been compared using two-factor analysis-of-variance (ANOVA) analysis. This statistical test is used to determine whether the differences in means of two or more groups at two or more levels are statistically significant; in this case, between the 'healthy' and 'unhealthy' trees, between the blocks, and the interaction effect between them. The post-hoc Tukey honest significant difference (HSD) test is applied following ANOVA—it is a comparison test that determines which pairs of means are responsible for the statistically significant test result. A significance level of $\alpha = 0.05$ has been used to determine whether the result is statistically significant or not.

2.2.4. Tree-Level Analysis

Different blocks were used for the individual tree analysis, as several of the blocks were not available in the hyperspectral dataset. Block 11 was selected for analysis for two reasons; firstly, it has the least visible blurring in the thermal dataset, and also it has a wide variety of trees with different canopy sizes.

As described in Section 2.2.2, the tree canopies have been identified based on NDVI. The next step was to select trees that can be used to extract more extensive information. In this case, trees were selected randomly from the hyperspectral dataset, and were matched manually to the trees in the thermal dataset. This was done by starting at the edges of the trees in each block and then counting vertically and the horizontally to identify the correct tree. A shapefile was created, and points were placed on the identified trees. A total of 100 trees was used for this block. As the trees are quite noticeable in each dataset, it is expected that this method was quite accurate at detection, although in some cases the tree may be misidentified.

Fourteen vegetation indices (VI) have been selected for further analysis (See Appendix A Table A1). These VIs can be split into general groups that focus on different regions of in the visible light spectrum, such as green, red, and the red-edge position. For each test tree, the correlation between each vegetation index and temperature and height has been derived, and R squared values have been calculated. A visual check has also been carried out.

3. Results

3.1. Block and Tree Identification Results

The results of the method which identifies the areas of entire tree canopies can be seen in Figure 3. The ITC segment tool was run on two different height rasters (spatial resolutions of 1 m and 2 m) derived from the LiDAR dataset. Table 3 shows the number of tree points identified using the two different height raster maps; these values have been compared with tree numbers provided by the farm manager. For the 1 m resolution data these values exceeded 90% in every block. Fewer trees are detected using the 2 m resolution data: 73–90% were detected when comparing the number of tree polygons to the number of tree points in each block.

Table 3. Number (#) of trees within the test blocks based on the farm managers records compared to results of the Individual Tree Crown (ITC) segment analysis for 1 m and 2 m height rasters.

Block No.	Recorded # of Trees	# of Trees (1 m Resolution)	# of Trees (2 m Resolution)
5	1188	1137	1043
7	1213	1197	1084
13	1417	1277	1131
16	1749	1572	1242
17	1422	1302	1013
19	1691	1617	1440

When comparing Figure 3a,b, it can be seen that in the denser parts of the plantation, using the 1m resolution LiDAR data has a higher detection rate of tree canopies.

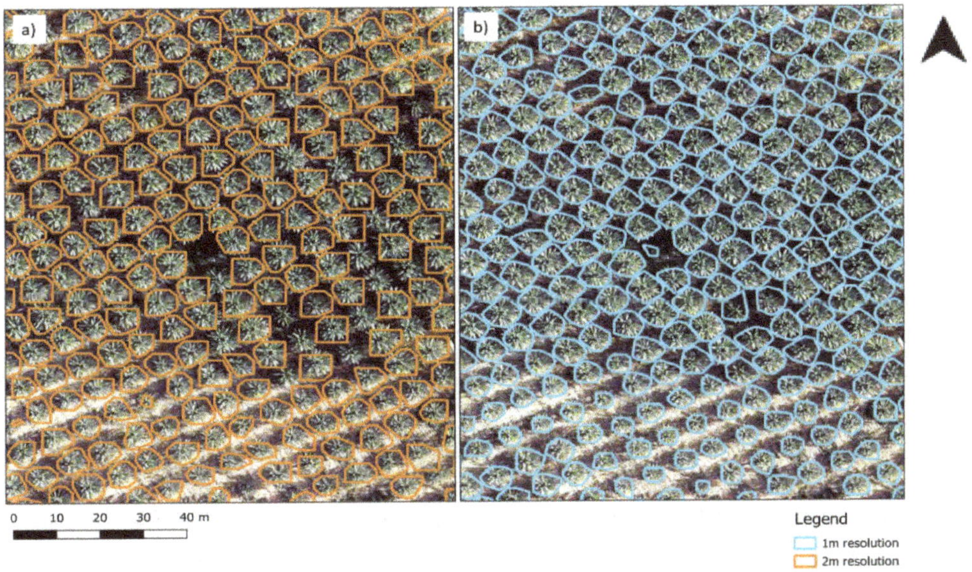

Figure 3. Section of an area of Block 5 in the farm. Canopy polygons derived using the ITC segment tool plotted in relation to the RGB data, for (**a**) the 2-m spatial resolution height raster (orange), and (**b**) 1-m spatial resolution (blue).

The second method used the NDVI image that was extracted from the hyperspectral dataset. Two different red bands were used to calculate the NDVI, and the comparisons can be seen in Figure 4. The outcome for both NDVI calculations is similar, although trees are identified more accurately in denser canopies when band 750 is used; for Block 16, for example, the number of trees recorded by the farm manager was 1749, compared to 1605 and 1601 for NDVI 750 and NDVI 800 respectively. For Block 19, the method using NDVI 750 identified 1661 trees, compared to 1656 (NDVI 800), and 1691 (trees recorded by farm manager). This indicates that in this case NDVI 750 is slightly more effective at identifying trees. Misidentification tends to be on the edges where the tree canopies are denser and therefore less distinct.

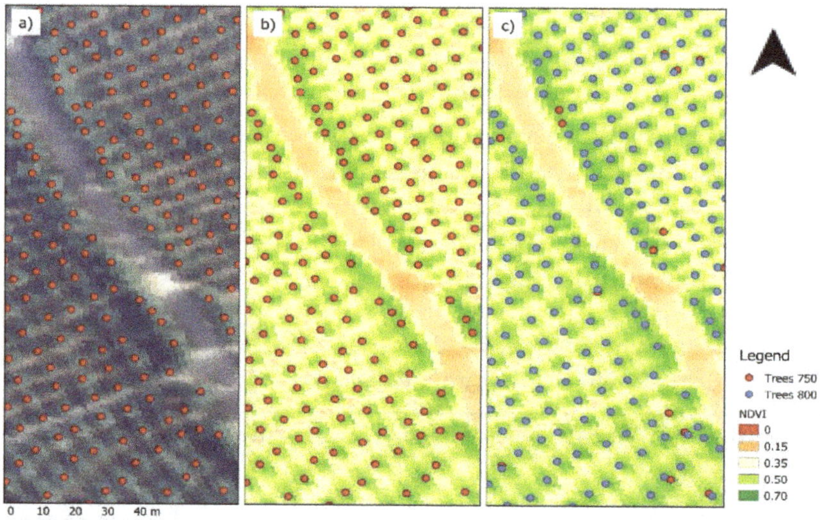

Figure 4. Tree points (red) derived using NDVI with band 750, overlaying (**a**) the hyperspectral image, and (**b**) the NDVI-derived using band 750, and (**c**) tree points derived using band 800 (blue) in comparison NDVI 750 points, overlying NDVI-derived using band 800.

For further analysis, the datasets with the highest detection rates were used. For analysis of the thermal and LiDAR datasets, the canopy polygons derived from the 1 m height rasters were used. The tree points resulting from the NDVI using spectral band 750 were chosen for the hyperspectral data analysis.

3.2. Block-Level Analysis

For block-level analysis, thermal data and height rasters were analysed. Six test blocks were chosen based on their attributes which have been discussed in the methods section. As the same six test blocks are not available for the hyperspectral dataset, this dataset has not been included in block-level analysis.

Boxplots of the thermal and height values for the six selected test blocks are presented in Figures 5 and 6. This illustrates the spread of heights across the test blocks, also in relation to all blocks in the farm. These boxplots illustrate the 25th, 50th and 75th percentiles, extreme values, and outlier values for the canopy pixels.

Figure 5 shows boxplots of the canopy temperature pixels of the test blocks. When treating block 5 as an outlier, it appears that two of the assumed healthy blocks have a mean temperature that is 0.5 °C lower than the unhealthy blocks. This is a positive result as the hypothesis was that unhealthy blocks would appear warmer than the healthy blocks. However, the mean temperature of block 5 is 0.5 °C higher than the mean temperature of the unhealthy test blocks. This could be because block 5 has been inaccurately classified as healthy. Feedback from the farm manager has indicated that the soil quality in this region is very poor so the trees there do not grow as well.

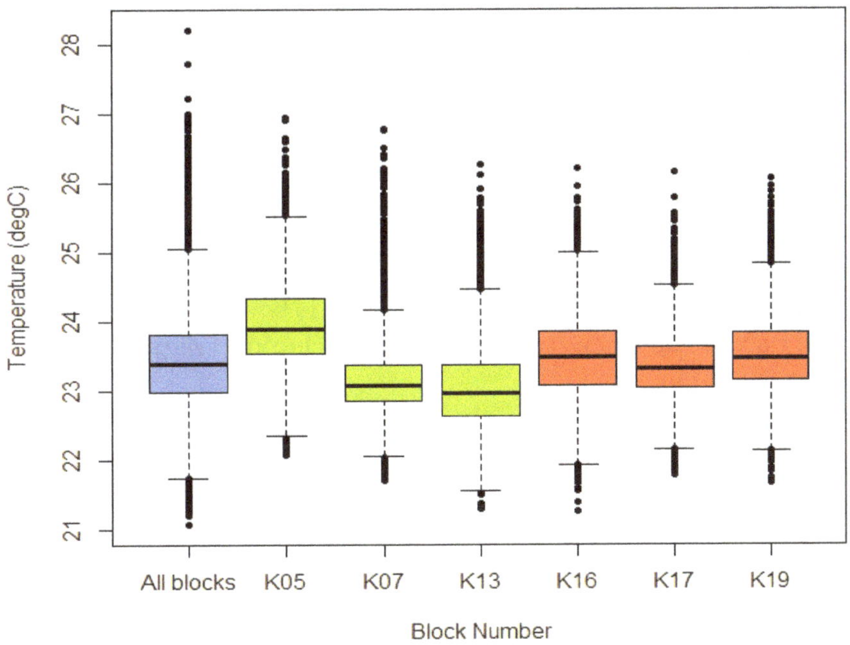

Figure 5. Boxplots of temperature for pixels defined as canopy using a buffer of 3 m and a height threshold of 4 m. The boxplot that is coloured blue shows the temperature for all blocks. The three test blocks in green (5, 7 and 13) are assumed healthy, and the three boxplots coloured red (16, 17 and 19) are unhealthy.

Figure 6 shows the outcome of the height analysis. The blocks classified as healthy appear to have similar heights to all blocks. Interestingly, the unhealthy blocks appear to deviate from the mean heights across all blocks. This is especially true for Block 19, which contains shorter trees than the other blocks. Block 16 appears to contain taller trees than the other blocks, and Block 17 has a wider spread within the upper and lower quartiles.

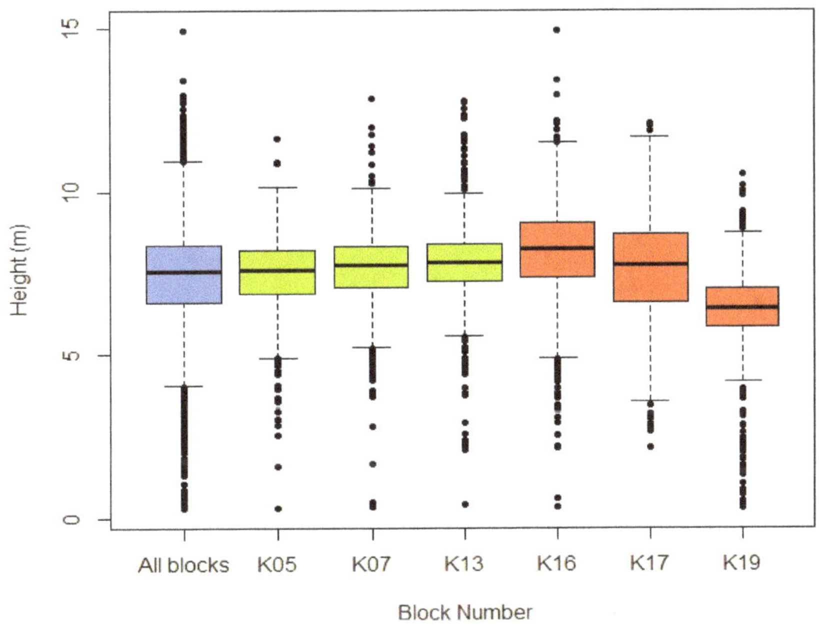

Figure 6. Boxplots for height of the blocks. The blue block shows the values for all blocks in the model farm. The healthy blocks are shown in green and the unhealthy blocks are shown in red.

The results of the boxplot analysis show some interesting trends, that are even more revealing when compared with the spatial plots of the blocks.

When looking at the spatial variation in the canopy pixels of the blocks (Figure 7), we can identify the areas that contribute to the varying values seen in the boxplots (Figure 5). Block 5 is evidently hotter than the other areas, especially in the south of the block, but also across the entire block as well. This is interesting, as the sparseness of the trees in the south of the block is similar to that in the sparse areas of the other blocks. It is difficult to determine the reason for this temperature difference. This area also corresponds to the high height area ratio (HAR) values seen in Figure 8. The unhealthy blocks clearly have more areas that are hotter. When comparing to the RGB data, these hotter areas tend to lie where the trees are sparser and thus more bare ground is exposed. There are also areas in the healthy blocks that have higher temperatures—this is because there is a high variability in all blocks and even those classified as healthy have unhealthy looking regions.

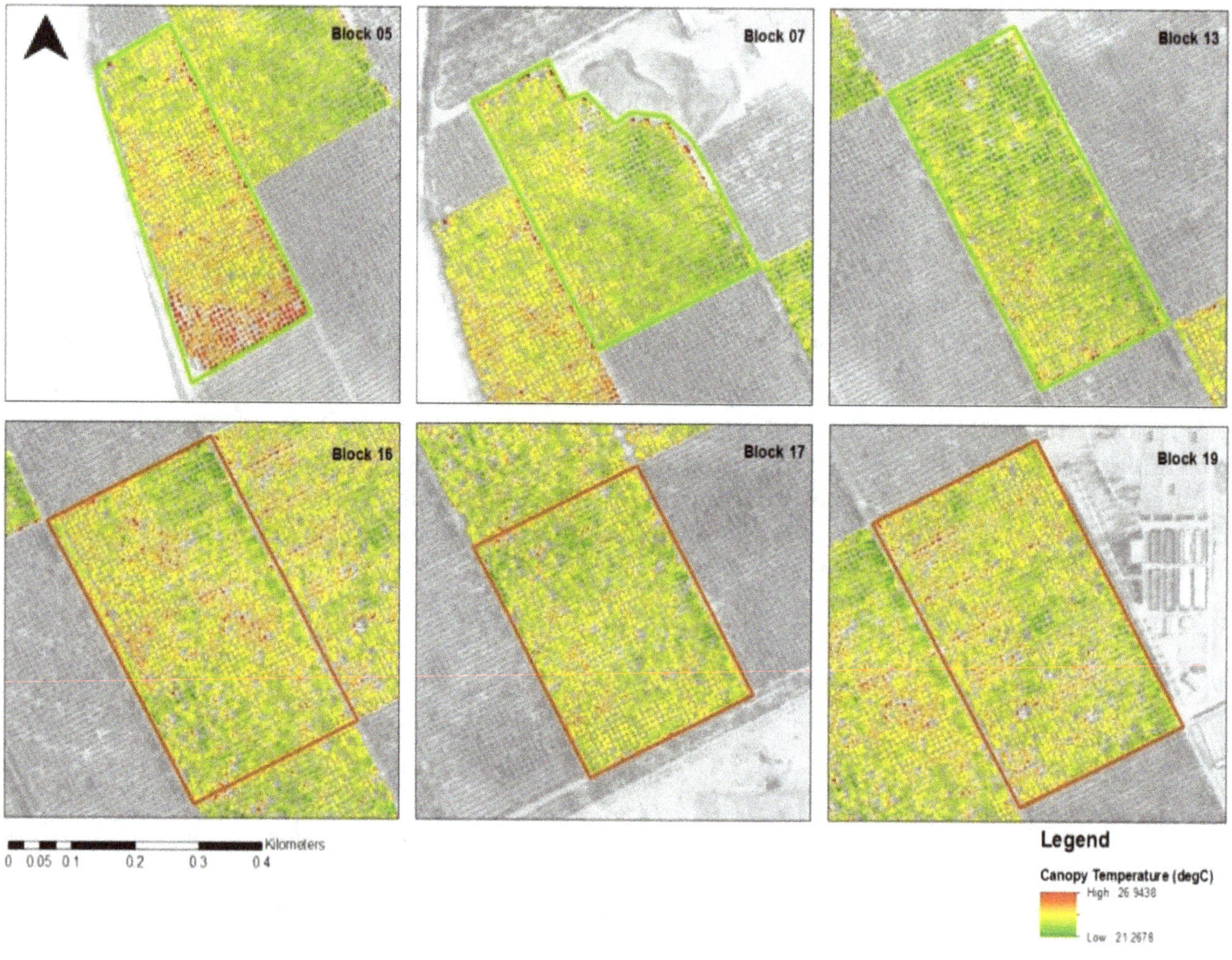

Figure 7. Temperature of the extracted canopy pixels representing individual trees for the test blocks.

Figure 8 shows the values for HAR in each block, where height is divided by canopy area. The healthy test blocks (5, 7 and 13) appear to have mostly values between 0.15 and 0.2 and Blocks 16 and 17 have more tree polygons with a larger height area ratio value, although one area of Block 5 contains trees with larger HAR values. This is to be expected as it indicates that there are more trees with abnormally small canopies, which one would expect in an unhealthier block. However, Block 19 has low HAR values, and in fact looks healthier than some of the "healthy" blocks. The trees in this block are shorter than in the other blocks, and the canopy area is similar, and thus it has the lowest mean height area ratio of all the blocks.

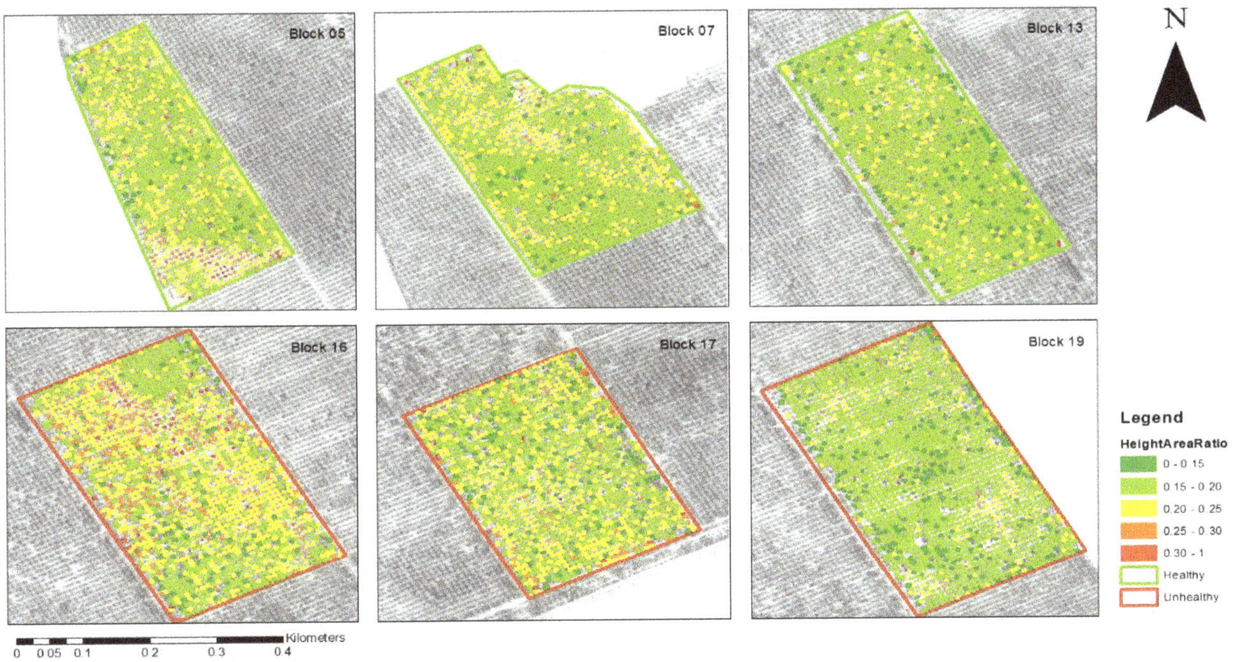

Figure 8. The height area ratios of the derived tree polygons representing individual trees for the test blocks.

Based on Figures 7 and 8, it is evident that within the blocks there is a lot of variation. This is especially true for Block 5, where there is a clear distinction in temperature and height between the north and south of the block.

ANOVA analysis was conducted on 10 randomly selected (within a height threshold; see Section 2.2.3) 'healthy' trees and 10 'unhealthy' trees per block, in order to evaluate whether the two factors of assumed health and the block containing the tree have an effect on tree temperature, and also whether there is interaction between the two (Table 4). There is a main effect of both health (probability (p) value = 0.000611) and block number (p value = 1.06E-14), and there is no interaction effect (p value = 0.834). The statistically significant outcome for block number indicates that the differences in the mean temperature of all the trees in a block are large enough to be considered as different. When comparing the mean temperature of all healthy trees and all unhealthy trees there is a statistically significant difference.

Table 4. Results of ANOVA analysis for temperature in relation to block number and the assumed block health.

	Df	Sum Sq	Mean Sq	F Value	Pr (>F)	Significance [1]
Block No.	5	20.515	4.103	21.161	1.06E-14	*** .
Health	1	2.417	2.41	12.465	0.000611	***
Block No.: Health	5	0.407	0.081	0.419	0.834311	
Residuals	108	20.94	0.194			

[1] Significance levels: 0 '***'; 0.001 '**'; 0.01 '*'; 0.05 '.'; 0.1 '·'; 1.

The results of post-hoc Tukey's HSD test reveal which mean differences contribute most to the statistical significance of the ANOVA results. A subset of results in Table 5 show six comparisons between healthy and unhealthy trees for each test block. When looking at the p value, we see that they are all greater than a significance level $\alpha = 0.05$, indicating no statistical significance. This suggests that when comparing the means of healthy and unhealthy trees within each block there are no statistically significant differences between these means. The differences in means of healthy and unhealthy trees range from 0.1 °C to 0.45 °C. This leads to the conclusion that the significant results seen in Table 4 are a result of comparing trees from different blocks. However, in all cases, we do see that the unhealthy trees have a mean temperature higher than the healthy trees, which does match the hypothesis that unhealthy trees have higher temperatures due to stressors [6,17].

Table 5. Selection of Tukey's HSD comparing mean temperatures of selected healthy (n = 10) and unhealthy trees (n = 10) per block.

	Mean Temperature (°C)			p Value
	Unhealthy	Healthy	Difference	
Block 5	24.37052	24.10113	0.26939	0.967108
Block 7	23.30163	23.04954	0.25209	0.979962
Block 13	23.02733	22.77498	0.25235	0.979803
Block 16	23.80391	23.4071	0.39681	0.682276
Block 17	23.40484	23.31995	0.08489	0.999999
Block 19	23.68493	23.23745	0.44748	0.50232

3.3. *Individual Tree-Level Analysis*

Individual tree analysis has been carried out in order to compare the temperature and height values per tree to the results of the vegetation analysis. Block 11 has been selected for further analysis as there are a wider variety of trees, such as those with wider and smaller canopies. The 100 randomly selected test trees are shown in Figure 9, overlaid on the RGB image, the Chlorophyll Index Green (CIG-Table A1) vegetation index and thermal imagery. The hyperspectral image has geometrical distortion across the whole image, and it is especially notable in the northern region, and along the edges. However, when comparing with the visual (RGB) image, definite patterns are visible when comparing the vegetation index values and the tree canopy density that is visible in Figure 9a. In the south of the block, for example, there is more dense vegetation, and on the right side on the bottom there are patches of sparser vegetation. The regions of sparse vegetation in the top right and dense vegetation in the central region are also visible in the thermal imagery. The randomly selected test points cover a good range of these areas, with points in both the dense and sparse regions of canopy.

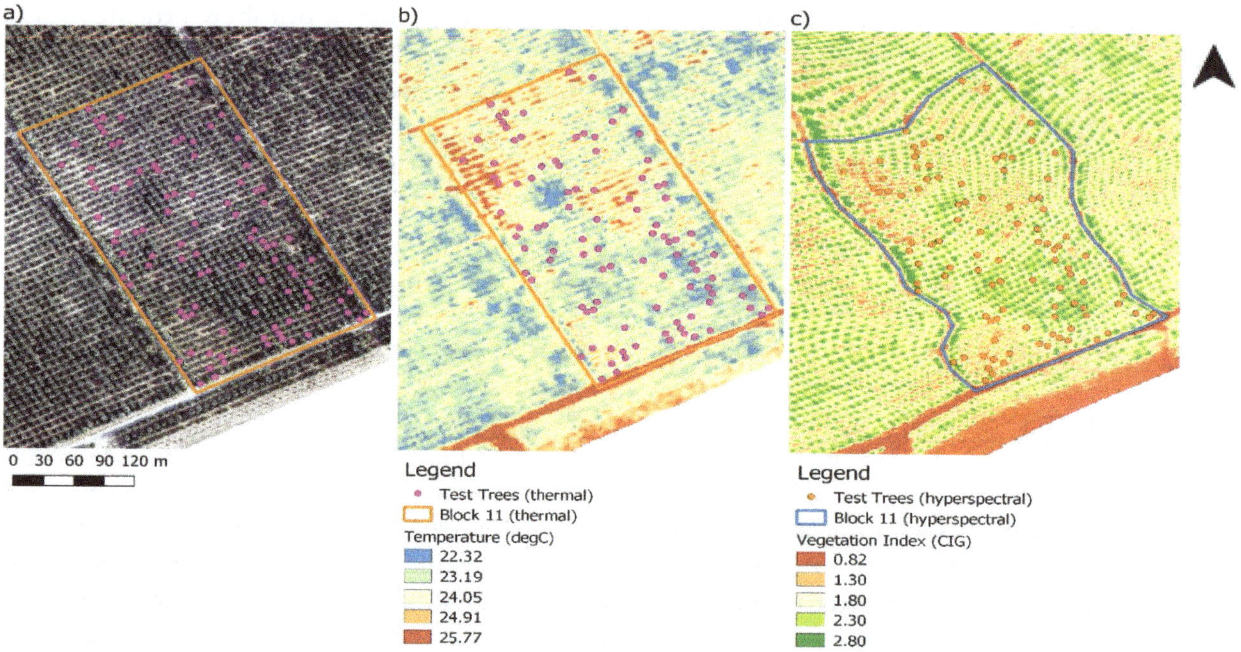

Figure 9. Illustrations of the selected test trees in Block 11 for (**a**) the visual RGB imagery, (**b**) the thermal imagery, and (**c**) the hyperspectral imagery represented as calculated Chlorophyll Index Green (CIG) (Table A1).

The correlation analysis between the vegetation indices and the height and temperature data show some interesting results. The R-squared values reach a maximum of 0.313 between the vegetation index red-edge position (REP) and temperature, and 0.253 between REP and height (Figure 10). The vegetation index Vogelmann red edge index (VOGI) also has a relatively strong correlation value with both temperature (R^2 = 0.227) and height (R^2 = 0.213) (Figure 10) and these indices both use bands in the red edge positions. Green NDVI (gNDVI) and Chlorophyll Index Green (CIG) also show relatively high correlation values (Table A2). These R-squared values are quite low in general, but as Figure 10 shows there is a trend between the two most highly correlated VIs (REP and VOGI) plotted against temperature and height for the 100 test trees. It is evident that temperature has a negative correlation with the VIs, indicating that at higher temperatures, these VI values tend to be lower. Height has the opposite correlation—taller trees have higher VI values.

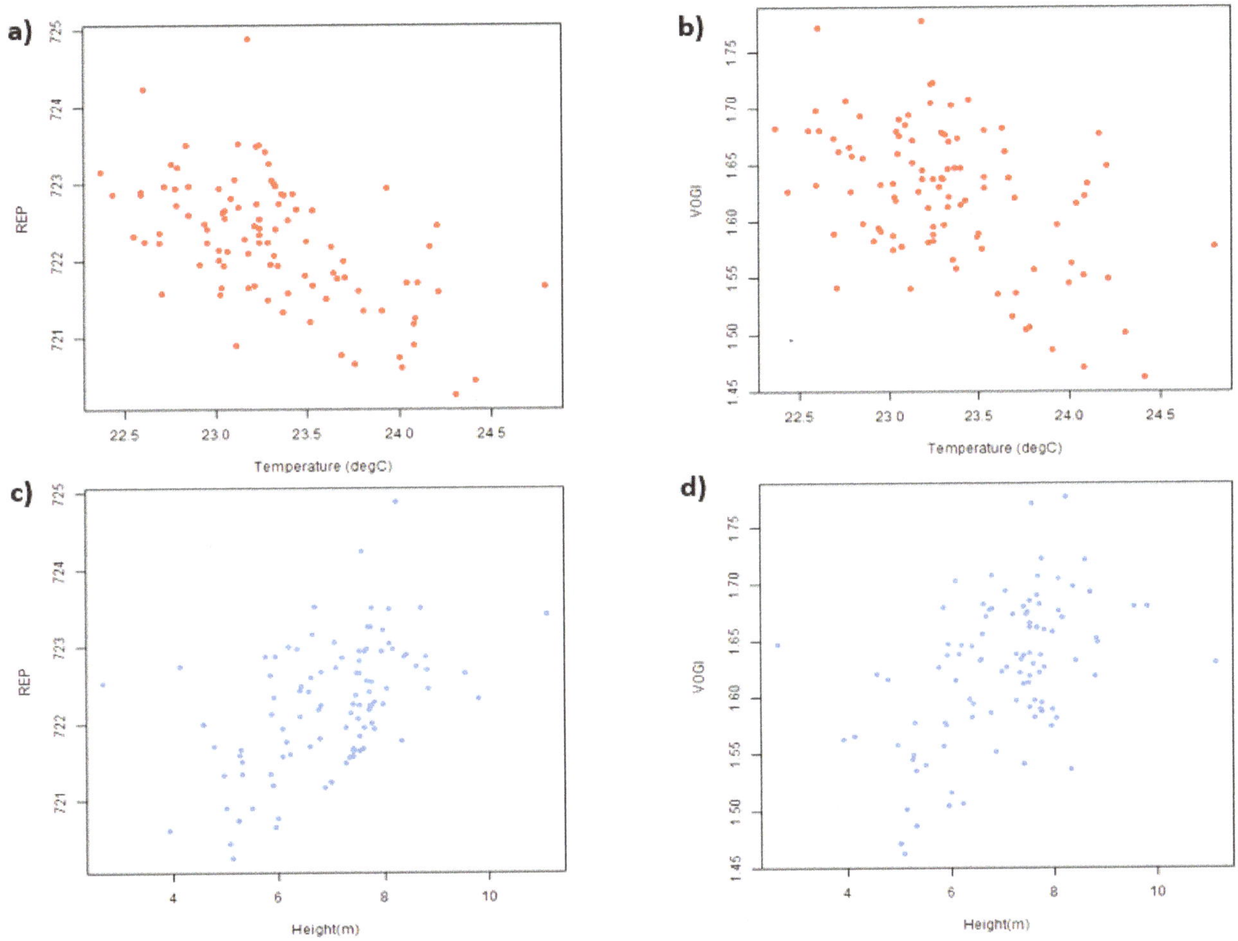

Figure 10. Correlation analysis for 100 test trees in Block 11: (**a**) and (**b**) show the correlation between temperature and two selected VIs, red-edge position (REP) and Vogelmann red edge index (VOGI), and (**c**) and (**d**) show the relationship between these indices and height.

In Figure 11 the temperature and height of the test trees are plotted overlaying the VOGI vegetation index image. It is evident that the regions with lower VI values tend to have higher temperatures and shorter trees; the opposite is generally true in regions with high VI values.

To get a clearer impression of the relation between the temperature and height values and VIs, the values have been visually displayed in relation to the VOGI vegetation index. It can be noted that the regions with lower VI values tend to have higher temperatures and shorter trees; the opposite is generally true in regions with high VI values.

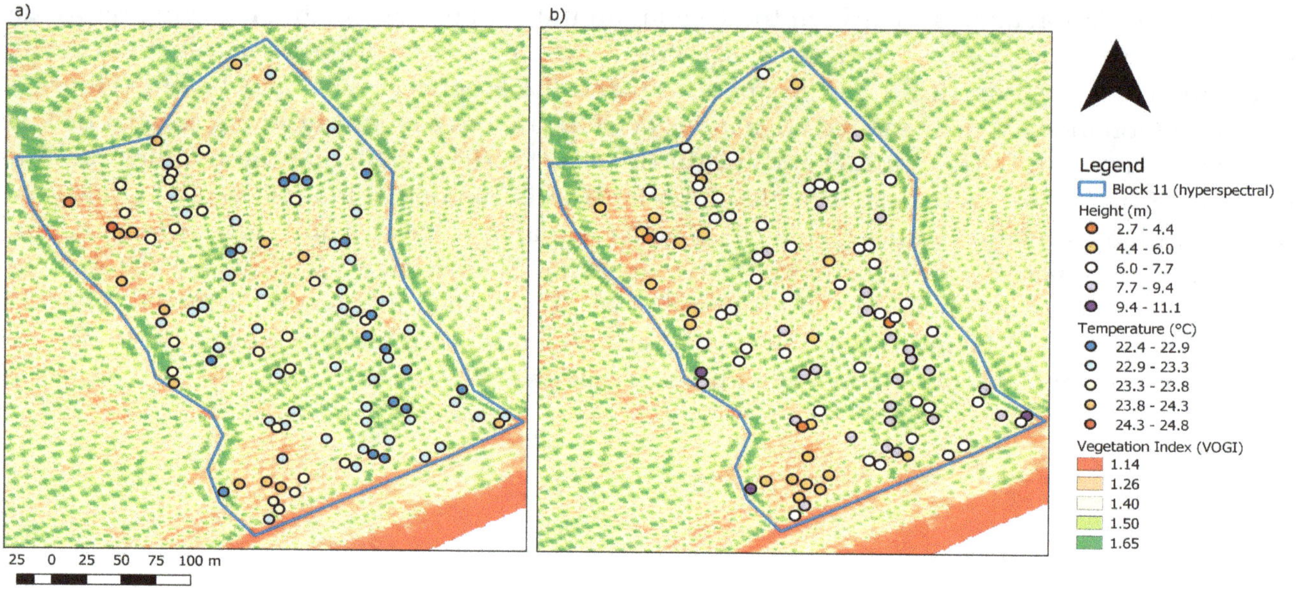

Figure 11. Illustration of the spatial patterns between tree-based point observations of (**a**) temperature and (**b**) height, with the vegetation index VOGI visualized as background map for comparison.

4. Discussion

This study explored various options for assessing vegetation health of date palm plantations using a combination of remote sensing data sources, specifically LiDAR, thermal, hyperspectral and visual RGB imagery. In order to differentiate between more general stressors and those that only affect individual palms, the plantations have been assessed on both a block and individual tree level. Although no direct field data was available for validation at tree level, exchange with the plantation manager provided support for the identified spatial patterns. The main contribution that this research can make is as a suggestion for methods to be used in further research. That being said, the preliminary results do show an ability to successfully differentiate between groups of trees and can be compared to earlier studies [6,16,17,19].

4.1. Exploring the Main Findings of the Study

On a block level, it is evident that the remote sensing images do highlight differences between the blocks; the results of the thermal box plots, for example, show up to a maximum of 1 °C difference between the blocks (Figure 5). These box plots show variation across the data—the spatial maps in Figures 7 and 8 illustrate how this variation is spatially distributed. Block 5 is a clear example, as it shows two very distinct areas in the north and the south of the block, where the trees are both much warmer and shorter. These inter-block variations can be seen to a lesser extent within each block. Therefore, it could be useful to add a further level to the current analysis, namely the development of management zones within the blocks that are based on tree characteristics.

ANOVA analysis was carried out to compare the temperature of individual healthy and unhealthy trees across each block, a method that has been applied in previous studies [6,17]. In the current study, the ANOVA results were statistically significant (Table 4) when comparing block number (p value = 1.06E-14) and tree health (p value = 0.000611) independently of one another. Tukey's HSD test (Table 5) shows the most significant differences come from comparing the trees in different blocks. There is no statistically significant result when comparing means of healthy trees in one block with means of unhealthy trees in that same block. The differences in means when comparing the unhealthy and healthy trees of one block range between 0.1 and 0.45 °C, which may not be practically significant to the extent that significant differences between trees can be identified based solely on temperature. When comparing the individual tree differences found in this study with those found by Cohen et al. (2012), who noted that trees under an 80% water deficit were warmer than the control treatment by approximately 1 °C following four months of the assigned water treatment, the differences in temperature seen in the current study seem notable [6]. Cohen et al. (2012) used comparable thermal images with a resolution of 1.8 m [6]. Golomb et al. (2015), using thermal images with a resolution of 0.5 m, also found statistically significant results when comparing mean Crop Water Stress Indices (CWSI) values of control trees and infested trees using ANOVA, with a statistical significance of $p = 0.001$ [17]. When looking at the absolute temperature difference, a preliminary study by Soroker et al. (2013) noted a mean temperature difference of 1.5 °C between trees infested with *R. ferrugineus* and control trees [16].

Spatial resolution for the thermal data was 1.8 m, which is comparable to the study of Cohen et al. (2012), but lower than for Golomb et al. (2015), where the pixel size was 0.5 by 0.5 m [6,17]. This spatial resolution could have several implications for the results. Based on the RGB images, a large palm tree canopy is approximately 6–8 m in diameter and is covered by approximately 9 to 16 pixels for 1.8 m resolution. However, due to the leaf structure of the date palm canopy there are areas where the background soil will be visible in the pixel, and temperature of the background soil will also be included in the pixel value—it is probable that only one or two pixels include only pure canopy. In studies looking for a large-scale pattern such as the effect of irrigation on an entire plantation, these changes could be visible at a resolution of 1.8 m because every tree in a block is affected by the lowered irrigation [6]. However, for analysis on an individual tree level this resolution may be too coarse to detect whether a tree has a markedly different temperature in comparison to the surrounding trees. Therefore, a maximum resolution of 0.5 m such as in the thermal images acquired by Golomb et al. (2015) might be required to be able to effectively assess individual tree stressors [17].

Considering the results of the hyperspectral data assessment several conclusions can be drawn for individual tree analysis based on the vegetation index analysis. In comparison with other studies, VOGI was one of the most indicative VIs of palm stress in the study by Shafri et al. (2012) on oil palms [19]. gNDVI was also found to identify *R. ferrugineus* infestation of date palms [18]. These were also two of the most indicative VIs when looking at correlations on an individual tree level in this study (Figure 10), indicating that VIs which focus on red-edge and green bands may be most suitable for date palm health analysis.

4.2. Added Value of Combined Datasets and New Indicators

Evaluating the combination of remote sensing data sources recorded at the same timestep makes this study unique in the sphere of date palm management. Several studies have shown that the combination of remote sensing sources can provide multivariate indicators with improved characterisation of the vegetation health status [20,26].

From a methodological perspective, the ability to combine datasets has proven to be very relevant when identifying trees. Using the tree objects derived from the LiDAR data to select the canopy pixels for temperature analysis provided an alternative to the method proposed in Cohen et al. (2012), which utilised the watershed algorithm to extract canopy pixels [6]. The LiDAR data again proved useful for temperature data analysis as the pixels below a certain height could be removed to reduce the chance of soil pixels being included in the analysis. A drawback is that an extra sensor is required, which will add costs to the data acquisition process. The high-resolution RGB data was compared with the outcome of the canopy extraction method to check whether the trees were delineated well (Figure 3).

Correlation analysis between thermal and hyperspectral datasets showed promising relationships (Figure 10). The expected trends are visible, with negative correlation between temperature and several of the VIs. A positive correlation was observed for height and the VIs. Whilst the R-squared values between the variables are quite low, there is a clear spatial pattern between the indicators and VIs as seen in Figure 10. This suggests that these data, especially thermal and hyperspectral imagery, could be combined to identify stressed trees. The temperatures of the trees range from 22.5 °C to 24.5 °C (Figure 10), a difference of 2 °C. Cohen et al. (2012) were able to illustrate differences in temperature of stressed and healthy palms of 1 °C [6]. This provides an indication that using thermal data could prove useful for diagnosis of individual tree health. Based on the correlation results, the hyperspectral data should be useful for reinforcing this diagnosis.

Based on the results of the tree point extraction using NDVI, it can be concluded that this method has high potential for identifying tree objects. Over 90% of trees were identified based on a comparison of field measured trees and NDVI, assuming accurate tree detection (Section 3.1). Misidentifications occur along the edges of the block where the trees are often close together and individual tree canopies are not distinct. As these results are based on distorted data one can assume that, when using a geometrically corrected dataset, identification accuracy will be higher. Using this identification method could prove to be an alternative to expensive LiDAR techniques. It should be noted that one of the reasons this method works well is because the canopies are distinct in a data palm plantation; however, in denser canopies such as the edges of date palm blocks, identification may be less accurate.

In this study, the height area ratio (HAR) index was used in order to assess the health of the blocks. When conducting analysis on the height of the trees, it became apparent that there were large differences in height of trees in certain blocks. For example, Block 19 had lower trees than the other blocks (Figure 6). From this information, one could conclude that either Block 19 is less healthy than the other blocks, or that the trees are younger in that block. Therefore, adding canopy area as a variable when considering tree health adds another consideration to the analysis. By determining the ratio between height and canopy area, one can determine whether a tree has an abnormally small canopy in relation to height—this could be an indication that the tree is experiencing growth limitations. In a study by Shendryk et al. (2016), a width height ratio was included amongst hyperspectral and LiDAR indicators to determine the most "useful" indicator for assessing tree health using Principal Component Analysis (PCA) [20]. Of the indicators derived from the recorded tree height, the width height ratio was the most important in their study. The conclusion that taller trees have smaller canopies in comparison to shorter trees indicates that using HAR alone does not give an accurate representation of tree health, as the growth of the tree affects this value for reasons other than health. Therefore, using an index that normalises height area ratio based on the height of a tree could prove more meaningful when making conclusions about tree health.

4.3. Data Limitations

As has been discussed in the Section 2.2.1 on preprocessing, there are some issues with data quality, especially for the hyperspectral dataset used in this study. Both the geometric distortions and the unknown scaling factor mean that the values used in the analysis might not be fully representative and certainly cannot be directly compared to other hyperspectral datasets that could be used in the future. For example, one step involved using bilinear interpolation for geo-correction, which resulted in multiple raster points with the same value because the hyperspectral image was effectively stretched to fit the corrected dataset.

Due to the distortions of the hyperspectral analysis, it was not possible to compare the trees on an individual level without manually matching the points. This meant that a relatively small sample size of 100 trees has been used as representative of over 10,000 trees. It is also possible that errors have been made when manually identifying the corresponding trees, as in some cases it was difficult to count rows and columns accurately within the distorted data.

Additionally, there are issues within the thermal dataset; specifically, for some limited areas blurring was noticed. This is due to the ortho-mosaicking technique used to merge the snapshot images. Trees are not identifiable in some regions, meaning that extracting the temperature per tree will result in values that are not an accurate representation of tree temperature. This was considered when selecting a block for individual tree analysis carried out in Section 2.2.4, as it was important that the blurring effect was minimal.

Finally, without ground truth data, it is not possible to determine a measure of actual tree health. This means that all assumptions made about tree health are based on indirect field observations and the relative health indicated by the RGB imagery (Table 2). The remote sensing datasets have been compared in various ways to determine whether the expected relations for healthy and unhealthy trees can be seen, and in some cases the initial assumptions were correct. However, as was seen in Figure 5, Block 5 had a much higher temperature than expected despite being considered as a healthy block. This suggests that there were factors not considered in classifying the blocks, and further field information was required. Following the analysis, further field information was provided, and the farm manager indicated that Block 5 had relatively poor soil quality in the southern region. Validation with field data is an essential step for future studies to explore the methods discussed in this paper further. As a result of the limited field data, especially on an individual tree level, the conclusions that can be made are limited to identifying that there are differences in date palms or date palm groups, and these differences cannot be attributed to a specific stressor.

4.4. Future Developments

Overall, this study reiterates the findings of earlier research indicating that different remote sensing datasets could prove useful for management of date palm plantations [6,16,17,19]. Thermal and hyperspectral data are the most promising, especially since identifying trees using NDVI is comparable to the results of LiDAR analysis. For future studies the opportunities of high-resolution RGB imagery to identify variation in palm canopy geometry could be investigated by using a combination of RGB based vegetation indices [27] and object-based image analysis.

Remote sensing data will prove most useful when provided as part of an integrated plantation management platform, where field data is being collected as the flight is taken, and ideally field data will be available over an extended period of time. The imagery could provide useful insights at all steps of an integrated system, including; field registration, field surveys, data acquisition, and monitoring. For example, the step of tree identification using NDVI or LiDAR would be useful for field registration. A topic that could be interesting to explore would be identifying within-block parcels with differing characteristics to be classified as management zones, which would allow for the adoption of precision agriculture techniques. Regular monitoring using a combination of satellite and aerial, or even drone imagery [28], potentially using the methods as explored in this article, could add value to the current monitoring systems that are in place. Based on a time-series analysis approach, changes in vegetation reflectance properties can be indicative of palm health, production and disease infestation processes over the growing season. For example, Jimenez-Brenes et al. [29] showed how multi-temporal UAV-based images can be adopted to support pruning management in olive tree orchards. The ground monitoring system should also be well designed and could include ground sensors [30] and standardised methods of recording tree characteristics. Based on these spatial-temporal data, spatial analytics and geostatistical functions can be adopted to evaluate infestation processes of pests and diseases both at farm and regional scale-level [31]. A design based on the DateGIS platform approach could prove useful, as it aims to integrate imagery data with a ground data recording system where farm employees can update information about specific trees in the field [32].

5. Conclusions

This research explored the options of using multiple types of high-resolution remote sensing imagery in order to assess its potential for detecting various health aspects of date palm plantations. The factors which affect health were considered on two levels; those that impact large areas of the plantation such as a block, and those that affect the individual trees. Based on the spatial variation within blocks, it is suggested to also consider a third level in future studies; namely, areas within blocks with similar characteristics. Despite issues with the hyperspectral and thermal data preprocessing and quality, the methods developed in this study provide new options for future analysis and indicate that remote sensing data could aid plantation management and supplement precision agriculture techniques. Specifically, using a combination of high resolution thermal and hyperspectral imagery can give an indication of individual tree health, and by using these indicators together it could be possible to define a status for each tree. There are many new developments to be made in this area of research—combining remote sensing with detailed field data could provide an early indication of *R. ferrugineus* infestation; and comparing flights at different time steps could provide insight into the rates of change of date palm health. Overall, this study adds to the currently small body of research regarding using remote sensing imagery for date palm plantation management by using a combination of data sources and by providing suggestions for future research in this topic.

Author Contributions: Conceptualization: M.M., L.K., L.B.; Methodology: M.M. and L.K.; Software: M.M.; Formal analysis: M.M..; Writing—Original draft preparation: M.M..; Writing—Review and editing: L.K and L.B.; Visualisation: M.M.; Supervision: L.K. and L.B.; Project administration: L.B.; Funding acquisition: L.B.

Acknowledgments: The authors want to acknowledge Mohamed Ali Bob, farm manager of the Al Mohamadia date farm (www.mohamadia.com.sa/en/) for his valuable support on date farm management and ground truthing feedback, and TAQNIA ETSC for provisioning and operating of the Diamond DA42 MPP GEOSTAR aircraft. Further, the authors want to acknowledge Harm Bartholomeus (Wageningen University and Research) for his support on LiDAR data preprocessing, Juha Suomalainen (Finnish Geospatial Research Institute) for support with the preprocessing to generate a mosaic for the thermal dataset, and Daniel Iordache (VITO) for support in preprocessing of the hyperspectral dataset.

Appendix A

Table A1. Formulae for all hyperspectral vegetation indices used for the hyperspectral analysis. R_n refers to the band number used.

Vegetation Index	Formula	Reference
SRI	$\frac{R_{800}}{R_{670}}$	[19,33]
NDVI 800	$\frac{R_{800}-R_{660}}{R_{800}+R_{660}}$	[19,34]
MSR	$\frac{R_{800}/R_{670}-1}{(R_{800}/R_{670})^{-1}+1}$	[19,35]
NDVI 750	$\frac{R_{750}-R_{660}}{R_{750}+R_{660}}$	[36]
VOGI	$\frac{R_{740}}{R_{720}}$	[19,37]
CIR	$\frac{R_{780}}{R_{710}}-1$	[38,39]
gNDVI	$\frac{R_{801}-R_{550}}{R_{801}+R_{550}}$	[18,40,41]
CIG	$\frac{R_{780}}{R_{550}}-1$	[38]
PRI	$\frac{R_{531}-R_{570}}{R_{531}+R_{570}}$	[42,43]
Optimum NDVI	$\frac{R_{738}-R_{610}}{R_{738}+R_{610}}$	[19]
SIPI	$\frac{R_{800}-R_{445}}{R_{800}+R_{680}}$	[18,44]
MCARI/OSAVI [705,750]	$\frac{[R_{750}-R_{705}-0.2(R_{750}-R_{550})](R_{750}/R_{705})}{(1+0.16)(R_{750}-R_{705})/(R_{750}+R_{705}+0.16)}$	[45–47]
TCARI/OSAVI [705,750]	$\frac{3[R_{750}-R_{705}-0.2(R_{750}-R_{550})(R_{750}/R_{705})]}{(1+0.16)(R_{750}-R_{705})/(R_{750}+R_{705}+0.16)}$	[46,47]
REP	$700+40\frac{R_{670}+R_{780}/2-R_{700}}{R_{740}-R_{700}}$	[48]

Appendix B

Table A2. Results of R-squared analysis of the vegetation indices contained in Table A1 and their relations to temperature and height.

Vegetation Index	Temperature (R-Squared)	Height (R-Squared)
REP	0.313	0.253
VOGI	0.227	0.213
gNDVI	0.206	0.189
CIG	0.196	0.204
CIR	0.196	0.202
TCOS750	0.191	0.103
SIPI	0.137	0.161
SRI	0.119	0.147
NDVI800	9.106	0.146
NDVI750	0.088	0.136
MSR	0.08	0.14
NDVI730	0.071	0.112
MCOS750	0.054	0.134
PRI	0.001	0.014

References

1. Stringer, L.C. Reviewing the links between desertification and food insecurity: From parallel challenges to synergistic solutions. *Food Secur.* **2009**, *1*, 113–126. [CrossRef]
2. Godfray, H.C.J.; Beddington, J.R.; Crute, I.R.; Haddad, L.; Lawrence, D.; Muir, J.F.; Pretty, J.; Robinson, S.; Thomas, S.M.; Toulmin, C. Food security: The challenge of feeding 9 billion people. *Science* **2010**, *327*, 812–818. [CrossRef] [PubMed]
3. Arias, E.; Hodder, A.J.; Oihabi, A. FAO support to date palm development around the world: 70 years of activity. *Emir. J. Food Agric.* **2016**, *28*, 1–11.
4. Al-Dosary, N.M.N.; Al-Dobai, S.; Faleiro, J.R. Review of the management of red palm weevil *Rhynchophorus ferrugineus* olivier in date palm *Phoenix dactylifera* L. *Emir. J. Food Agric.* **2016**, *28*, 34–44. [CrossRef]
5. Shabani, F.; Kumar, L.; Nojoumian, A.H.; Esmaeili, A.; Togheyani, M. Projected future distribution of date palm and its potential use in alleviating micronutrient deficiency. *J. Sci. Food Agric.* **2015**, *96*, 1132–1140. [CrossRef] [PubMed]
6. Cohen, Y.; Alchanatis, V.; Prigojin, A.; Levi, A.; Soroker, V.; Cohen, Y. Use of aerial thermal imaging to estimate water status of palm trees. *Precis. Agric.* **2012**, *13*, 123–140. [CrossRef]
7. Elshibli, S.; Elshibli, E.M.; Korpelainen, H. Growth and photosynthetic CO_2 responses of date palm plants to water availability. *Emir. J. Food Agric.* **2016**, *28*, 58–65. [CrossRef]
8. Tripler, E.; Ben-Gal, A.; Shani, U. Consequence of salinity and excess boron on growth, evapotranspiration and ion uptake in date palm (*Phoenix dactylifera* L., cv. Medjool). *Plant Soil* **2007**, *297*, 147–155. [CrossRef]
9. Askri, B.; Ahmed, A.T.; Abichou, T.; Bouhhlila, R. Effects of shallow water table, salinity and frequency of irrigation water on the date palm water use. *J. Hydrol.* **2014**, *513*, 81–90. [CrossRef]
10. Yaish, M.W.; Kumar, P.P. Salt tolerance research in date palm tree (*Phoenix dactylifera* L.), past, present, and future perspectives. *Front. Plant Sci.* **2015**, *6*, 348. [CrossRef]
11. Abraham, V.A.; Al Shuaibi, M.A.; Faleiro, J.R.; Abozuhairah, R.A.; Viyasagar, P.S.P.V. An integrated management approach to for red palm weevil *Rhynchophorus ferrugineus* Oliv. A key pest of date palm in the Middle East. *Agric. Sci.* **1998**, *3*, 77–84. [CrossRef]
12. Faleiro, J.R. A review of the issues and management of the Red Palm Weevil *Rhynchophorus ferrugineus* (Coleoptera: Rhynchophoridae) in coconut and date palm during the last one hundred years. *Int. J. Trop. Insect Sci.* **2006**, *26*, 135–154.
13. Ali, A.-S.A.; Hama, N.N. Integrated management for major date palm pests in Iraq. *Emir. J. Food Agric.* **2016**, *28*, 24–33. [CrossRef]

14. Gutiérrez, A.; Ruiz, V.; Moltó, E.; Tapia, G.; del Mar Téllez, M. Development of a bioacoustic sensor for the early detection of Red Palm Weevil (*Rhynchophorus ferrugineus* Olivier). *Crop Prot.* **2010**, *29*, 671–676. [CrossRef]
15. Dembilio, O.; Jacques, J.A. *Sustainable Pest Management in Date Palm: Current Status and Emerging Challenges*, 1st ed.; Springer International Publishing: Basel, Switzerland, 2015; pp. 13–36.
16. Soroker, V.; Suma, P.; la Pergola, A.; Cohen, Y.; Cohen, Y.; Alchanatis, V.; Golomb, O.; Goldshtein, E.; Hetzroni, A.; Galazan, L.; et al. Early detection and monitoring of Red Palm Weevil: Approaches and challenges. In Proceedings of the Palm Pest Mediterranean Conference, Nice, France, 19–18 July 2013.
17. Golomb, O.; Alchanatis, V.; Cohen, Y.; Levin, N.; Cohen, Y.; Soroker, V. *Precision Agriculture '15*, 1st ed.; Wageningen Academic Publishers: Wageningen, The Netherlands, 2015; pp. 643–650.
18. Bannari, A.; Khurshid, K.S.; Staenz, K.; Schwarz, J. Potential of Hyperion EO-1 hyperspectral data for wheat crop chlorophyll content estimation. *Can. J. Remote Sens.* **2008**, *34*, 139–157. [CrossRef]
19. Shafri, H.Z.M.; Hamdan, N.; Anuar, M.I. Detection of stressed oil palms from an airborne sensor using optimized spectral indices. *Int. J. Remote Sens.* **2012**, *33*, 4293–4311. [CrossRef]
20. Shendryk, I.; Broich, M.; Tulbure, M.G.; McGrath, A.; Keith, D.; Alexandrov, S.V. Mapping individual tree health using full-waveform airborne laser scans and imaging spectroscopy: A case study for a floodplain eucalypt forest. *Remote Sens. Environ.* **2016**, *187*, 202–217. [CrossRef]
21. Alhammadi, M.S.; Glenn, E.P. Detecting date palm trees health and vegetation greenness change of the eastern coast of the United Arab Emirates using SAVI. *Int. J. Remote Sens.* **2008**, *29*, 1745–1765. [CrossRef]
22. Vidyasagar, P.S.P.V.; Al Saihati, A.; Al Mohanna, O.E.; Subbei, A.I.; Mohsin, A.M.A. Management of Red Palm Weevil *Rhynchophorus ferrugineus* oliv., a serious pest of date palm in Al Qatif, Kindgdom of Saudi Arabia. *J. Plant. Crop.* **2000**, *28*, 35–43.
23. Aleid, S.M.; Al-Kharyi, J.M.; Al-Bahrany, A.M. *Date Palm Genetic Resources and Utilization: Volume 2: Asia and Europe*, 1st ed.; Springer: Dordrecht, The Netherlands, 2015; pp. 49–95.
24. InfraTec. *VarioCam HD Head: User Manual*; InfraTec: Dresden, Germany, 2016; Available online: https://www.infratec.at/downloads/en/thermography/manuals/infratec-manual-variocam-hd-head.pdf (accessed on 11 January 2019).
25. Westoby, M.J.; Brasington, J.; Glasser, N.F.; Hambrey, M.J.; Reynolds, J.M. 'Stucture-from-Motion' photogrammetry: A low-cost, effective tool for geoscience applications. *Geomorphology* **2012**, *179*, 300–314. [CrossRef]
26. Dalponte, M.; Reyes, F.; Kandare, K.; Gianelle, D. Delineation of Individual Tree Crowns from ALS and Hyperspectral data: A comparison among four methods. *Eur. J. Remote Sens.* **2015**, *48*, 365–382. [CrossRef]
27. Hamuda, E.; Glavin, M.; Jones, E. A survey of image processing techniques for plant extraction and segmentation in the field. *Computelectron. Agric.* **2016**, *125*, 184–199. [CrossRef]
28. Gevaert, C.M.; Suomalainen, J.; Tang, J.; Kooistra, L. Generation of spectral-temporal response surfaces by combining multispectral satellite and hyperspectral UAV imagery for precision agriculture applications. *IEEE J.-Stars* **2015**, *8*, 3140–3146. [CrossRef]
29. Jiménez-Brenes, F.M.; López-Granados, F.; de Castro, A.I.; Torres-Sánchez, J.; Serrano, N.; Peña, J.M. Quantifying pruning impacts on olive tree architecture and annual canopy growth by using UAV-based 3D modelling. *Plant Methods* **2017**, *13*, 55. [CrossRef] [PubMed]
30. Massimo, P.; Alberto, R.A.; Roberto, M.; Khalid, A.; Ali, A. Devices to detect red palm weevil infestation on palm species. *Precis. Agric.* **2018**, *19*, 1049. [CrossRef]
31. Al-Kindi, K.M.; Kwan, P.; Andrew, N.R.; Welch, M. Modelling spatiotemporal patterns of dubas bug infestations on date palms in northern Oman: A geographical information system case study. *Crop Prot.* **2017**, *93*, 113–121. [CrossRef]
32. DateGIS Demo—An Integrated Geo-Information System (GIS) for Precision Agriculture of Date Palm Farming. Available online: https://business.esa.int/projects/dategis-demo (accessed on 9 December 2018).
33. Jordan, C.F. Derivation of Leaf Area Index from quality of light on the forest floor. *Ecology* **1996**, *90*, 663–666. [CrossRef]
34. Baret, F.; Guyot, G. Potentials and limits of vegetation indices for LAI and APAR assessment. *Remote Sens. Environ.* **1991**, *35*, 161–173. [CrossRef]
35. Chen, J.M. Evaluation of vegetation indices and a modified simple ratio for boreal applications. *Can. J. Remote Sens.* **1996**, *22*, 229–242. [CrossRef]

36. Apan, A.; Held, A.; Phinn, S.; Markley, J. Detecting sugar cane 'orange rust' disease using EO-1 Hyperion imagery. *Int. J. Remote Sens.* **2004**, *25*, 489–498. [CrossRef]
37. Vogelmann, J.E.; Rock, B.N.; Moss, D.M. Red edge spectral measurements from sugar maple leaves. *Int. J. Remote Sens.* **1993**, *14*, 1563–1575. [CrossRef]
38. Gitelson, A.A.; Gritz, Y.; Merzlyak, M.N. Relationships between leaf chlorophyll content and spectral reflectance and algorithms for non-destructive chlorophyll assessment in higher plant leaves. *J. Plant Physiol.* **2003**, *160*, 271–282. [CrossRef] [PubMed]
39. Clevers, J.G.P.W.; Kooistra, L. Using Hyperspectral Remote Sensing Data for Retrieving Canopy Chlorophyll and Nitrogen Content. *IEEE J.-Stars* **2012**, *5*, 574–583. [CrossRef]
40. Gitelson, A.A.; Kaufman, Y.J.; Merzlyak, M.N. Use of a green channel in remote sensing of global vegetation from EOS-MODIS. *Remote Sens. Environ.* **1996**, *58*, 289–298. [CrossRef]
41. Blackburn, G.A. Relationships between spectral reflectance and pigment concentrations in stacks of deciduous broadleaves. *Remote Sens. Environ.* **1999**, *70*, 224–237. [CrossRef]
42. Gamon, J.A.; Peñuelas, J.; Field, C.B. A narrow-waveband spectral index that tracks diurnal changes in photosynthetic efficiency. *Remote Sens. Environ.* **1992**, *14*, 35–44. [CrossRef]
43. Barton, C.V.M. Advances in remote sensing of plant stress. *Plant Soil* **2012**, *354*, 41–44. [CrossRef]
44. Peñuelas, J.; Baret, J.; Filella, I. Semi-empirical indices to assess carotenoids/chlorophyll a ratio from leaf spectral reflectance. *Photosynthetica* **1995**, *31*, 221–230.
45. Daughtry, C.S.T.; Walthall, C.L.; Kim, M.S.; Brown de Colstoun, E.; McMurtrey, J.E., III. Estimating corn leaf chlorophyll concentration from leaf and canopy reflectance. *Remote Sens. Environ.* **2000**, *74*, 229–239. [CrossRef]
46. Haboudane, D.; Miller, J.R.; Tremblay, N.; Zarco-Tejada, P.J.; Dextraze, L. Integrated narrowband vegetation indices for prediction of crop chlorophyll content for application to precision agriculture. *Remote Sens. Environ.* **2002**, *81*, 416–426. [CrossRef]
47. Wu, C.; Niu, Z.; Tang, Q.; Huang, W. Estimating chlorophyll content from hyperspectral vegetation indices: Modelling and validation. *Agric. For. Meteorol.* **2008**, *148*, 1230–1241. [CrossRef]
48. Clevers, J.G.P.W.; de Jong, S.M.; Epema, G.; van der Meer, F.; Bakker, W.H.; Skidmore, A.K.; Addink, E.A. MERIS and the red-edge position. *Int. J. Appl. Earth Obs. Geoinf.* **2001**, *3*, 313–320. [CrossRef]

12

Automatic Parameter Tuning for Adaptive Thresholding in Fruit Detection

Elie Zemmour *, Polina Kurtser and Yael Edan

Department of Industrial Engineering and Management, Ben-Gurion University of the Negev, Beer-Sheva 8410501, Israel; kurtser@post.bgu.ac.il (P.K.); yael@bgu.ac.il (Y.E.)

* Correspondence: eliezem@post.bgu.ac.il

Abstract: This paper presents an automatic parameter tuning procedure specially developed for a dynamic adaptive thresholding algorithm for fruit detection. One of the major algorithm strengths is its high detection performances using a small set of training images. The algorithm enables robust detection in highly-variable lighting conditions. The image is dynamically split into variably-sized regions, where each region has approximately homogeneous lighting conditions. Nine thresholds were selected to accommodate three different illumination levels for three different dimensions in four color spaces: RGB, HSI, LAB, and NDI. Each color space uses a different method to represent a pixel in an image: RGB (Red, Green, Blue), HSI (Hue, Saturation, Intensity), LAB (Lightness, Green to Red and Blue to Yellow) and NDI (Normalized Difference Index, which represents the normal difference between the RGB color dimensions). The thresholds were selected by quantifying the required relation between the true positive rate and false positive rate. A tuning process was developed to determine the best fit values of the algorithm parameters to enable easy adaption to different kinds of fruits (shapes, colors) and environments (illumination conditions). Extensive analyses were conducted on three different databases acquired in natural growing conditions: red apples (nine images with 113 apples), green grape clusters (129 images with 1078 grape clusters), and yellow peppers (30 images with 73 peppers). These databases are provided as part of this paper for future developments. The algorithm was evaluated using cross-validation with 70% images for training and 30% images for testing. The algorithm successfully detected apples and peppers in variable lighting conditions resulting with an F-score of 93.17% and 99.31% respectively. Results show the importance of the tuning process for the generalization of the algorithm to different kinds of fruits and environments. In addition, this research revealed the importance of evaluating different color spaces since for each kind of fruit, a different color space might be superior over the others. The LAB color space is most robust to noise. The algorithm is robust to changes in the threshold learned by the training process and to noise effects in images.

Keywords: adaptive thresholding; fruit detection; parameter tuning

---

1. Introduction

Fruit detection is important in many agricultural tasks such as yield monitoring [1–8], phenotyping [9–11], precision agriculture operations (e.g., spraying [12] and thinning [13–15]), and robotic harvesting [16–18]. Despite intensive research conducted in identifying fruits, implementing a real-time vision system remains a complex task [17,18]. Features like shape, texture, and location are subject to high variability in the agricultural domain [18]. Moreover, fruits grow in unstructured environments with highly-variable lighting conditions [19,20] and obstructions [21] that influence detection performance. Color and texture are fundamental characteristics of natural images and play an important role in visual perception [22].

Nevertheless, despite the challenges, several algorithms have been developed with impressive detection rates of over 90–95%. However, these detection rates were achieved only for specific fruit (apples, oranges, and mangoes) [13,23,24]. These crops are known for their high ratio of fruits per image allowing easier acquisition of large quantities of data. Other crops such as sweet peppers and rock melons [23] with a lower fruit-to-image ratio yield lower results of 85–90% [19,23], even with the employment of cutting edge techniques such as deep learning [25]. Additionally, the crops with high detection rates with these results are very distinct from their background in terms of color, a central feature for a color-based-only detection. The only recent research made to detect green crops is on weed detection [26,27], but these green crops are held against a brown background. In grape clusters' detection [28], a rate of about 90% accuracy was achieved. Some work in the development of cucumber harvesters [29] has been done, but the success rate of harvesting was not distinguished from detection success rates and therefore cannot be reported.

In this research, we focus on the detection of three different types of challenging crops: red apples (a high ratio of fruits per image; however, we used a very small dataset http://icvl.cs.bgu.ac.il/lab_projects/agrovision/DB/Sweeper05/#/scene), green grapes ("green-on-green" dataset [28]), and yellow sweet peppers (a low fruit-to-image ratio http://icvl.cs.bgu.ac.il/lab_projects/agrovision/DB/Sweeper06/#/scene).

The adaptive thresholding algorithm presented in this paper is based on previous work [19] that was developed for sweet peppers' detection for a robotic harvester. A set of three thresholds was determined for each region of the image according to its lighting setting. Preliminary results of the same algorithm for an apple detection problem have been previously presented [30].

The current paper advances previous research [30] with several new contributions: (1) a new parameter tuning procedure developed to best-fit the parameters to the specific database; (2) the application and evaluation of the adaptive thresholding algorithm for different color spaces; (3) application of the algorithm to different types of fruits along with intensive evaluation and sensitivity analyses; (4) comparing the contribution of the new developments (Items 1–2) to previous developments.

2. Literature Review

2.1. Detection Algorithms in Agriculture

While this paper does not aim to be a review paper, in addition to the many recent reviews (e.g., [16,18,25]), a summary of previous results helps place the outcomes of this paper into context (Table 1).

As can be seen in the table, most algorithms focus on pixel-based detection (e.g., segmentation). This is indeed a common method in fruit detection (e.g., [31–34]). Many segmentation algorithms have been developed [35] including: K-means [36], mean shift analysis [37], Artificial Neural Networks (ANN) [38], Support Vector Machines (SVM) [39], deep learning [25], and several others.

A common challenge facing agriculture detection research is the lack of data [20], due to the harsh conditions for image acquisition and the tedious related ground truth annotation [40]. Current advanced algorithms (e.g., deep learning) require collecting many data. Therefore, to date, the best detection results are provided for crops with high fruit-to-image ratios (e.g., apples, oranges, and mangoes) and fruits that grow in high density, and hence, each image provides many data. Some research [26,41] aimed to cope with the need for large quantities of highly-variable data by pre-training a network on either non-agricultural open access data [26] or by generating synthetic data [41]. Both methods have shown promising results.

In this paper, we present an alternative direction, focusing on the development of algorithms based on smaller datasets. This research focuses on segmenting objects in the image using an adaptive thresholding method. Observing the histogram of the image color implies that a threshold can be determined to best differentiate between the background and the object distributions [42].

The threshold is computed by finding the histogram minimum (Figure 1) separating two peaks: the object and the background. However, the global minimum between the distributions is very hard to determine in most cases [43].

Currently, most optimal thresholding algorithms determine the threshold only in a one-dimensional space, for example in the RGB space, either R, or G, or B, or a linear combination of their values (e.g., grayscale transformation) [44]. In the transformation from three color dimensions into one, information is lost. In this research, a three-dimensional thresholding algorithm based on [19] was applied and evaluated also for additional color spaces (RGB, HSI, LAB, and NDI color spaces); a threshold is determined for each dimension in the color space.

There are two common adaptive thresholding algorithm concepts: (1) global thresholding, in which for each image, a different threshold is determined according to specific conditions for the entire image that is then transformed into a binary image; (2) local thresholding, in which the image is divided into sections and a different threshold is calculated for each section; the sections are then combined to a binary image. There are several methods that utilize dynamic local thresholding algorithms [45,46]. A common approach is to use multi-resolution windows that apply a bottom-up method, merging pixels while a criterion is met [45,46]. Another approach is the top down method, where the image is divided into subregions according to specific criteria. The top-down approach reduces execution speed and improves generalization [47] and was therefore used in this research.

The previously-developed algorithm by Vitzrabin et al., [19], which this research is based on, dynamically divides the image into several regions, each with approximately the same lighting conditions. The main contribution of the adaptive local 3D thresholding is a very high True Positive Rate (TPR) and a Low False Positive Rate (FPR) in the fruit detection task in an unstructured, highly-variable, and dynamic crop environment. Another contribution is the ability to change in real time the task objective to enable fast adaption to other crops, varieties, or operating conditions requiring small datasets and fast training. The algorithm's adaptation to the desired ratio between TPR and FRP makes it specifically fit for robotic harvesting tasks for which it was originally designed; it contributes to a better success rate in robotic operations in which at first, FPR should be minimum (to reduce cycle times), and when approaching the grasping operation itself [19], TPR should be maximized (to increase grasping accuracy). This can be applicable towards other fruit detection tasks (e.g., same as above for spraying, thinning and in yield detection first maximizing TPR for deciding on harvesting timing and then minimizing TPR for accurate marketing estimation).

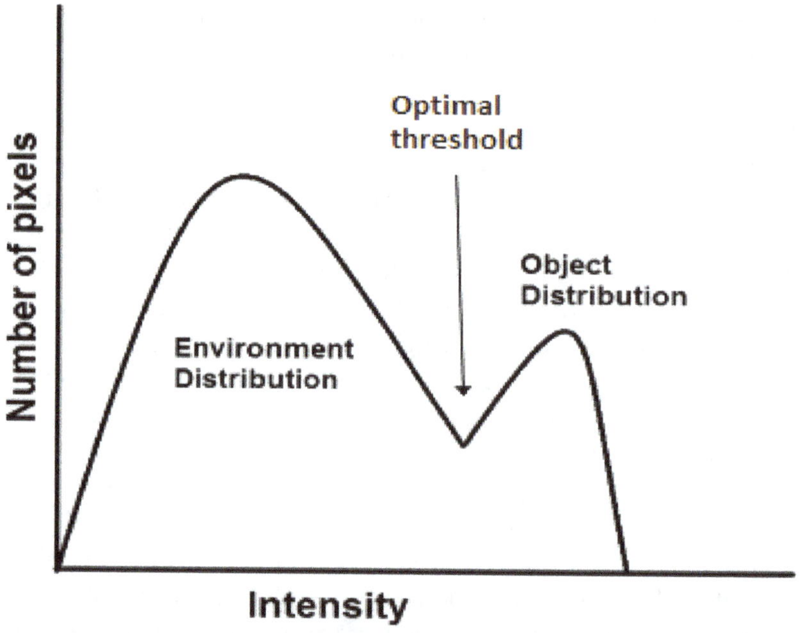

Figure 1. Optimal threshold in the bimodal histogram.

Table 1. Summary of previously-published results.

Paper	Crop	Dataset Size	Detection Level	Algorithm Type	FPR %	TPR %	F %	A %	P %	R %
Arad et al., 2019 [20]	Peppers	156 img	PX W	AD DL	NA	NA	NA	NA	65(95) 84	94(95) NA
Ostovar et al., 2018 [34]	Peppers	170 img	PX	AD	NA	NA	NA	91.5	NA	NA
Chen et al., 2017 [24]	Apples Oranges	1749 (21 img) 7200 (71 img)	PX	DL	5.1 3.3	95.7 96.1	95.3 * 96.4*	NA	NA	NA
McCool et al., 2017 [26]	Weed	Pre-train:10^6 img tuning and testing: 60 img	PX	D-CNN	NA	NA	NA	93.9	NA	NA
Milioto et al., 2017 [27]	Weed	5696 (867 img) 26,163 (1102 img)	PX	CNN	NA	NA	NA	96.8 99.7	97.3 96.1	98.1 96.3
Sa et al., 2016 [23]	Sweet pepper Rock melon Apple Avocado Mango Orange	122 img 135 img 64 img 54 img 170 img 57 img	W	DL	NA	NA	82.8 84.8 93.8 93.2 94.2 91.5	NA	NA	NA
Vitzrabin et al., 2016 [19]	Sweet pepper	479 (221 img)	PX	AD	4.6	90.0	92.6 *	NA	NA	NA
Song et al., 2014 [11]	Pepper plants	1056 img	W	NB+SVM	NA	NA	65	NA	NA	NA
Nuske et al., 2011 [6]	Grapes	2973 img	PX	K-NN	NA	NA	NA	NA	63.7	98
Berenstein et al., 2010 [28]	Grapes	100 img	PX	Morphological	NA	NA	NA	90	NA	NA
Zheng et al., 2009 [37]	Vegetation	20 img 80 img	PX	Mean-Shift	NA	NA	NA	95.4 95.9	NA	NA
Our results	Sweet pepper Apples Grapes	73 (30 img) 113 (9 img) 1078 (129 img)	PX	AD	0.81 2.59 33.35	99.43 89.45 89.48	99.31 93.17 73.52	NA	NA	NA

DL = Deep Learning; PX = Pixels' segmentation; AD = Adaptive threshold; NB = Naive Bias; W = Window detection; F = F-score; A = Accuracy; P = Precision; R = Recall; * Calculated F-score based on reported TPR and FPR according to Equation (4)

2.2. Color Spaces

Images can be represented by different color spaces (e.g., RGB, HSI, LAB, and NDI), each one emphasizing different color features [22]. RGB is the most common color space representing each pixel in the image in three color channels as acquired: red, green, and blue. HSI represents every color with three components: hue (H), saturation (S), and intensity (I), also known as HSV [37]. The LAB color space is an approximation of human vision [36] and presents for each pixel the L* (Lightness) from black to white, a* from green to red, and b* from blue to yellow. An additional color space commonly employed in the agriculture field [19] is the Normalized Difference Index (NDI) space. The NDI is used to differentiate between fruit and background [48] since it helps to overcome changes in illumination and shading due to its normalization technique [49]. Each dimension in the NDI space is the normalized difference index between two colors in the RGB space, resulting in three dimensions (Equation (1)). These operations are applied for all pixel locations in the image, creating a new image with this contrast index. These equations yield NDI values ranging between −1 and +1.

$$NDI_1 = \frac{R-G}{R+G}; NDI_2 = \frac{R-B}{R+B}; NDI_3 = \frac{B-G}{B+G} \quad (1)$$

3. Materials and Methods

3.1. Databases

The algorithm was evaluated on three databases representing three different fruit colors: red (apples), green (grapes), and yellow (peppers) and different types of fruits (high image and low image ratios) for two environmental settings (greenhouse and field) in different illumination conditions. Images were acquired with different cameras. Each image was processed by a human labeler who performed manual segmentation of the image into targets and background (Figures 2 and 3) by visually analyzing the image and marking all the pixels considered as a fruit, in accordance with the common protocols used in the computer vision community [50].

3.1.1. Apples

The orchard apples database included 113 "Royal Gala" apples in 9 images acquired from an orchard in Chile in March 2012 under natural growing conditions with a Prosilica GC2450C (Allied Vision Technologies GmbH, Stadtroda, Germany) camera with 1536 × 2048 resolution; the camera was attached to a pole. The images were captured in daylight; half of the images were acquired under direct sunlight, and half of the images were acquired in the shade.

Figure 2. Apple (**top**) and grape (**bottom**) RGB image (**left**) and ground truth (**right**) examples

3.1.2. Grapes

The images were acquired in a commercial vineyard growing green grapes of the "superior" variety. An RGB camera (Microsoft NX-6000) with 600 × 800 resolution was manually driven, at mid-day, along a commercial vineyard in Lachish, Israel, during the summer season of 2011, one month before harvest time. The images were captured from five different growing rows. A set of 129 images was acquired and included 1078 grape clusters.

3.1.3. Peppers

The dataset included 30 images of 73 yellow peppers acquired in a commercial greenhouse in Ijsselmuiden, Netherlands, using a 6 degree of freedom manipulator (Fanuc LR Mate 200iD/7L), equipped with an iDS Ui-5250RE RGB camera with 600 × 800 resolution. Two different datasets were created by marking the images twice. The first dataset included only peppers with high visibility (denoted as "high visibility peppers"; this was done for 10 images of 25 yellow peppers). In the second dataset, all peppers were marked including peppers in dark areas that were less visible in the image (denoted as "including low-visibility peppers", done for all 30 images) (Figure 3).

Figure 3. Peppers tagging example. **Top**: RGB image (**left**) and ground truth (**right**) example in high visibility. **Bottom**: RGB image (**left**) and labeled image (**Right**). "High-visibility peppers" marked in red and "low-visibility peppers" marked in blue.

3.1.4. Performance Measures

Metrics included the TPR (True Positive Rate, also noted as hit), FPR (False Positive Rate, also noted as false alarms), and the F-score (the harmonic mean of precision and recall [51]. The TPR metric (Equation (2)) states the number of correctly-detected objects relative to the actual number of objects, while the FPR metric calculates the number of false objects detected relative to the actual number of objects (Equation (3)). The F-score (Equation (4)) balances between TPR and FPR equally.

$$TPR = \frac{N_{TDF}}{N_F} \tag{2}$$

where N_{TDF} is the number of pixels detected correctly as part of the fruit and N_F is the actual number of pixels that represent the fruit.

$$FPR = \frac{N_{FDF}}{N_B} \tag{3}$$

where N_{FDF} is the number of pixels falsely classified as fruit and N_B is the number of pixels that represent the background.

$$F(TPR, FPR) = \frac{2 * (TPR * (1 - FPR))}{TPR + (1 - FPR)} \tag{4}$$

3.2. Analyses

The following analyses were conducted for the three databases, apples, grapes, and peppers, using 70% of the data for training and 30% for testing [52]. This rate was chosen to make the algorithm performances more rigid since the number of images in each DB was relativity small. In addition, to ensure robust results, each split into training and testing was randomly performed 5 times, and all detection results reported are an average of the 5 test sets.

- Tuning parameters: Parameters were computed for each database with procedures defined in Section 4.3 and compared to previous predefined parameters

- Color spaces' analyses: Algorithm performances were tested on all databases for four different color spaces: RGB, HSI, LAB, and NDI.

- Sensitivity analysis: Sensitivity analyses were conducted for all the databases and included:
    1. Noise: Noise was created by adding to each pixel in the RGB image a random number from the mean normal distribution for noise values up to 30%. The artificial noise represents the algorithms' robustness toward other cameras with more noise, or when capturing images with different camera settings. Noise values of 5%, 10%, 20%, and 30%, were evaluated.
    2. Thresholds learned in train process: Thresholds were changed by $\pm$5%, $\pm$10%, and $\pm$15% according to the threshold in each region.
    3. Stop condition: The selected STD value was changed by 5% and 10% to test the robustness of the algorithm to these parameters.
    4. Training vs. testing: The algorithm performances were evaluated while using different percentages of DB images for the training and testing processes.

- Morphological operation (erosion and dilation) contribution: Performances were tested for imaging with and without a morphological operations process.

4. Algorithm

4.1. Algorithm Flow

The overall flow of the algorithm is outlined in Figure 4, and it is as follows. The RGB images were the inputs for the training process. Some areas in the images contained more illumination than others, depending on the position of the light source and shading caused by leaves, branches, and the covering net when it existed. To overcome this issue, the algorithm divided each image into multiple sub-images, with approximately homogeneous illumination conditions (Figure 5). These sub-images were categorized into three illumination conditions: low, medium, and high. The illumination level was obtained by calculating the average on the grayscale sub-images. The grayscale image showed values between zero (completely dark) and 255 (completely white). In the previous algorithm [19], the sub-images were categorized into groups using levels selected empirically as 10, 70, and 130, corresponding to low-, medium-, and high-level images based on manual image analyses. The high value was set as 130 in order to filter overexposed areas in the images. In the current algorithm, a tuning parameter process (detailed in Section 4.3) was developed to determine these three values.

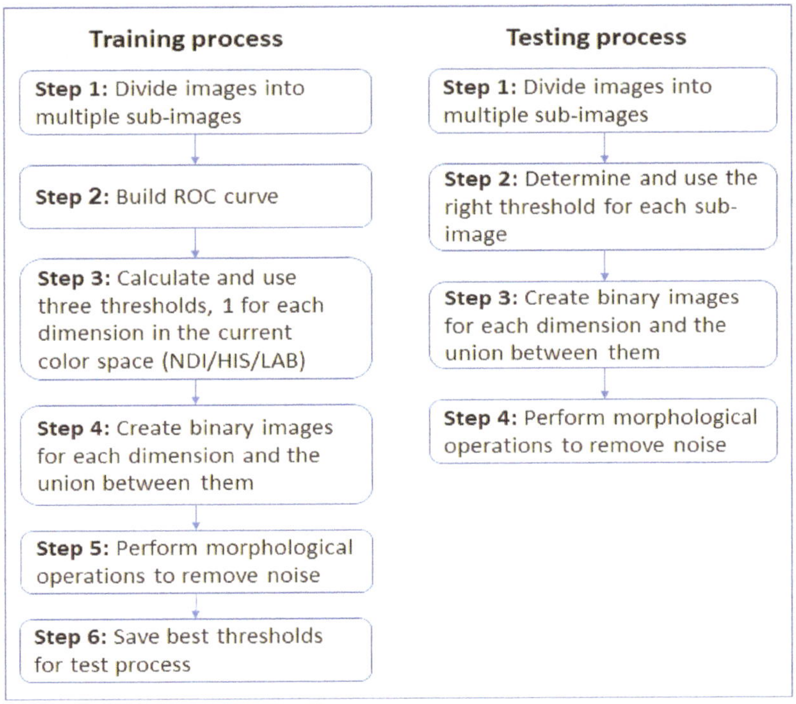

Figure 4. Algorithm flowchart.

Figure 5. Image split into sub-images: visualization.

The algorithm then created a 3D color space image (transformed the RGB image to NDI, HSI, and LAB space or used the RGB space directly). For each color dimension, a binary image (mask) was created, where each pixel that represents the fruit received a value of one and all other pixels received a value of zero. Finally, the algorithm created an ROC (Receiver Operator characteristics Curve) representing TPR as a function of FPR [53] including all nine thresholds learned from the training process. Figure 6 presents an example of nine ROC curves computed for three sub-images with different Light levels (L1, L2, L3) in the NDI color space. In this example, the sub-image with Light Level 2 (L2) in the first NDI dimension obtained the best performances (high TPR and low FPR).

In the test process, the algorithm received RGB images from the camera in real time, transformed the representation to the relevant color space (HSI/LAB/NDI), and created a binary image by applying the thresholds as follows: three thresholds, one for each dimension, were calculated from the nine thresholds learned by linear interpolation between two of the three illumination regions (low, medium, and high) selected as closest to the calculated illumination level for the specific sub-image from the grayscale image and using Equation (5).

$$T = \frac{T(LL[i]) * (currentLL - LL[i])(LL[i+1] - currentLL)}{LL[i+1] - LL[i]} \tag{5}$$

where LL is an array of the light level values for each group: LL = [low,medium,high]), and i is the light level index representing the group that is the closest to the current image light level from below.

For example, if the current light level is 40 and the thresholds in the train process for the low, medium, and high light levels were 10, 70, and 130, the threshold would be calculated in the following way (Equation (6)):

$$T = \frac{T(10)(40-10) + T(70)(70-40)}{70-10} \tag{6}$$

The end of the process results in binary images where white areas (pixels with a value of one) in the binary image represent the fruits and the black areas (pixels with a value of zero) represent the background (see Figures 2 and 3). In total, the algorithm created 7 binary images, 3 images corresponding to the three color space dimensions and 4 binary images corresponding to the intersections between the first 3 binary images. For example, the intersection between Binary Images 1 and 2 resulted in a binary image $1 \cap 2$ that contained white pixels only, where the same pixels were white in both Images 1 and 2 (Figure 7).

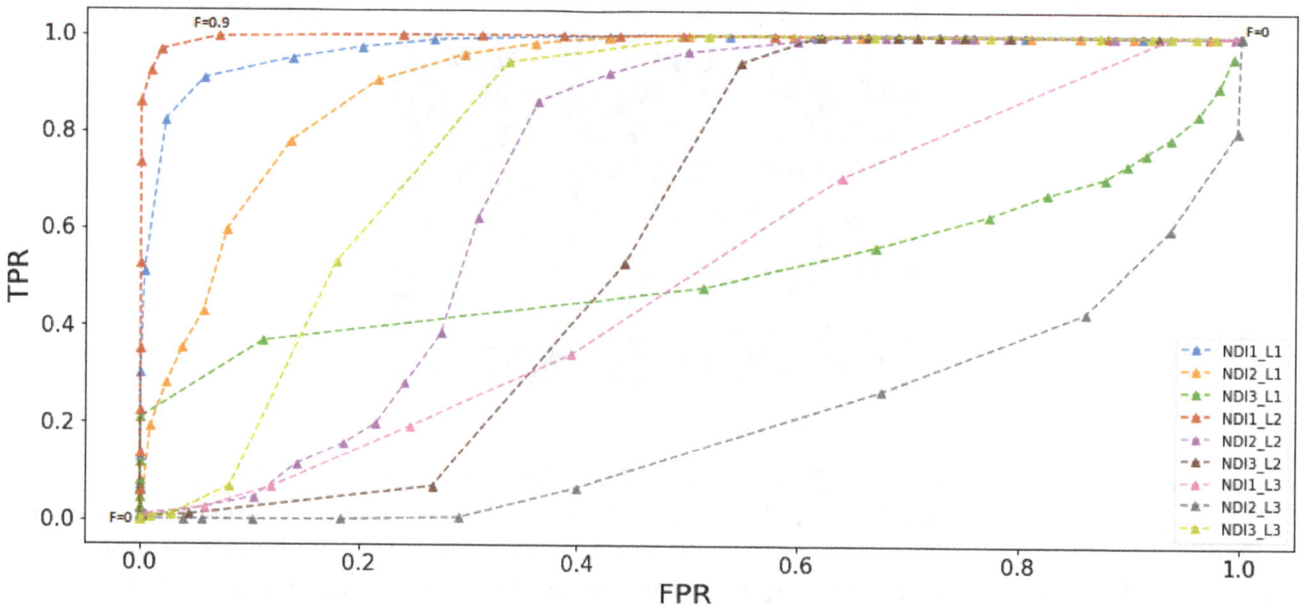

Figure 6. Nine ROC curves: 3 dimensions × 3 light levels $NDI_i - L_j$; i represents the color space dimension; j represents the illumination level.

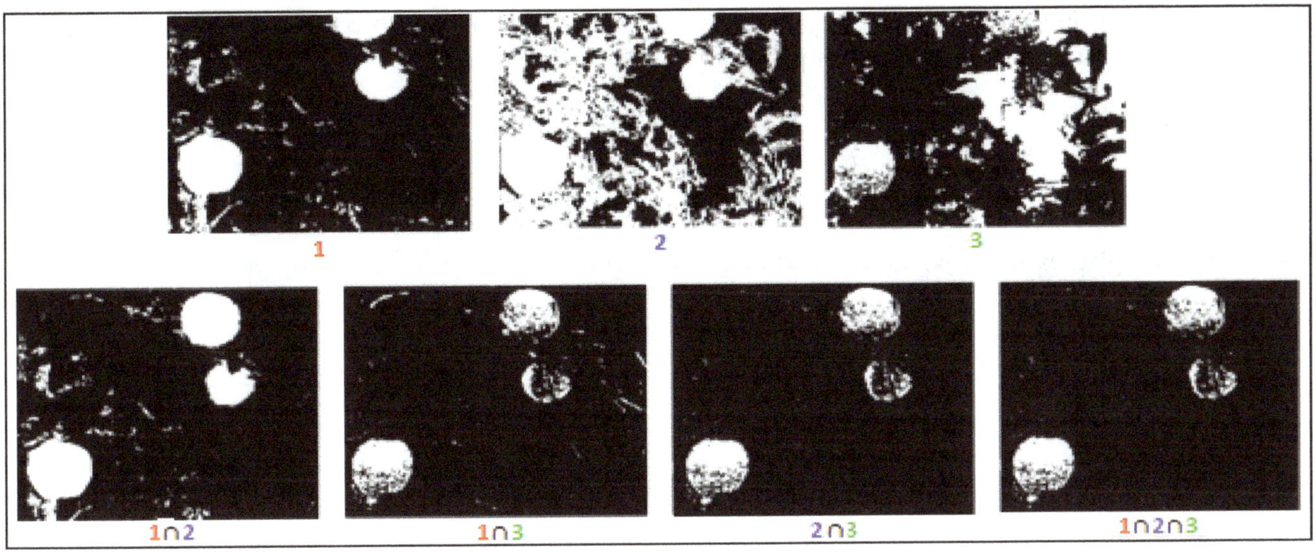

Figure 7. Use of dimension intersection to increase performance.

4.2. Morphological Operations: Erosion and Dilation

The algorithm result is a binary image with major fruit detected and small clusters of pixels that were wrongly classified as fruits (e.g., Figure 8). In addition, some fruits were split between several clusters (e.g., Figure 8). To overcome these problems, several morphological operations were performed based on previous research that indicated their contribution [19]: erosion followed by dilation with a neighborhood of 11 × 11-pixel squares. The square function was used since there was no pre-defined knowledge about the expected fruit orientation. To connect close clusters, the closing morphological operation was then applied by dilation followed by erosion implemented with a 5 × 5-pixel square neighborhood.

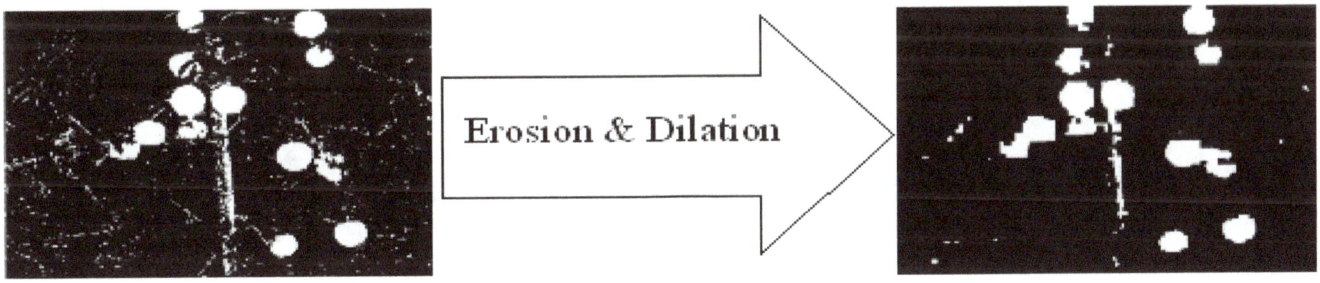

Figure 8. Morphological operation.

4.3. Parameter Tuning

The algorithm used several parameters that influence the algorithm performances: T1, T2, STD, classification rule Direction (D1/D2). The following parameter tuning procedure (Figure 9) was developed and should be performed when exploring images from a new database or when exploring a new color space or new operating conditions (cameras, illumination). The parameters, as detailed below, are: light level thresholds, stop splitting condition, classification rule direction.

4.3.1. Light Level Thresholds (T1, T2)

The algorithm split the images into sub-images set to 1% pixels of the entire image. Then, the algorithm computed the light level of each sub-image by calculating the average pixels values of the grayscale sub-image. Finally, the algorithm grouped the sub-images into three light level categories (see Figure 10) using two thresholds as presented in Equation (7).

$$i = \begin{cases} Low & 0 < x < T1 \\ Medium & T1 < x < T2 \\ High & x > T2 \end{cases} \tag{7}$$

where i is the light level index, as detailed above in Equation (5).

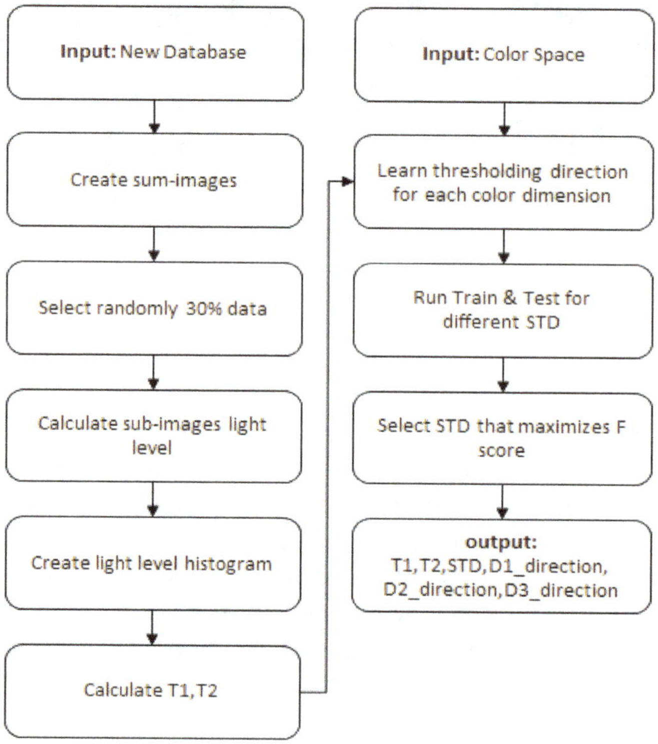

Figure 9. Parameter tuning process.

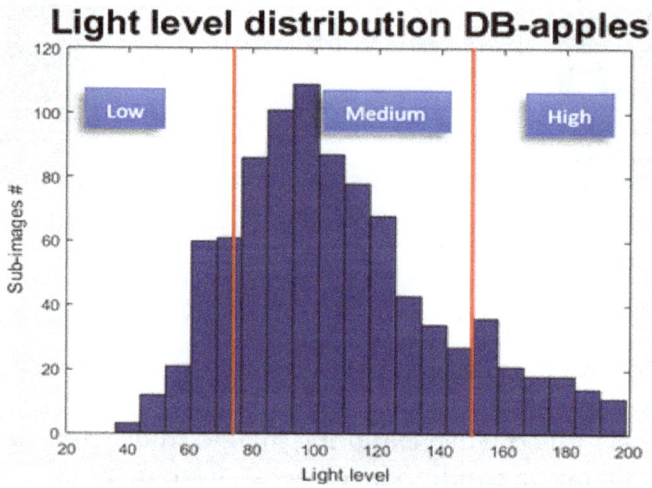

Figure 10. Sub-images level of light distribution.

Research was done to identify the PDF function of the data distributions of each database through a χ^2 goodness of fit test. However, since these tests did not reveal significant results [54], the thresholds were selected as follows: T1 and T2 were chosen so that 15% of the data would be categorized as low, 15% as high, and 70% as medium.

Note that as described in the algorithm flow, the algorithm used a third threshold. Sub-images above that threshold were ignored in the training process since they were almost completely white.

4.3.2. Stop Splitting Condition (STD)

The algorithm split an image into sub-images until the sub-image achieved a predefined Standard Deviation (STD) value. This approach assumes that a larger sub-image contains a higher STD value. To test this assumption, STD was calculated for different sizes of sub-images for the different databases. The stop condition value (STD minimum value) was determined by maximizing the F-score (Equation (4)).

4.3.3. Classification Rule Direction (D1, D2, D3)

As detailed in the Introduction section, as part of the thresholding process, an intensity value was determined to differentiate between objects and background pixels. In order to execute the thresholding process, the algorithm must receive as input the classification rule direction (the algorithm must automatically determine if the intensity of the background distribution is higher or lower than the intensity of the object distribution in each color dimension).

This information was learned as part of the tuning procedure. A simple heuristic rule was used as follows based on the assumption that the images contained more background pixels than objects: (1) execute $image > Threshold$; (2) if the pixels categorized as background represent less than 70% of the image, reverse the thresholding direction $images < Threshold$.

5. Results and Discussion

5.1. Sub-Image Size vs. STD Value

For images with a size of 300 × 300 or lower, splitting an image into small sub-images (small S) decreases the average STD of the sub-images (in all three databases; Figure 11). Although a direct decrease in very large images is not noted, we still can conclude that splitting a large image to 300 × 300 or lower will decrease the average STD.

Figure 11. Sub image size vs. average STD.

5.2. *Tuning Process*

This section presents the tuning procedure results, including thresholds derived to categorize the sub-images into light level groups, as well as the recursive stop condition that achieved the best result for each database.

5.2.1. Light Level Distribution

The light level distribution was computed for each database (Figure 12) along with T1 and T2 (Table 2). The variation in the light distributions between the different databases are described in Table 3. The variance of light in the grape databases was significantly higher than in both the apple and the pepper databases, the pepper database being significantly darker and highly skewed. Therefore, for each database, the selected T1 and T2 were significantly different, implying the importance of the tuning procedure.

Table 2. T1 and T2 values determined for each database.

Measure\DB	Apples	Grapes	Peppers
T1	84	49	18
T2	140	130	47

Table 3. Descriptive statistics of the different light distributions.

DB\Measure	Mean	Std	Skewness	Kurtosis	Median
Apples	118.46	28.04	0.4	−0.17	116.31
Grapes	88	37.9	0.68	−0.13	81.06
Peppers	32.09	18.92	3.16	15.36	26.93

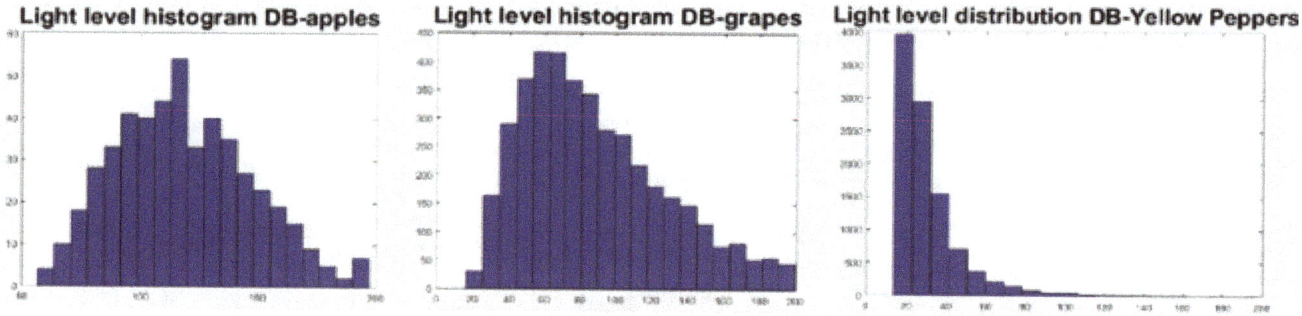

Figure 12. Light level distribution computed for each database.

5.2.2. Stop Splitting Condition

Using a low STD value as a stop condition increased the performance (Figure 13). This happens since smaller sub-images contain less illumination differences. However, small STD values can create also too small sub-images, which may not contain fruit and background pixels in the same frame. In these cases, the algorithm cannot learn a threshold that could differ between them. Additionally, results revealed that when using high STD values, the performances remained constant. This happens since beyond a certain value, the algorithm did not split the image even once.

As part of the parameter tuning process, the STD value was selected by testing the performances of a range of STD [0, 100]. For each STD value, the algorithm ran five iterations where it randomly selected P% of the images, from the selected images; it used 70% for training and 30% testing. The final selected STD values are presented in Table 4 for each database and color space (using P = 30% and 50%).

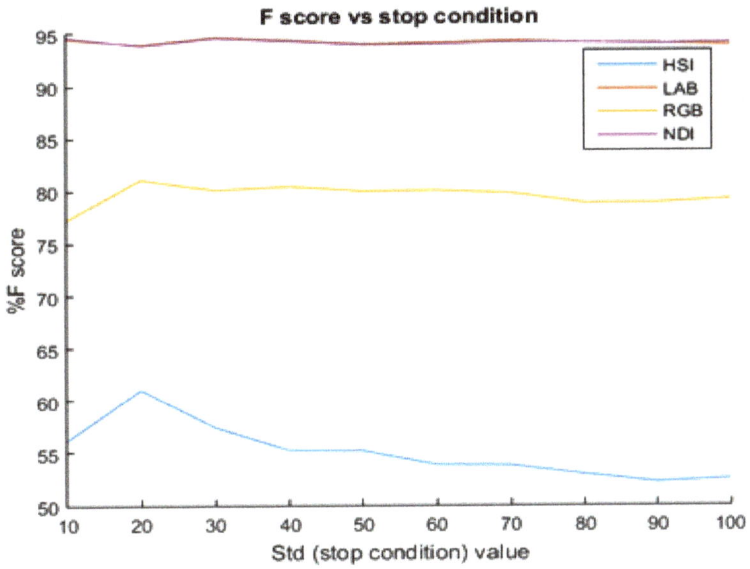

Figure 13. F-score vs. increasing STD value as the stop condition for the recursive function on the apple DB.

5.2.3. Classification Rule Direction

As shown in Table 5, the direction of the classification rule in the thresholding process can be different for each color dimension; therefore, this must be learned as part of the tuning procedure.

Table 4. STD value chosen for each database and color space. D, Direction.

DB		Apples				Grapes				Peppers			
Color space		HSI	LAB	NDI	RGB	HSI	LAB	NDI	RGB	HSI	LAB	NDI	RGB
STD (P = 30%)		20	30	10	20	10	20	60	20	100	10	10	10
STD (P = 50%)		20	10	10	30	20	20	70	20	100	20	10	10
Classification rule direction	D1	>	<	>	>	<	>	>	>	<	>	>	>
	D2	>	>	>	<	>	<	>	>	>	>	>	>
	D3	>	>	>	<	<	>	<	<	>	>	<	<

5.3. Color Space Analyses

In this section, algorithm performance results are presented for each color space followed by a table representing the best color space performances including the performances for all color space dimensions' combinations.

5.3.1. Apples

Results (Figure 14) revealed that NDI and LAB color spaces resulted in similar best performances. In Table 5, the preferences for each dimension in the NDI color space and the performances when using the intersection between them is shown. The NDI first dimension (see Equation (1)) represents the difference between the red and green colors in the image. The objects in this database were red apples, and most of the background was green leaves; therefore, as expected, the first NDI obtained the best F of 93.17%. In the LAB color space, results (Table 5) revealed that the second dimension (A) yielded the best F-score of 93.19.

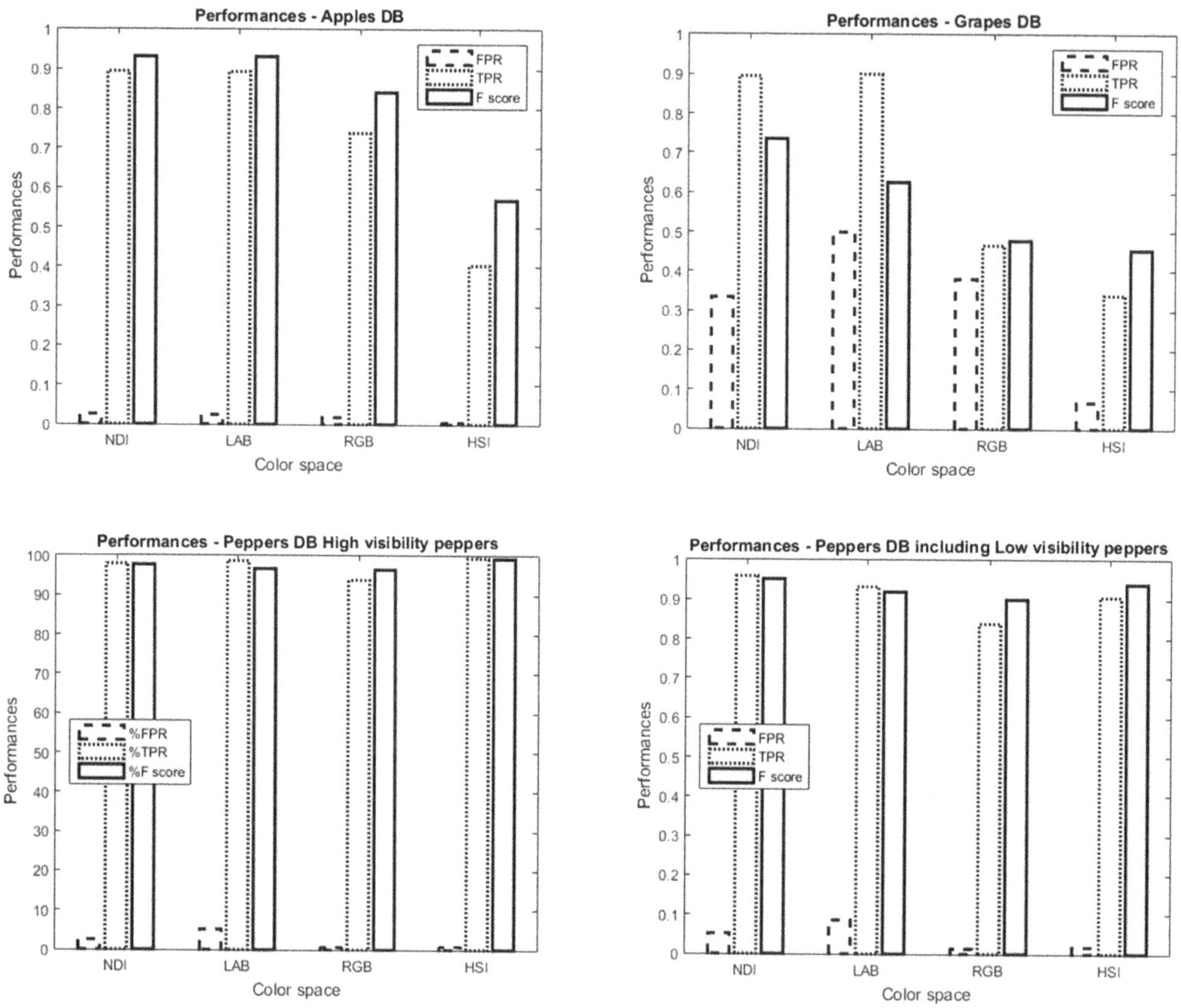

Figure 14. Color space performances for (**top**) apples (**left**), grapes (**right**), peppers (**bottom**) with high visibility (**left**), and peppers with low visibility (**right**).

5.3.2. Grapes

The NDI color space obtained the best result for grapes (Figure 14) with an F-score of 73.52%. The second-best color space was the LAB with an F-score of 62.54%. The best NDI results were obtained using the second dimension (Table 5).

5.3.3. Peppers

High visibility: Figure 14 indicates that the HSI color space obtained the best results with relatively low FPR (0.81%) and very high TPR (99.43%), resulting in a high F-score (99.31%). The second-best color space was NDI with FPR = 2.48% and TPR = 97.96% (F = 97.72%). The best HSI result was obtained using the combination of the first and the second dimensions (Table 5).

Including low visibility: Figure 14 indicates that the NDI color space obtained the best results with relatively low FPR (5.24%) and very high TPR (95.93) resulting in a high F-score (95.19%). Although for the "high visibility" peppers HSI obtained the best performances, when trying to detect peppers in dark areas that were less visible, NDI showed better results. The best NDI result was obtained using the intersection between the first and the second dimensions (Table 5).

Table 5. Performances of each color space for each dimension and intersection for all datasets.

DB	Color Space	Measure	Dimension 1	2	3	1∩2	1∩3	2∩3	1∩2∩3
Apples	NDI	% FPR	**2.59**	40.91	31.38	1.64	1.32	2.48	0.48
		% TPR	**89.45**	83.53	68.39	78.52	64.82	54.65	54.1
		% F	**93.17**	67.85	67.8	86.75	77.6	69.67	69.69
	LAB	% FPR	33.58	**2.45**	77.08	1.78	28.55	1.55	1.07
		% TPR	61.26	**89.34**	85.26	56.59	52.95	76.27	48.8
		% F	56.79	**93.19**	35.85	69.02	54.61	85.37	62.58
Grapes	NDI	% FPR	35.86	33.35	52.9	4.86	5.5	**32.35**	4.09
		% TPR	44.52	89.48	89.99	38.53	37.27	**87.5**	36.7
		% F	47.19	73.52	58.05	50.12	48.93	**73.2**	48.65
Peppers High Visibility	HSI	% FPR	18.27	1.52	2.16	**0.81**	0.64	0.17	0.14
		% TPR	99.96	99.43	86.42	**99.43**	86.41	86.23	86.23
		% F	89.77	98.95	91.41	**99.31**	92.12	92.24	92.25
Peppers Inc. Low Visibility	NDI	% FPR	66.57	**5.24**	9.23	1.42	1.24	4.57	0.99
		% TPR	85.61	**95.93**	92.49	82.2	78.64	92.33	78.61
		% F	46.96	**95.19**	91.24	88.91	86.59	93.51	86.67

5.4. Sensitivity Analysis

5.4.1. Noise

Analysis showed that the algorithm was robust to noise in the image up to 15% in the apple and pepper databases (Figure 15). The grape images were more sensitive to noise, and performance dropped when noise values of 5% were added. Although better F-score values were obtained for NDI and HSI for grapes and peppers, we can see that the LAB color space yielded more robust performance when adding noise to the images.

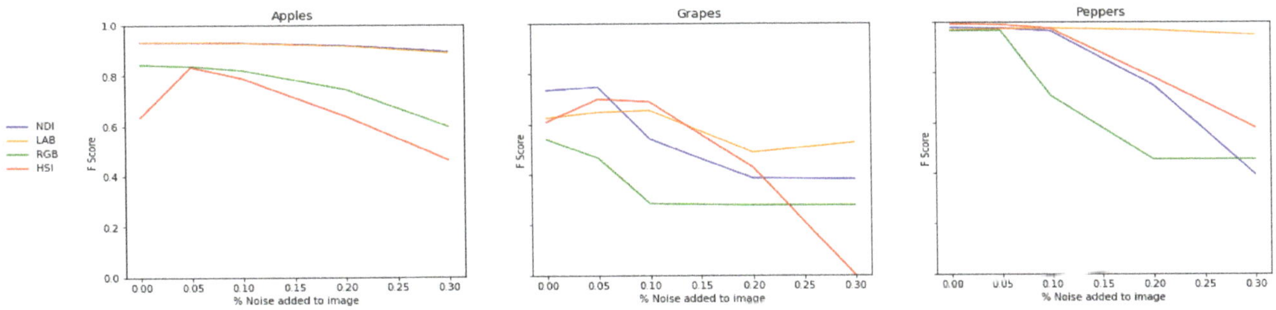

Figure 15. Sensitivity analysis: adding noise to the image.

5.4.2. Thresholds Learned in the Training Process

As expected, TPR decreased when the threshold values changed. The algorithm was relatively robust to the change in the thresholds for apples and peppers. Performance in the grape images was more sensitive to threshold changes, and yielded a significant decrease in TPR when increasing the threshold value (Table 6).

Table 6. Threshold values changed by ±5%, ±10%, and ±15% according to the threshold in each region.

		Changes in Threshold						
DB	**Measure**	**−15%**	**−10%**	**−5%**	**0%**	**5%**	**10%**	**15%**
Apples	**% FPR**	3.58	3.43	3.3	2.59	3.06	2.93	2.81
	% TPR	91.47	91.28	91.07	89.45	90.75	90.57	90.44
Grapes	**% FPR**	21.59	18.4	15.53	33.35	11.00	9.23	7.72
	% TPR	78.02	72.63	66.42	89.48	50.99	43.63	36.24
Peppers	**% FPR**	0.98	0.91	0.86	0.81	0.78	0.7	0.65
	% TPR	99.25	99.22	99.2	99.43	99.12	99.07	99.04

5.4.3. Stop Condition

The algorithm showed more robustness to apple and pepper images than grapes (Figure 16).

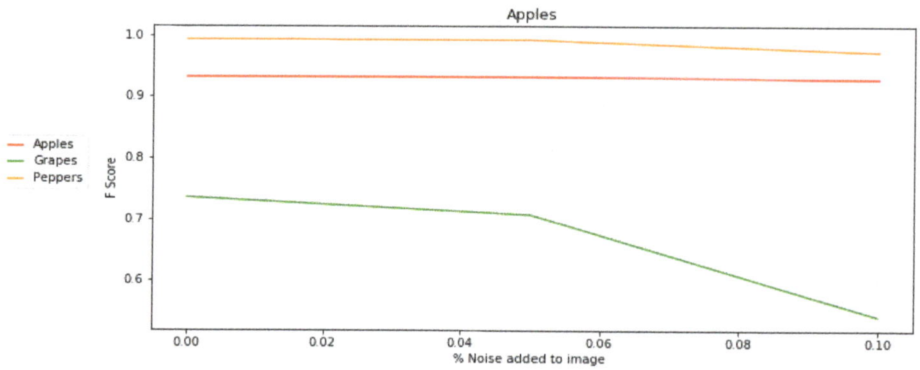

Figure 16. Sensitivity analysis: adding noise to STD stop condition.

5.4.4. Training/Testing

The expectation was that more training images would lead to better performance until over fitting was accommodated. There was a clear increase in TPR; however, FPR increased as well at 80% and 90% training (Table 7).

The tuning process resulted (Table 8) in increased performances for both the grape and pepper databases with a 6.22% and 0.84% increase, respectively. The results for the apple database were similar with only a 0.1% increase, as expected (since this was similar to the database from which the previous parameters were derived).

Table 7. Performances vs. different % images database as the training set.

		Training %: For a Dataset Size of 129 Images								
DB	**Measure**	**10**	**20**	**30**	**40**	**50**	**60**	**70**	**80**	**90**
Grapes	% FPR	32.81	37.08	28.54	36.16	31.83	29.1	29.63	40.51	40.8
	% TPR	88.79	89.62	87.01	88.44	87.58	82.55	87.14	94.53	95.85
	% F	73.35	70.2	75.19	69.41	72.49	70.24	73.92	72.55	72.54

5.5. Morphological Operations

The morphological operations process increased the F-score by 2.85%, 8.59%, and 2.71% for the apple, grape, and pepper databases respectively, (Figure 17).

Table 8. Parameter tuning contribution to algorithm performances.

DB	Measure	Performances Using Previous Params	Performances Using Tuning Process
Apples	% FPR	2.53	2.59
	% TPR	89.23	89.45
	% F	93.08	93.17
Grapes	% FPR	18.63	33.35
	% TPR	63.7	89.48
	% F	67.3	73.52
Peppers	% FPR	1	0.81
	% TPR	97.97	99.43
	% F	98.47	99.31

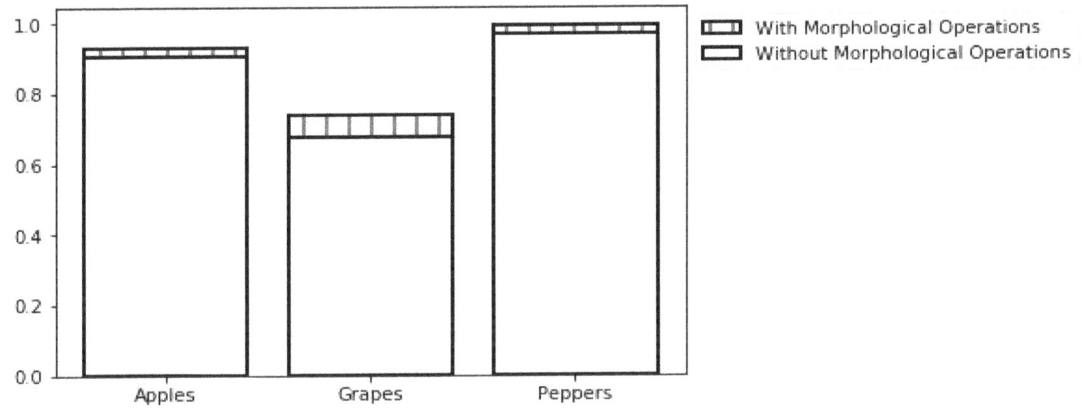

Figure 17. Sensitivity analysis: adding noise to the STD stop condition.

6. Conclusions and Future Work

The algorithm successfully detected apples and peppers (Table 1) in variable lighting conditions resulting in an F-score of 93.17% and 99.31%, respectively, which is one of the best detection rates achieved to date in fruit detection to the best of our knowledge. The average F-score across all datasets was 88.8 (Table 1). Previous research achieved the lowest F-score (65) with the method of [11] for red and green pepper plants, while oranges obtained the highest F-score (96.4) with that of [24]. Previous reported results (Table 1) revealed a 91.5 and 92.6 F-score for peppers in [19,34], respectively, versus our method, which resulted in an F-score of 99.43. For apple images, our method obtained similar F-score performances as in previous work (~93), even though the dataset was much smaller (64 vs. 9 images).

The high F-score was mostly due to low FPR values (except for grapes). In addition, our method achieved high performances using a relatively small dataset.

The algorithm resulted in less impressive results in the grape database of 73.52% due to the difficulties in differentiating between green fruits and green background (leaves). In this case, additional features (e.g., morphological operations fitted for grapes; see [28]) should be used to increase performance. However, this requires the development of specially-tailored features. It is important to note that these results cannot be compared to the weed detection results presented in Table 1, since the background of the green obkects was the ground on which it grew and not the green leaves. Different color spaces yielded the best results for each fruit variety, implying that the color space must be analyzed and fitted to the specific fruit. The LAB color space was more robust to noise in images and hence should be used when images are of low quality. The algorithm was robust to changes in the threshold learned by the training process and to noise effects in images. Morphological operations such as erosion and dilation can improve performance in agriculture images and hence should be utilized. The tuning process developed in this paper enabled the previous algorithm [30] to adapt automatically to changing conditions/objectives (i.e., to detect other fruit with different colors and other outdoor conditions) and, hence, should be used for improved target detection in highly-variable illumination conditions. Finally, this work has presented the feasibility of color-based algorithms solving the challenges that advanced machine learning algorithms face such as small training sets (small number of images and/or small number of fruits per image). This work has shown that for challenging color conditions (e.g., green on green for grapes), additional features should be considered for improved fruit detection.

Author Contributions: E.Z.: Formal algorithm development and analysis, Investigation, Methodology, Software, Writing—Original draft, Writing—review & editing; P.K.: Formal analysis, Methodology, Supervision, Validation; Writing—review & editing; Y.E.: Methodology, Supervision, Validation, Writing—original draft; Writing—review & editing.

References

1. Kalantar, A.; Dashuta, A.; Edan, Y.; Gur, A.; Klapp, I. Estimating Melon Yield for Breeding Processes by Machine-Vision Processing of UAV Images. In Proceedings of the Precision Agriculture Conference, Montpellier, France, 8–11 July 2019.
2. Bargoti, S.; Underwood, J.P. Image segmentation for fruit detection and yield estimation in apple orchards. *J. Field Robot.* **2017**, *34*, 1039–1060. [CrossRef]
3. Stein, M.; Bargoti, S.; Underwood, J. Image based mango fruit detection, localisation and yield estimation using multiple view geometry. *Sensors* **2016**, *16*, 1915. [CrossRef] [PubMed]
4. Dorj, U.O.; Lee, M.; Yun, S.S. An yield estimation in citrus orchards via fruit detection and counting using image processing. *Comput. Electron. Agric.* **2017**, *140*, 103–112.
5. Wang, Q.; Nuske, S.; Bergerman, M.; Singh, S. Automated crop yield estimation for apple orchards. In *Experimental Robotics*; Springer: Berlin, Germany, 2013; pp. 745–758.
6. Nuske, S.; Achar, S.; Bates, T.; Narasimhan, S.; Singh, S. Yield estimation in vineyards by visual grape detection. In Proceedings of the IEEE/RSJ International Conference on Intelligent Robots and Systems, San Francisco, CA, USA, 25–30 September 2011; pp. 2352–2358.
7. Zaman, Q.; Schumann, A.; Percival, D.; Gordon, R. Estimation of wild blueberry fruit yield using digital color photography. *Trans. ASABE* **2008**, *51*, 1539–1544. [CrossRef]
8. Oppenheim, D. Object recognition for agricultural applications using deep convolutional neural networks. Master's Thesis, Ben-Gurion University of the Negev, Beer-Sheva, Israel, 2018.
9. Mack, J.; Lenz, C.; Teutrine, J.; Steinhage, V. High-precision 3D detection and reconstruction of grapes from laser range data for efficient phenotyping based on supervised learning. *Comput. Electron. Agric.* **2017**, *135*, 300–311. [CrossRef]
10. Pound, M.P.; Atkinson, J.A.; Wells, D.M.; Pridmore, T.P.; French, A.P. Deep learning for multi-task plant phenotyping. In Proceedings of the IEEE International Conference on Computer Vision, Venice, Italy, 22–29 October 2017; pp. 2055–2063.
11. Song, Y.; Glasbey, C.; Horgan, G.; Polder, G.; Dieleman, J.; Van der Heijden, G. Automatic fruit recognition and counting from multiple images. *Biosyst. Eng.* **2014**, *118*, 203–215. [CrossRef]
12. Oberti, R.; Marchi, M.; Tirelli, P.; Calcante, A.; Iriti, M.; Tona, E.; Hočevar, M.; Baur, J.; Pfaff, J.; Schütz, C.; et al. Selective spraying of grapevines for disease control using a modular agricultural robot. *Biosyst. Eng.* **2016**, *146*, 203–215. [CrossRef]
13. Payne, A.B.; Walsh, K.B.; Subedi, P.; Jarvis, D. Estimation of mango crop yield using image analysis-segmentation method. *Comput. Electron. Agric.* **2013**, *91*, 57–64. [CrossRef]
14. Wang, Z.; Verma, B.; Walsh, K.B.; Subedi, P.; Koirala, A. Automated mango flowering assessment via refinement segmentation. In Proceedings of the International Conference on Image and Vision Computing New Zealand (IVCNZ), Palmerston North, New Zealand, 21–22 November 2016; pp. 1–6.

15. Wouters, N.; De Ketelaere, B.; Deckers, T.; De Baerdemaeker, J.; Saeys, W. Multispectral detection of floral buds for automated thinning of pear. *Comput. Electron. Agric.* **2015**, *113*, 93–103. [CrossRef]
16. Bac, C.W.; van Henten, E.J.; Hemming, J.; Edan, Y. Harvesting robots for high-value crops: State-of-the-art review and challenges ahead. *J. Field Robot.* **2014**, *31*, 888–911. [CrossRef]
17. Kapach, K.; Barnea, E.; Mairon, R.; Edan, Y.; Ben-Shahar, O. Computer vision for fruit harvesting robots–state of the art and challenges ahead. *Int. J. Comput. Vis. Robot.* **2012**, *3*, 4–34. [CrossRef]
18. Gongal, A.; Amatya, S.; Karkee, M.; Zhang, Q.; Lewis, K. Sensors and systems for fruit detection and localization: A review. *Comput. Electron. Agric.* **2015**, *116*, 8–19. [CrossRef]
19. Vitzrabin, E.; Edan, Y. Adaptive thresholding with fusion using a RGBD sensor for red sweet-pepper detection. *Biosyst. Eng.* **2016**, *146*, 45–56. [CrossRef]
20. Arad, B.; Kurtser, P.; Barnea, E.; Harel, B.; Edan, Y.; Ben-Shahar, O. Controlled Lighting and Illumination-Independent Target Detection for Real-Time Cost-Efficient Applications. The Case Study of Sweet Pepper Robotic Harvesting. *Sensors* **2019**, *19*, 1390. [CrossRef] [PubMed]
21. Barth, R.; Hemming, J.; van Henten, E. Design of an eye-in-hand sensing and servo control framework for harvesting robotics in dense vegetation. *Biosyst. Eng.* **2016**, *146*, 71–84. [CrossRef]
22. Arivazhagan, S.; Shebiah, R.N.; Nidhyanandhan, S.S.; Ganesan, L. Fruit recognition using color and texture features. *J. Emerg. Trends Comput. Inf. Sci.* **2010**, *1*, 90–94.
23. Sa, I.; Ge, Z.; Dayoub, F.; Upcroft, B.; Perez, T.; McCool, C. Deepfruits: A fruit detection system using deep neural networks. *Sensors* **2016**, *16*, 1222. [CrossRef] [PubMed]
24. Chen, S.W.; Shivakumar, S.S.; Dcunha, S.; Das, J.; Okon, E.; Qu, C.; Taylor, C.J.; Kumar, V. Counting apples and oranges with deep learning: A data-driven approach. *IEEE Robot. Autom. Lett.* **2017**, *2*, 781–788. [CrossRef]
25. Kamilaris, A.; Prenafeta-Boldú, F.X. Deep learning in agriculture: A survey. *Comput. Electron. Agric.* **2018**, *147*, 70–90. [CrossRef]
26. McCool, C.; Perez, T.; Upcroft, B. Mixtures of lightweight deep convolutional neural networks: Applied to agricultural robotics. *IEEE Robot. Autom. Lett.* **2017**, *2*, 1344–1351. [CrossRef]
27. Milioto, A.; Lottes, P.; Stachniss, C. Real-time blob-wise sugar beets vs weeds classification for monitoring fields using convolutional neural networks. *ISPRS Ann. Photogramm. Remote Sens. Spat. Inf. Sci.* **2017**, *4*, 41. [CrossRef]
28. Berenstein, R.; Shahar, O.B.; Shapiro, A.; Edan, Y. Grape clusters and foliage detection algorithms for autonomous selective vineyard sprayer. *Intell. Serv. Robot.* **2010**, *3*, 233–243. [CrossRef]
29. Van Henten, E.; Van Tuijl, B.V.; Hemming, J.; Kornet, J.; Bontsema, J.; Van Os, E. Field test of an autonomous cucumber picking robot. *Biosyst. Eng.* **2003**, *86*, 305–313. [CrossRef]
30. Zemmour, E.; Kurtser, P.; Edan, Y. Dynamic thresholding algorithm for robotic apple detection. In Proceedings of the IEEE International Conference on the Autonomous Robot Systems and Competitions (ICARSC), Coimbra, Portugal, 26–28 April 2017; pp. 240–246.
31. Wang, J.; He, J.; Han, Y.; Ouyang, C.; Li, D. An adaptive thresholding algorithm of field leaf image. *Comput. Electron. Agric.* **2013**, *96*, 23–39. [CrossRef]
32. Jiang, J.A.; Chang, H.Y.; Wu, K.H.; Ouyang, C.S.; Yang, M.M.; Yang, E.C.; Chen, T.W.; Lin, T.T. An adaptive image segmentation algorithm for X-ray quarantine inspection of selected fruits. *Comput. Electron. Agric.* **2008**, *60*, 190–200. [CrossRef]
33. Arroyo, J.; Guijarro, M.; Pajares, G. An instance-based learning approach for thresholding in crop images under different outdoor conditions. *Comput. Electron. Agric.* **2016**, *127*, 669–679. [CrossRef]
34. Ostovar, A.; Ringdahl, O.; Hellström, T. Adaptive Image Thresholding of Yellow Peppers for a Harvesting Robot. *Robotics* **2018**, *7*, 11. [CrossRef]
35. Zhang, Y.J. A survey on evaluation methods for image segmentation. *Pattern Recognit.* **1996**, *29*, 1335–1346. [CrossRef]
36. Shmmala, F.A.; Ashour, W. Color based image segmentation using different versions of k-means in two spaces. *Glob. Adv. Res. J. Eng. Technol. Innov.* **2013**, *1*, 30–41.
37. Zheng, L.; Zhang, J.; Wang, Q. Mean-shift-based color segmentation of images containing green vegetation. *Comput. Electron. Agric.* **2009**, *65*, 93–98. [CrossRef]
38. Al-Allaf, O.N. Review of face detection systems based artificial neural networks algorithms. *Int. J. Multimed. Its Appl.* **2014**, *6*, 1. [CrossRef]

39. Sakthivel, K.; Nallusamy, R.; Kavitha, C. Color Image Segmentation Using SVM Pixel Classification Image. *World Acad. Sci. Eng. Technol. Int. J. Comput. Electr. Autom. Control Inf. Eng.* **2015**, *8*, 1919–1925.
40. Kurtser, P.; Edan, Y. Statistical models for fruit detectability: Spatial and temporal analyses of sweet peppers. *Biosyst. Eng.* **2018**, *171*, 272–289. [CrossRef]
41. Barth, R.; IJsselmuiden, J.; Hemming, J.; Van Henten, E.J. Data synthesis methods for semantic segmentation in agriculture: A Capsicum annuum dataset. *Comput. Electron. Agric.* **2018**, *144*, 284–296. [CrossRef]
42. Park, J.; Lee, G.; Cho, W.; Toan, N.; Kim, S.; Park, S. Moving object detection based on clausius entropy. In Proceedings of the IEEE 10th International Conference on Computer and Information Technology (CIT), Bradford, West Yorkshire, UK, 29 June–1 July 2010; pp. 517–521.
43. Hannan, M.; Burks, T.; Bulanon, D. A real-time machine vision algorithm for robotic citrus harvesting. In Proceedings of the 2007 ASAE Annual Meeting, American Society of Agricultural and Biological Engineers, Minneapolis, MN, USA, 17–20 June 2007; p. 1.
44. Bulanon, D.M.; Burks, T.F.; Alchanatis, V. Fruit visibility analysis for robotic citrus harvesting. *Trans. ASABE* **2009**, *52*, 277–283. [CrossRef]
45. Gunatilaka, A.H.; Baertlein, B.A. Feature-level and decision-level fusion of noncoincidently sampled sensors for land mine detection. *IEEE Trans. Pattern Anal. Mach. Intell.* **2001**, *23*, 577–589. [CrossRef]
46. Kanungo, P.; Nanda, P.K.; Ghosh, A. Parallel genetic algorithm based adaptive thresholding for image segmentation under uneven lighting conditions. In Proceedings of the IEEE International Conference on Systems Man and Cybernetics (SMC), Istanbul, Turkey, 10–13 October 2010; pp. 1904–1911.
47. Hall, D.L.; McMullen, S.A. *Mathematical Techniques in Multisensor Data Fusion*; Artech House: Norwood, MA, USA, 2004.
48. Woebbecke, D.M.; Meyer, G.E.; Von Bargen, K.; Mortensen, D.A. Plant species identification, size, and enumeration using machine vision techniques on near-binary images. In *Optics in Agriculture and Forestry*; SPIE-International Society for Optics and Photonics: Bellingham, WA, USA, 1993; Volume 1836, pp. 208–220.
49. Shrestha, D.; Steward, B.; Bartlett, E. Segmentation of plant from background using neural network approach. In Proceedings of the Intelligent Engineering Systems through Artificial Neural Networks: Proceedings Artificial Neural Networks in Engineering (ANNIE) International Conference, St. Louis, MO, USA, 4–7 November 2001; Volume 11, pp. 903–908.
50. Deng, J.; Dong, W.; Socher, R.; Li, L.J.; Li, K.; Fei-Fei, L. Imagenet: A large-scale hierarchical image database. In Proceedings of the IEEE Computer Vision and Pattern Recognition, Miami Beach, FL, USA, 20–25 June 2009.
51. Goutte, C.; Gaussier, E. A probabilistic interpretation of precision, recall and F-score, with implication for evaluation. In *European Conference on Information Retrieval*; Springer: Berlin, Germany, 2005; pp. 345–359.
52. Guyon, I. *A Scaling Law for the Validation-Set Training-Set Size Ratio*; AT&T Bell Laboratories: Murray Hill, NJ, USA, 1997; pp. 1–11.
53. Siegel, M.; Wu, H. Objective evaluation of subjective decisions. In Proceedings of the IEEE International Workshop on Soft Computing Techniques in Instrumentation, Measurement and Related Applications, Provo, UT, USA, 17 May 2003.
54. Zemmour, E. Adaptive thresholding algorithm for robotic fruit detection. Master's Thesis, Ben-Gurion University of the Negev, Beer-Sheva, Israel, 2018.

13

Delineation of Agricultural Drainage Pipe Patterns using Ground Penetrating Radar Integrated with a Real-Time Kinematic Global Navigation Satellite System

Barry Allred [1,*], DeBonne Wishart [2], Luis Martinez [1], Harry Schomberg [3], Steven Mirsky [3], George Meyers [3], John Elliott [4] and Christine Charyton [4]

1 U.S. Dept. of Agriculture, Agricultural Res. Serv., Soil Drainage Res. Unit, Columbus, OH 43210, USA; Luis.Martinez@ars.usda.gov
2 C.J. International Center for Water Resources Management, Central State University, Wilberforce, OH 45384, USA; dwishart@centralstate.edu
3 U.S. Dept. of Agriculture, Agricultural Res. Serv., Beltsville Agric. Res. Cntr., Beltsville, MD 20705, USA; Harry.Schomberg@ars.usda.gov (H.S.); Steven.Mirsky@ars.usda.gov (S.M.); George.Meyers@ars.usda.gov (G.M.)
4 Veselka Farms, Pataskala, OH 43062, USA; elliott.327@osu.edu (J.E.); Christine.Charyton@gmail.com (C.C.)
* Correspondence: Barry.Allred@ars.usda.gov

Abstract: Better methods are needed for mapping agricultural drainage pipe systems. Prior research on small test plots indicates that ground penetrating radar (GPR) is oftentimes capable of detecting buried drainage pipes; however, the feasibility of employing this geophysical technique in larger field areas has not been adequately evaluated. Ground penetrating radar integrated with a Real-Time Kinematic (RTK) Global Navigation Satellite System (GNSS) may be an effective and efficient means of mapping drain lines within agricultural fields. Therefore, GPR-RTK/GNSS was tested in three agricultural settings; with Site 1 and Site 2 located in Beltsville, MD, USA and Site 3 near Columbus, OH, USA. Soils at the three sites ranged from silty clay loam to loamy sand. A GPR unit with 250 MHz antennas was used to detect drainage pipes, and at Sites 1 and 2, a physical GNSS base station was utilized, while a virtual base station was employed at Site 3. The GPR-RTK/GNSS configurations used in this study delineated a complex rectangular drainage pipe system at Site 1, with one set of drainage pipes oriented southwest-northeast and a second oriented southeast-northwest. At Site 2, a herringbone drain line pattern was outlined, and at Site 3, random drain lines were found. When integrated with RTK/GNSS, spiral or serpentine GPR transects (or spiral/serpentine segments of a GPR transects) were utilized to provide insight on drain line directional trends. Consequently, given suitable field conditions, GPR integrated with RTK/GNSS can be a valuable tool for farmers and drainage contractors needing to map subsurface drainage systems.

Keywords: Drainage Pipe Mapping; Ground Penetrating Radar (GPR); Real-Time Kinematic Global Navigation Satellite System (RTK/GNSS)

---

1. Introduction

A 1985 economic survey showed that the states comprising the Midwest U.S. (Illinois, Indiana, Iowa, Ohio, Minnesota, Michigan, Missouri, and Wisconsin) by that year had approximately 12.5 million ha that contained subsurface drainage systems, with cropland accounting for the large majority of areas possessing these buried drainage pipe networks [1]. In the more than 30 years that have elapsed since the 1985 survey, a very sizeable amount of additional agricultural drainage pipe has been installed throughout the Midwest U.S. In order to improve crop yields, farmers within this region often need to repair drain lines that are not functioning properly or install new drain lines between the old ones to improve soil water removal efficiency. Whether for system repairs or efficiency improvements, locations of the preexisting drain lines are required; however, in most cases, a map of the original subsurface drainage system installation is no longer available.

Present methods of locating drain lines involve the use of hand-held tile probes or heavy trenching equipment. Finding drainage pipes with a tile probe (Figure 1a) is both time consuming and extremely tedious. With a tile probe, it is sometimes difficult at a particular spot beneath the ground surface to distinguish a clay tile drainage pipe from the large stones that are commonly present in glacially derived soils of the Midwest U.S. A tile probe also easily penetrates through a corrugated plastic tubing (CPT) drainage pipe, thereby making the presence of the CPT drainage pipes difficult to ascertain, while at the same time potentially causing pipe collapse problems. Locating drainage pipes with heavy trenching equipment (Figure 1b) is effective, but causes costly pipe damage that has to be repaired before the trench is backfilled. Consequently, an effective, efficient, and noninvasive (i.e., nondestructive) method is needed for mapping subsurface drainage systems in agricultural fields.

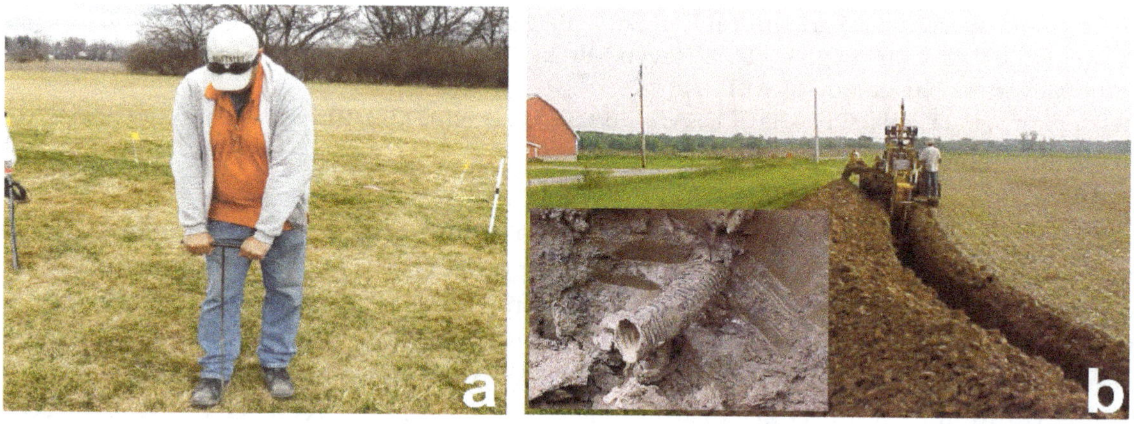

Figure 1. Present methods for finding drainage pipes: (**a**) hand-held tile probe; (**b**) heavy trenching equipment with inset photo showing an example of the drainage pipe damage caused during this excavation operation.

In a preliminary study [2], an unmanned aircraft system (UAS) with a thermal infrared camera exhibited promise in regard to drainage pipe mapping at an agricultural field in central Ohio, USA; however, the technology is now only just beginning to be rigorously evaluated for this particular application. Prior research has demonstrated the feasibility of using ground penetrating radar (GPR) in agricultural settings to locate buried clay tile and CPT drainage pipe over a range of soil types and shallow hydrologic conditions [3–5]. Additional research has shown that GPR, under certain circumstances, can isolate the location of drainage pipe obstructions [6]. The strength of the GPR drainage pipe response for specific soil moisture conditions was found to depend on the orientation of the GPR antennas relative to the drain line directional trend [7].

All prior GPR drainage pipe detection research focused on testing the feasibility of this geophysical technology on small agricultural experiment plots. The practicality of employing GPR for determining drain line patterns in larger farm fields has not been adequately assessed. Ground penetrating radar integrated with a Real-Time Kinematic (RTK) Global Navigation Satellite System (GNSS) may be an effective, efficient, and noninvasive (i.e., nondestructive) means of mapping drain lines within agricultural fields. Consequently, GPR-RTK/GNSS was evaluated at three agricultural fields. The overall process involved mapping the latitude and longitude coordinates of points in the field beneath which drainage pipes were potentially detected and then using this map to provide insight on drain line patterns. The guiding hypothesis for this research can be stated; "GPR integrated with RTK/GNSS is an effective, efficient, and noninvasive tool for delineating drain line patterns in agricultural fields." The results of this study confirmed this hypothesis by successfully demonstrating GPR-RTK/GNSS mapping of complex subsurface drainage systems at three agricultural field sites.

2. Materials and Methods

2.1. Site Descriptions

Testing of GPR integrated with an RTK/GNSS for drainage pipe mapping was carried out at three agricultural field sites. Two of the sites are located in Beltsville, MD, USA, and the third site is near Columbus, OH, USA. Aerial images of the three sites, with soil map overlays, are shown in Figure 2. Note: Throughout this article, aerial images obtained via Google Earth (Google LLC., Mountain View, CA, USA). were used as a base map to overlay soils and GPR information. Soil map overlays were obtained via the SoilWeb Earth App (UC Davis, California Soil Resource Lab, Davis, CA, USA). Descriptions for each test site are provided as follows.

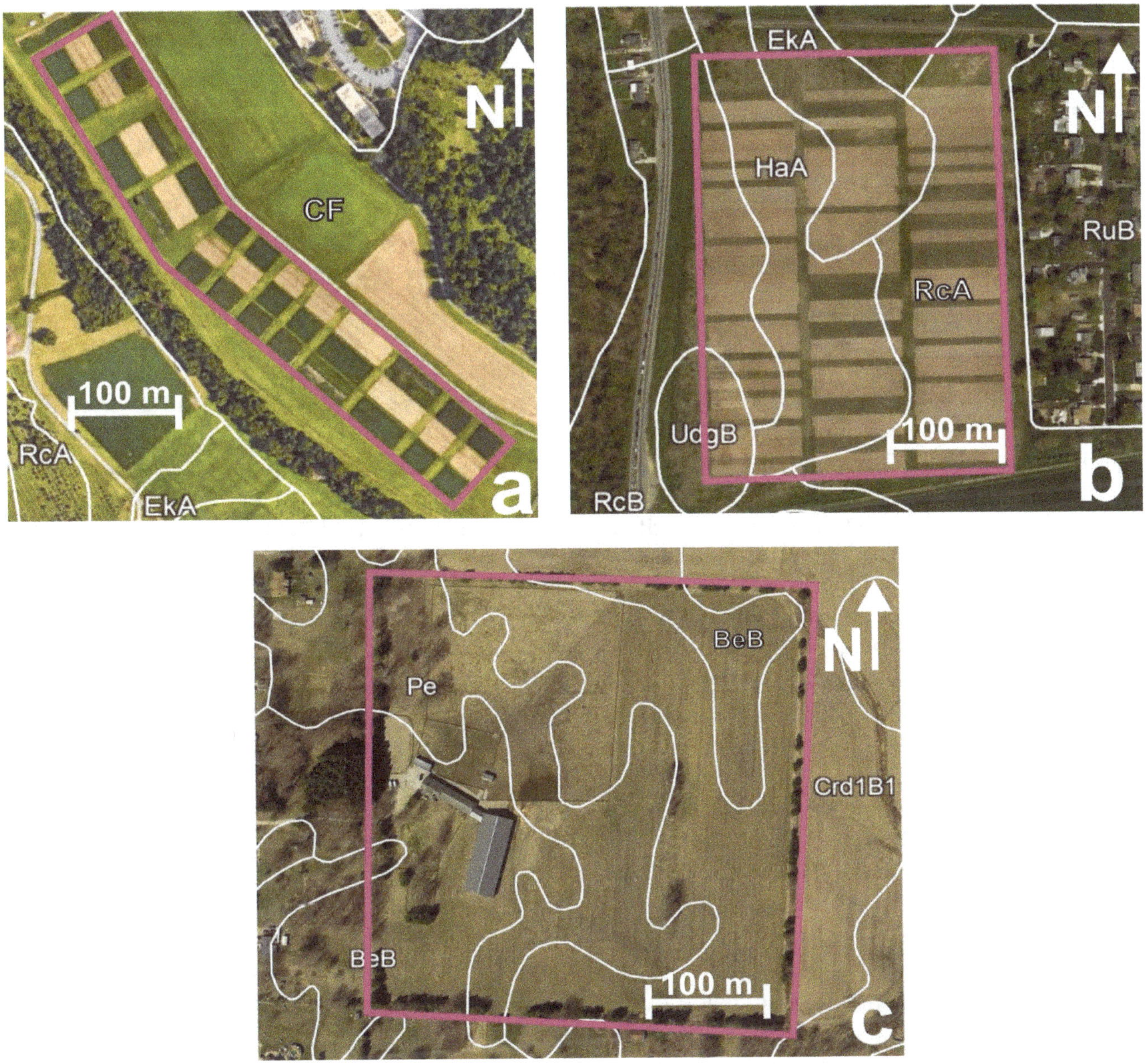

Figure 2. Aerial images with soil map overlays of the three agricultural sites that were investigated with GPR-RTK/GNSS: (**a**) Site 1; (**b**) Site 2; (**c**) Site 3. Purple lines mark site boundaries.

2.1.1. Site 1 (Latitude: 39.012555, Longitude: −76.940204)

Site 1 is located at the U.S. Department of Agriculture (USDA)—Agricultural Research Service (ARS)—Beltsville Agricultural Research Center (BARC). Site 1 is comprised of numerous small experiment plots (Figure 2a). The total area of this agricultural field is 5.6 ha. Cordorus and Hatboro soils, CF, cover Site 1. Both the Cordorus (*mesic Fluvaquentic Dystrudept*) and Hatboro (*mesic Fluvaquentic Endoaquept*) are classified as loams.

2.1.2. Site 2 (Latitude: 39.025426, Longitude: −76.899118)

Site 2 is also located at BARC and is approximately 4 km northeast of Site 1. Site 2, as with site 1, is comprised of numerous small experiment plots (Figure 2b). The total area of the Site 2 agricultural field is 11.3 ha. The soils that cover Site 2 include Elkton silt loam, EkA, (*mesic Typic Endoaquults*), Hammonton loamy sand, HaA, (*mesic Aquic Hapludults*), Russett—Christiana complex, RcA or RcB, (Russett loam—*mesic Aquic Hapludults*, Christiana—*mesic Aquic Hapludults*), and Udorthents, UdgB, (reclaimed gravel pits). Udorthents cover only a very small part of Site 2, just in the southwest corner.

2.1.3. Site 3 (Latitude: 40.029482, Longitude: −82.727897)

Site 3 is a horse boarding, riding, and training facility located near Columbus, Ohio. The northeastern portion of the site has fenced-in areas for horses to exercise and graze (Figure 2c). The building structures contain offices, boarding stalls, and an indoor riding area. A large hay field is present within southern and western portions of the site. The total area of Site 3 is 13.0 ha. The soils in this agricultural setting include Bennington silt loam, BeB, (*mesic Aeric Epiaqualfs*), Cardington silt loam, Crd1B1, (*mesic Aquic Hapludalfs*), and Pewamo silty clay loam, Pe, (*mesic Typic Argiaquolls*).

2.2. Equipment

A Sensors & Software Inc. (Mississagua, ON, Canada) Noggin® GPR system having 250 MHz antennas (Figure 3a,c) was employed to detect buried drainage pipes. The antennas used were shielded and had a frequency range from 125 to 375 MHz, with a center frequency of 250 MHz. Previous research [3–7] indicated that 250 MHz center frequency antennas worked best for finding clay tile and CPT drainage pipes in agricultural settings. Integration of RTK/GNSS with the GPR system allowed accurate latitude and longitude coordinates to be obtained at field locations where potential drainage pipes were detected. In order to obtain RTK/GNSS coordinates at Sites 1 and 2, a Topcon Corporation (Itabashi, Japan) GNSS (Global Positioning System—GPS and Globalnaya Navigazionnaya Sputnikovaya Sistema—GLONASS) FC-200 rover receiver (Figure 3a) and Topcon Corporation GNSS (GPS and GLONASS) HiPer XT base station receiver (Figure 3b) were utilized. At Site 3, a Topcon Corporation GRS-1 dual-frequency, GNSS (GPS and GLONASS) rover receiver and Topcon Corporation PG-S1 hand-held controller were employed together (Figure 3c) with a virtual base station set-up through the Ohio Department of Transportation's (ODOT) network of Continuously Operating Reference Stations (CORS) [8]. Time domain reflectometry (TDR) soil water content values were collected using a Spectrum Technologies, Inc. (East Plainfield, Illinois) Field Scout TDR-300 with 20 cm waveguides (Figure 3d). The TDR water content measurements were utilized to determine soil dielectric constant values [9], that in turn were used to calculate preliminary soil radar velocities [10,11], which were then employed to convert GPR two-way travel times to depth values.

2.3. Field Data Collection

Almost all of the GPR-RTK/GNSS data for Site 1 was collected on 5 April 2017, with just a few additional short GPR transects carried out on 24 January 2018. All GPR-RTK/GNSS data for Site 2 was obtained on 25 January 2018. For Site 3, linear GPR transects were completed on 6 November 2017, followed by the spiral/serpentine GPR transects on 5 February 2018. At each site, between four to six hours total was needed to complete the GPR-RTK/GNSS surveys. The GPR equipment settings included a 5 cm distance between signal trace measurement points along a transect (i.e., station interval = 5 cm). At each measurement point on the GPR transect, 16 signal traces were collected and averaged (i.e., stacking = 16). The GPR two-way travel time for each signal trace was set to provide a depth of investigation of 1.5 m (Site 3) or 2.0 m (Sites 1 and 2). The GPR signal trace two-way travel time needed to achieve 1.5 m or 2.0 m investigations depths were calculated from soil radar velocity values based on TDR measurements.

Figure 3. Equipment used in investigation: (**a**) Noggin GPR system (250 MHz antennas) with GNSS FC-200 rover receiver; (**b**) GNSS HiPer XT base station receiver; (**c**) Noggin GPR system (250 MHz antennas) with GNSS GRS-1 rover receiver and PG-S1 hand-held controller; (**d**) Field Scout TDR-300.

One main component of the Noggin® GPR unit is the Digital Video Logger (DVL), which is used to input equipment settings (station interval, stacking, radar velocity, depth of investigation, etc.) and store GPR-RTK/GPS data. The DVL also has a display screen that provides a real-time GPR cross-section view of the subsurface as data is being collected along a measurement transect (Figure 3a,c). Consequently, the DVL can provide an almost instant indication of a buried drainage pipe. Suspected locations of the buried drainage pipes can then be flagged in the field, with flagged locations that line-up, pointing to the presence and trend of a drain line. Additionally, spiral/serpentine GPR transects or segments of transects can be carried out, on the spot, based on DVL indications of a buried drainage pipe, directly for the purpose of confirming drain line presence and determining its directional trend.

Site maps showing GPR measurement transects (yellow lines—obtained through RTK/GNSS data) are provided in Figure 4. The physical GNSS base station locations at Sites 1 and 2 are marked with orange square symbols (Figure 4a,b). (Again, a virtual base station was employed at Site 3, so there is no orange square symbol depicted in Figure 4c) At Sites 1 and 2, there were essentially two sets of GPR transects, one that coincided or paralleled the longest side boundaries at the sites and then a second set that was oriented perpendicular to the first set. At Site 2, most of the GPR transects contained spiral path segments that provided insight on drain line directional trends (Figure 4b). At Site 3, linear GPR transects around the perimeter of the property and within the fenced-in grazing/exercise areas or hay field were used for initial indications of where drainage pipes might exist. These linear GPR transects at Site 3 were then followed-up with spiral or serpentine GPR transects that were again used to determine the trend and extent of individual drain lines.

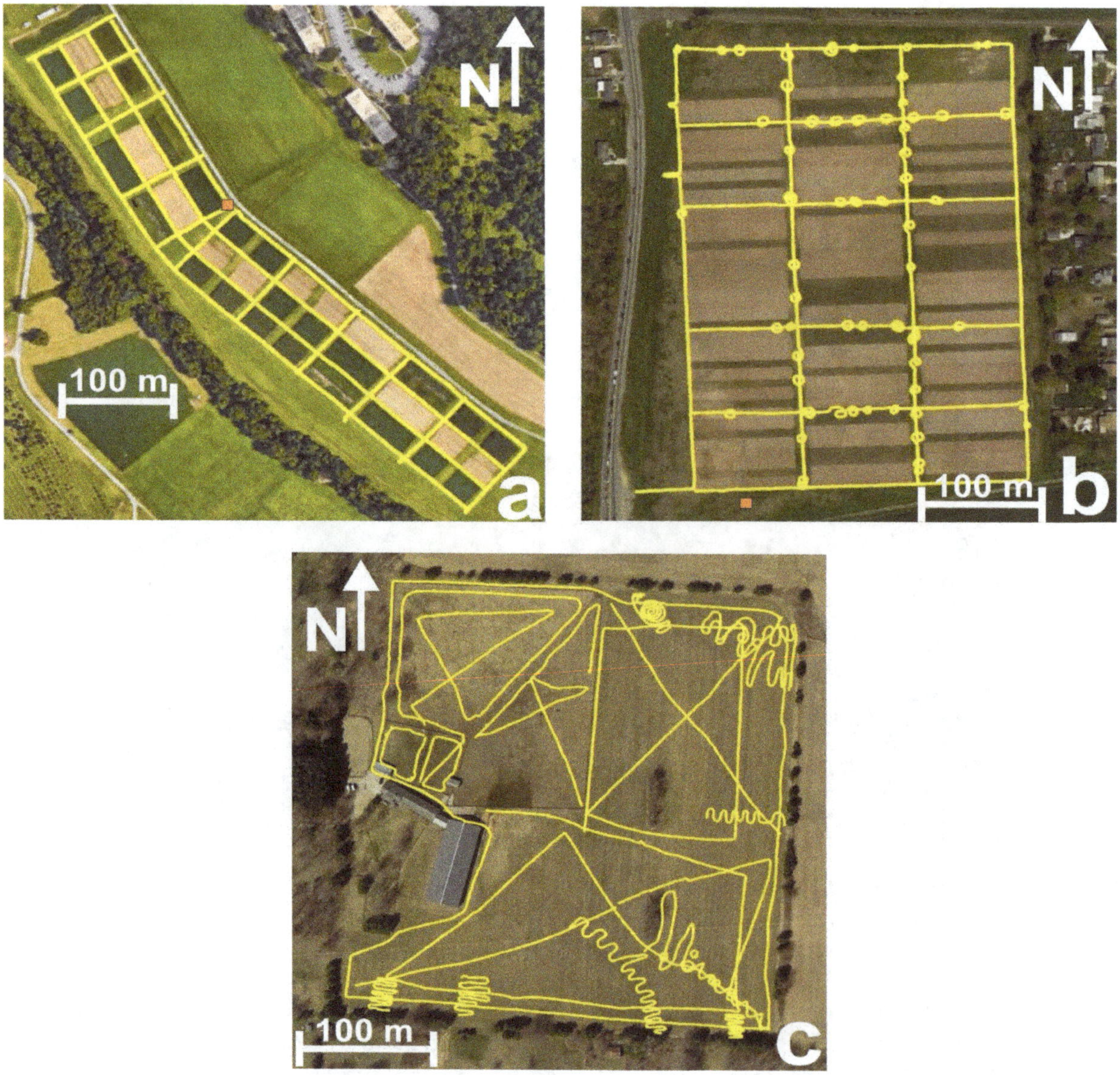

Figure 4. GPR measurement transects (yellow lines) and locations of physical RTK/GNSS base stations (orange squares): (**a**) Site 1; (**b**) Site 2; (**c**) Site 3.

2.4. Data Processing and Interpretation

The principle of GPR operation is conceptually simple and involves directing an electromagnetic radio energy (radar) pulse into the subsurface, followed by measurement of the elapsed time taken by the signal as it is travels downwards from the transmitting antenna, partially reflects off a buried feature, and is eventually returned to the surface, where it is recorded by a receiving antenna. For each point of measurement along a GPR transect, reflections from different depths produce a signal trace, which is a function of the radar wave amplitude (and energy) versus two-way travel time. (Note: As previously indicated, soil water content measurements can be used to convert two-way travel times to depth values.) Antenna frequency, soil moisture conditions, clay content, salinity, and the amount of iron oxide present all have a substantial influence on the distance beneath the surface to which the radar signal penetrates. Differences in the dielectric constant across a buried feature discontinuity govern the amount of radar energy that reflects off the buried feature and then returns to surface to be recorded by the receiving antenna. Previous research [5] shows that the GPR drainage pipe response (i.e., amount of radar energy reflected from a buried pipe) does not depend on the type of pipe (clay tile or CPT) but rather the difference between the dielectric constant of the soil surrounding the pipe versus the dielectric constant of the air and/or water inside the pipe.

A GPR profile (i.e., cross-section) of the subsurface beneath a GPR transect is generated by sequentially combining, side-by-side, one after another, the signal traces obtained at each measurement point along the transect. The horizontal axis on a GPR profile represents distance along the transect (in meters), while the vertical axis represents two-way radar signal travel time (in nanoseconds) and/or depth (in meters). With depths of investigation of 1.5 or 2.0 m, the GPR profiles generated for this study essentially depict GPR responses only within the soil profile.

Figure 5 depicts the two types of GPR profile drainage pipe responses. Where there is a somewhat modest to large angle (i.e., $15° < x° < 90°$) between the GPR transect orientation and drain line directional trend, the GPR response is that of an upside-down U-shaped feature (i.e., reflection hyperbola), and the position for the top of the drainage pipe corresponds with the apex of the reflection hyperbola (Figure 5a). The drainage pipe reflection hyperbola is horizontally compressed if the angle is closer to 90°, while alternatively, if the angle is closer to 15°, then the reflection hyperbola becomes spread out horizontally. It is important to note other buried features, such as large stones can produce a reflection hyperbola, but in map view, the location of these features are isolated, while mapped locations of actual drainage pipe reflection hyperbolas form a line. Where the GPR transect is essentially over top and along trend ($x° < 15°$) of a drain line, the GPR response formed is that a banded linear feature, with the position for the top of the drainage pipe corresponding to the top of the banded linear feature (Figure 5b). Obtaining the banded linear GPR drainage pipe response is fairly uncommon because this response requires, without any prior knowledge of drain line locations and directional trends, that a GPR transect just happens to be oriented over the top and along the trend of a drain line.

EKKO Project 5 software (Sensors & Software Inc., Mississagua, ON, Canada) was used to process and interpret the GPR-RTK/GNSS data acquired at Sites 1, 2, and 3, starting with the generation of GPR profiles from the GPR measurement transects. EKKO Project 5 reflection hyperbola curve fitting procedures, employed only in cases where the GPR transect was perpendicular to the drain line directional trend, allowed soil radar velocities to be refined at each site (from the original TDR measurements), and in turn, improve the accuracy of GPR profile depth scales. Soil radar velocities, refined via EKKO Project 5, were 0.072 m/ns on 5 April 2017 and 0.063 m/ns on 24 January 2018 at Site 1, 0.067 m/ns on 25 January 2018 at Site 2, and 0.060 m/ns on both 6 November 2017 and 5 February 2018 at Site 3. The only processing steps used to produce GPR profiles via EKKO Project 5 were (1) application of a signal saturation correction filter (i.e., Dewow) to remove slowly decaying low frequency noise and (2) utilization of a spreading and exponential calibrated compensation gain function to amplify potential GPR drainage pipe responses. An interpretation module in EKKO Project 5 allowed potential drainage pipe responses in a GPR profile to be marked (Figure 6). With GPR and RTK/GNSS data collected together, EKKO Project 5 was then employed to save a spreadsheet file of the latitudes, longitudes, and depths corresponding to the marked potential drainage pipe responses. Furthermore, because GPR and RTK/GNSS data were collected concurrently, EKKO Project 5 was then able to generate a KMZ file that stored a site map of the GPR transects along with the marked potential drainage pipe locations. Opening this KMZ file in Google Earth allowed GPR measurement transects and the possible pipe locations to be overlaid on an aerial image.

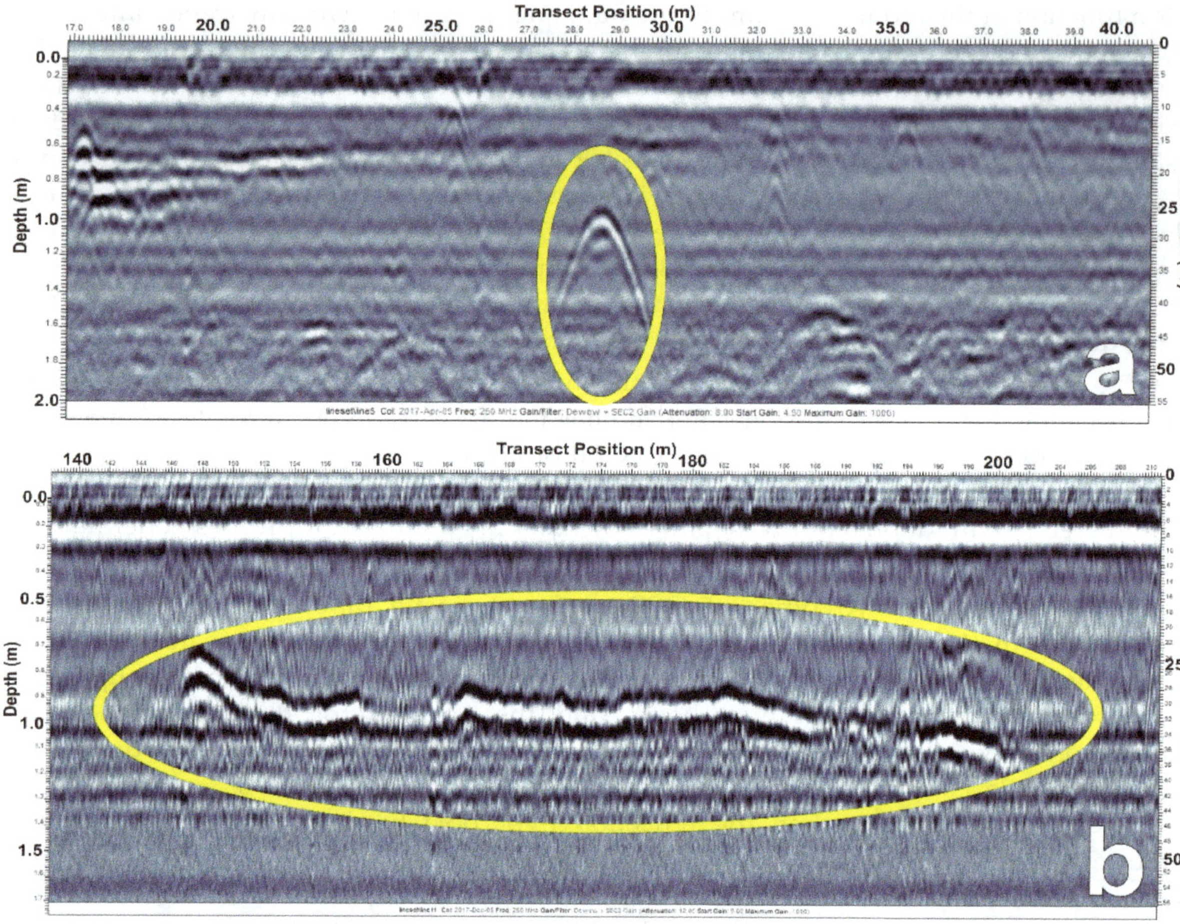

Figure 5. GPR profile drainage pipe responses; (**a**) horizontally compressed upside down U-shaped feature (i.e., reflection hyperbola) within yellow line oval results from the GPR transect being oriented perpendicular to drain line directional trend (from Site 1); (**b**) the banded linear feature within the yellow line oval results from the GPR transect being oriented the over top and along the trend of a drain line (from Site 3).

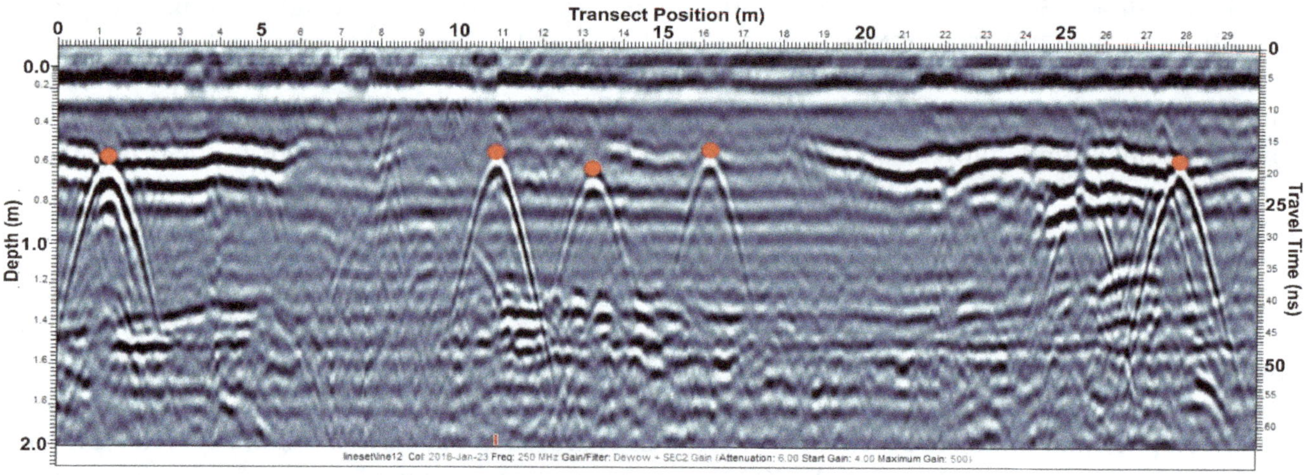

Figure 6. GPR profile from Site 1 with potential reflection hyperbola drainage pipe responses marked via EKKO Project 5. Because GPR and RTK/GNSS data are collected concurrently, the latitude, longitude, and depth corresponding to a potential drainage pipe response mark (red dot) could then all be saved (via EKKO Project 5) into spreadsheet and KMZ files.

3. Results

Locations on GPR transects where potential buried drainage pipes were detected are shown (with red dots) in Figure 7. At Site 1, to the west of the white dashed line in Figure 7a, the overall soil profile had a relatively homogeneous appearance regarding GPR response (as exhibited in GPR profiles). Conversely, to the east of the dashed line, the overall soil profile had a far more heterogeneous appearance regarding GPR response. The overall GPR appearance of the soil profile (i.e., homogeneous vs. heterogeneous) definitely impacted Site 1 drainage pipe detection. In the more homogeneous soils on the west side of Site 1, reflection hyperbola potential drainage pipe responses were fairly easy to isolate within GPR profiles (Figure 8a), while in the more heterogeneous soils on the east side of Site 1, isolating the reflection hyperbola potential drainage pipe responses from the surrounding clutter was much more difficult (Figure 8b).

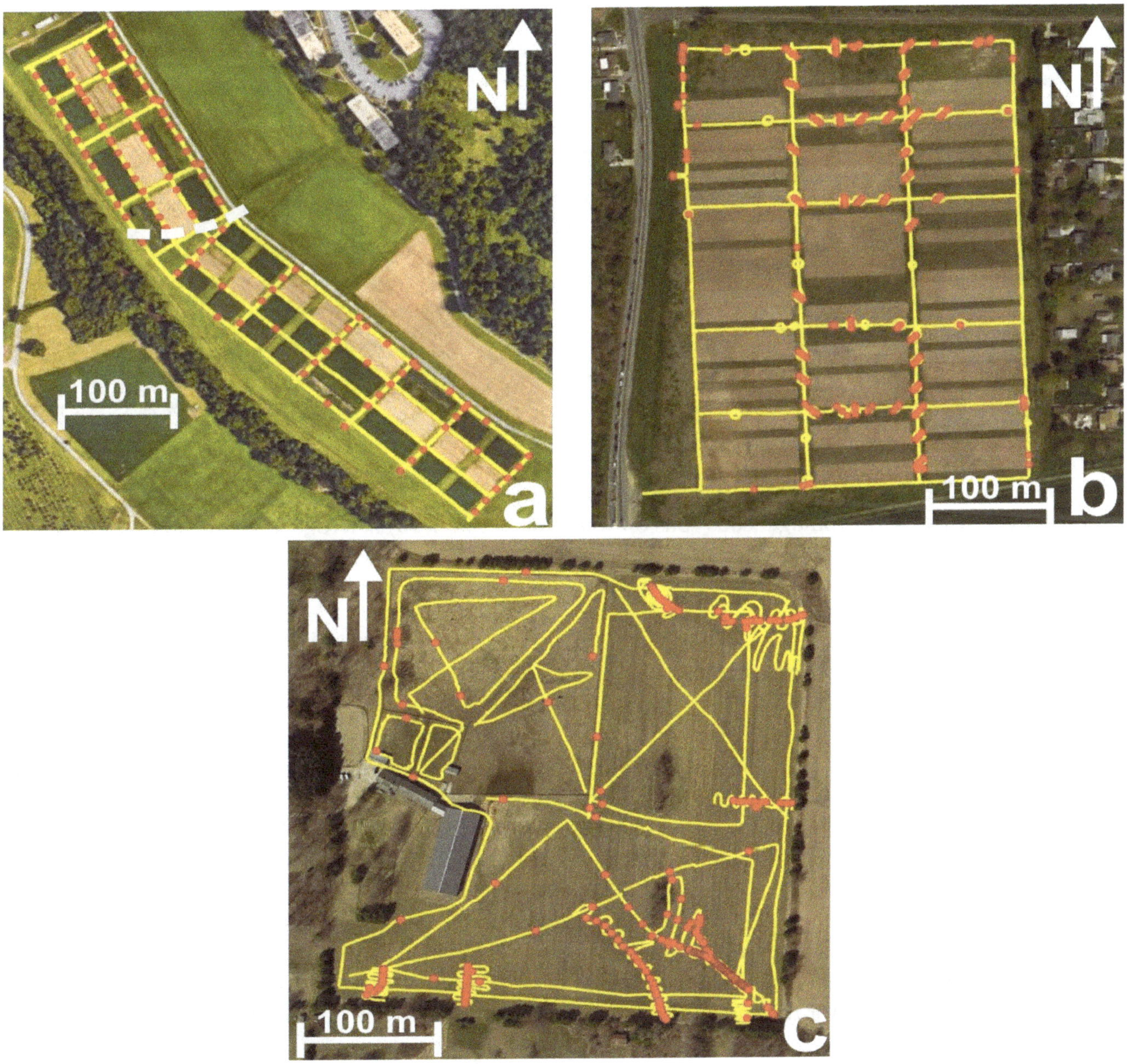

Figure 7. GPR measurement transects (yellow lines) and potential drainage pipe locations (red dots): (**a**) Site 1; (**b**) Site 2; (**c**) Site 3.

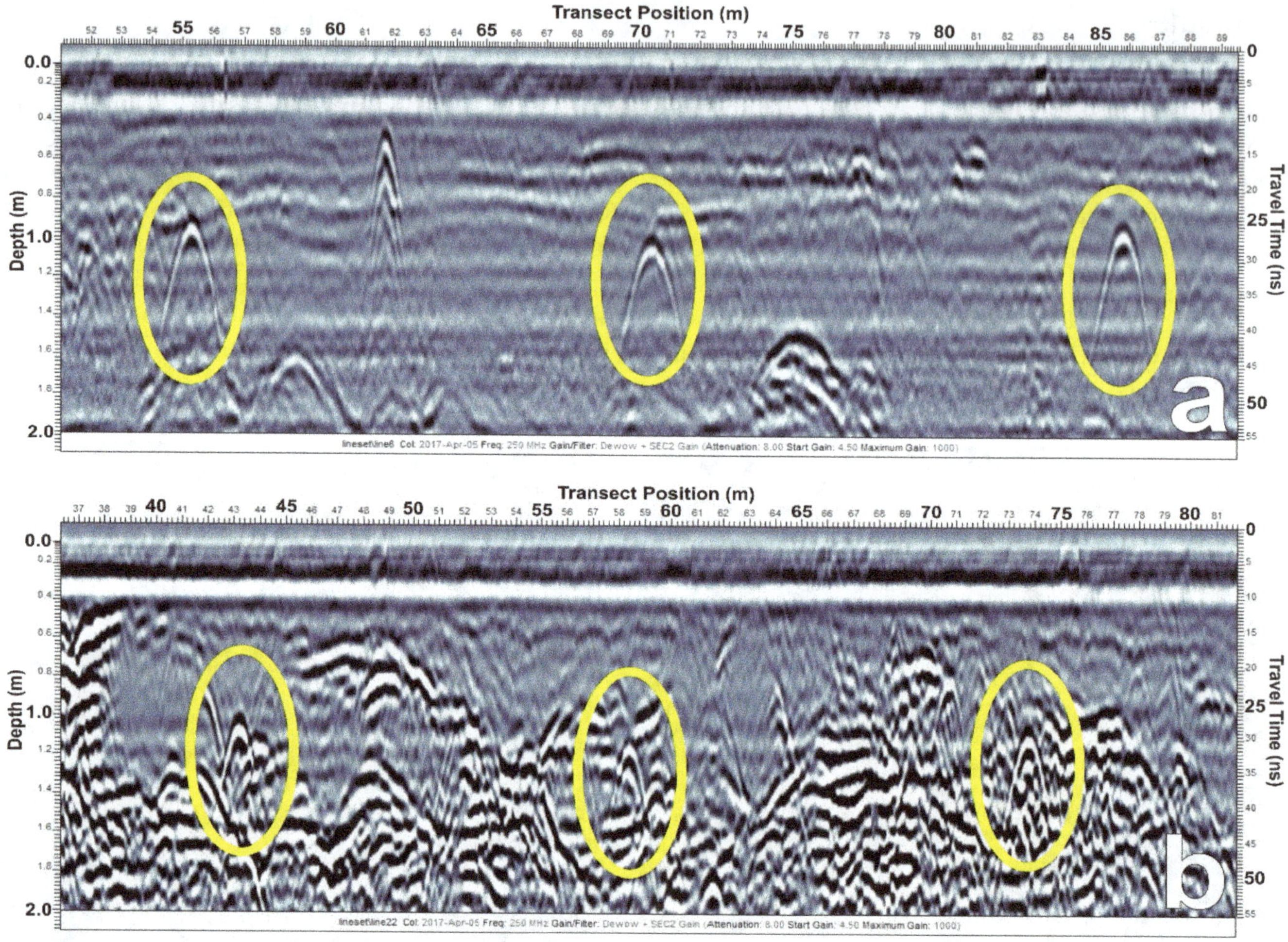

Figure 8. GPR profiles showing reflection hyperbola potential drainage pipe responses highlighted by yellow line ovals; (**a**) homogeneous soils on west side of Site 1; (**b**) heterogeneous soils on east side of Site 1. The reflection hyperbola potential drainage pipe responses are far easier to discern in the homogeneous soils as opposed to the heterogeneous soils.

Drain lines were interpreted to exist in places where mapped locations of potential drainage pipe responses clearly followed a line. Again, a potential drainage pipe location that is isolated likely implies that the corresponding GPR profile reflection hyperbola response is not due to a drainage pipe, but rather some solitary buried object, such as a large stone. The interpreted drain line patterns (blue lines) are depicted in Figure 9 for all three sites. The GPR-RTK/GNSS data collected in this study delineated a complex rectangular drainage pipe system at Site 1 (Figure 9a), with a set of southwest-northeast drain lines present on the west side of the site and a southeast-northwest drain line set present on the east side of the site. Figure 9b indicates that although not all the drainage pipe present at Site 2 was detected, GPR integrated with RTK/GNSS did find a sufficient number of drain lines to unambiguously determine that a herringbone subsurface drainage system had been installed within this agricultural field. Random drain lines were found at Site 3 (Figure 9c), which are likely to be very old and installed for the purpose of draining wet areas in the hay field.

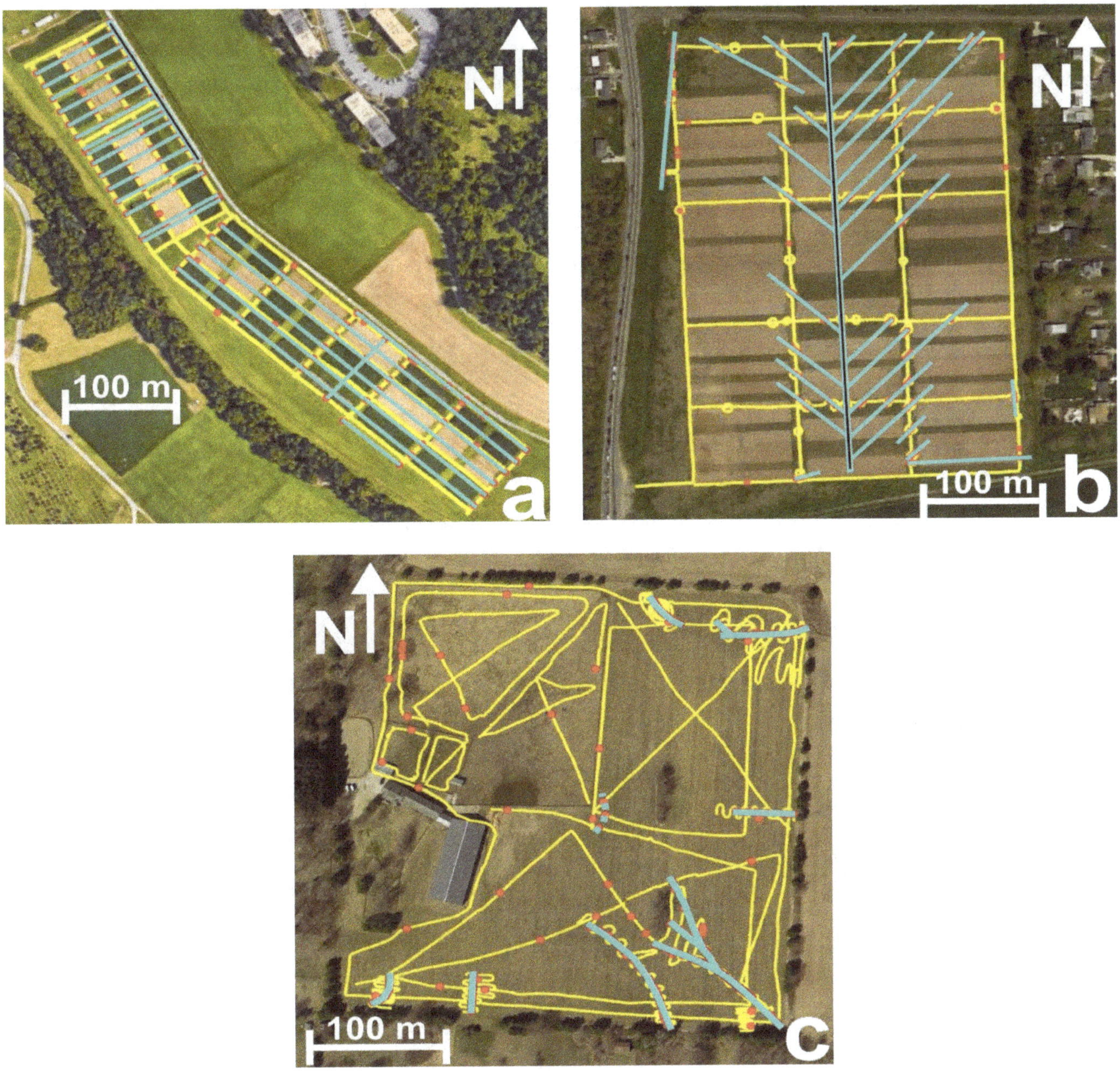

Figure 9. Interpreted drain line patterns (blue lines) overlaid on GPR measurement transects (yellow lines) and potential drainage pipe locations (red dots): (**a**) Site 1; (**b**) Site 2; and (**c**) Site 3. A thin black line within a thicker blue line indicates a main collector pipe.

4. Discussion

At Site 1, GPR profiles (Figure 8) showed soils to the west of the white dashed line in Figure 7a to have a relatively homogeneous GPR response, while soils to the east of the white dashed line exhibited a far more homogeneous GPR response. Prior research [5] has shown that fine-textured silty/clayey soils tend to have a homogeneous appearance in GPR profiles, while coarse-textured sandy soils tend to have a more heterogeneous, or cluttered, appearance in GPR profiles. This finding could indicate for Site 1 that there were generally fine-textured silty/clayey soils west of the white dashed line in Figure 7a and generally coarse-textured sandy soils east of the white dashed line. Textural analysis of soil cores obtained across Site 1 will be required in order to confirm this supposition.

When GPR is integrated with RTK/GNSS, spiral or serpentine GPR transects, or spiral/serpentine segments of GPR transects, can be employed where needed to determine drain line directional trends.

Initial linear GPR transects at Site 1 clearly indicated, just by themselves, based on DVL potential pipe locations flagged in the field, the drain line directional trends; therefore, spiral/serpentine GPR transects or spiral/serpentine segments of GPR transects were not needed to establish drain line patterns at this site. While not required at Site 1 to determine drain line patterns, spiral or serpentine GPR transects (or spiral/serpentine segments of a GPR transects) did provide valuable insight on drain line directional trends at Sites 2 and 3. Figure 10 provides clear examples from Sites 2 and 3 as to how spiral/serpentine GPR transect segments were employed to determine the direction of an agricultural field drain line. It is readily apparent that the drain line patterns at Sites 2 and 3 would have been almost impossible to discern without the integration of GPR with RTK/GNSS and the use of spiral/serpentine GPR transects or spiral/serpentine segments of GPR transects.

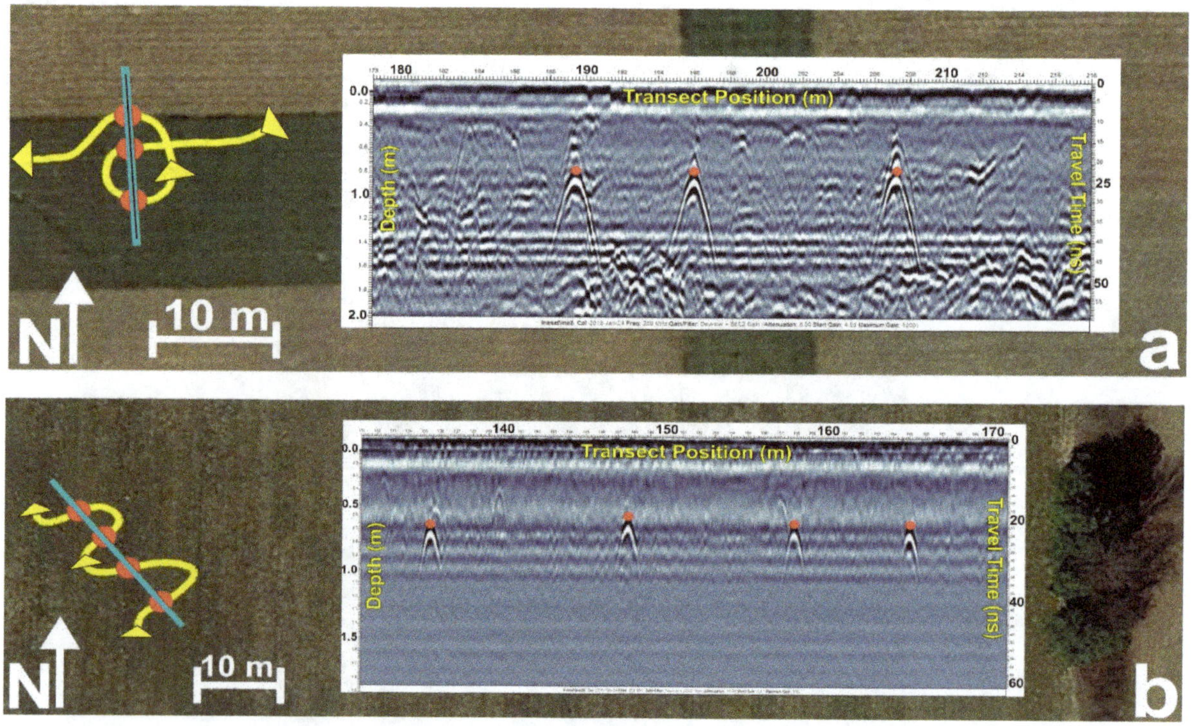

Figure 10. Spiral/serpentine segments of a GPR transect can be employed to determine drain line presence and direction: (**a**) spiral transect segment example from Site 2 (main collector pipe); (**b**) serpentine transect segment example from Site 3. Drainage pipe reflection hyperbolas interpreted with EKKO Project 5 (red dots on GPR profile insets) were mapped (red dots) along with transect segments paths (yellow lines with directional arrowheads). The mapped drainage pipe locations are in a line, thereby indicating the presence and direction of a buried agricultural drain line.

Field data collection using GPR integrated with RTK/GNSS to map subsurface drainage was quite efficient, requiring only four to six hours at each of the three sites. This study shows that a dense grid of GPR transects is not needed to determine drainage pipe patterns. Actually, all that is required in a particular agricultural field to find drain lines and determine their directional trends are a few select GPR transects (usually with incorporated spiral/serpentine segments). The efficiency of GPR-RTK/GNSS for subsurface drainage mapping could be improved further over what was obtained in this investigation by using an all-terrain vehicle (ATV) pulling a sled to mount all the GPR and RTK/GNSS equipment instead of mounting the equipment on the push cart shown in Figure 3a,c.

The GPR-RTK/GNSS drainage pipe mapping approach described is applicable for mapping subsurface drainage systems in agricultural fields up to tens of hectares in size. With suitable soil/hydrologic conditions and widespread adoption of this approach over large areas, it might be possible to gain insight on drainage practice intensity at small watershed scales. As indicated, the types of soils present can be important. Soils having very high clay contents oftentimes, but not always, severely limit radar signal penetration into the subsurface, thereby making GPR drainage pipe detection difficult [4,5]. Transient shallow hydrologic conditions also have a substantial impact. Soils that are not too dry or too wet are best for GPR drainage pipe detection, especially good are moist soils surrounding air-filled drainage pipes, which is a common situation occurring a day or more after a large rainfall event [5].

The integration of GPR with RTK/GNSS can therefore be a valuable tool for determining drainage pipe patterns in agricultural fields. The subsurface drainage system information obtained by GPR-RTK/GNSS could be very useful for retrofitting a subsurface drainage system for subirrigation or to improve soil water removal efficiency, with both scenarios requiring installation of new drain lines in between pre-existing drain lines [12–14], which in turn cannot be accomplished without knowing where the preexisting drain lines are located. Knowledge of drain line patterns obtained via GPR-RTK/GNSS could also be used for repairing part of a subsurface drainage system that is not functioning properly. Finding drainage pipes with GPR-RTK/GNSS is certainly an improvement over tile probe and trench excavation methods, because GPR-RTK/GNSS field data collection, compared to these other two methods, is much faster, is better at determining drain line directional trends, and causes no pipe damage. Consequently, in areas where GPR is capable of detecting buried drainage pipes, GPR-RTK/GNSS can supplant tile probe and trench excavation methods for mapping drain line patterns.

5. Summary and Conclusions

Ground penetrating radar integrated with a Real-Time Kinematic (RTK) Global Navigation Satellite System (GPR-RTK/GNSS) successfully determined drainage pipe patterns at three agricultural field sites. At one site, a rectangular pattern of drain lines was mapped, at the second site, a herringbone pattern was delineated, and at the third site, random drain lines were found. A novel field data collection approach was developed, taking advantage of GPR-RTK/GNSS, by which spiral or serpentine GPR measurement transects (or spiral/serpentine segments of GPR measurement transects) were employed to establish the presence and direction of drain lines. Using GPR-RTK/GNSS to map drain lines can therefore provide useful information needed for retrofitting or repair of agricultural subsurface drainage systems. Furthermore, in areas where GPR can detect drainage pipes, GPR-RTK/GNSS is a substantial improvement (faster, better able to resolve drain line trends, and causes no pipe damage) over the traditional tile probe and trench excavation methods for finding drainage pipes.

Author Contributions: Conceptualization, B.A., D.W., H.S., S.M., J.E.; Methodology, B.A., D.W., L.M.; Validation, B.A.; Formal Analysis, B.A.; Investigation, B.A., D.W., L.M., G.M.; Resources, H.S., S.M., J.E., C.C.; Writing-Original Draft Preparation, B.A.; Writing-Review & Editing, B.A., D.W., L.M., H.S., S.M., G.M., J.E., C.C.; Supervision, B.A.; Project Administration, B.A.

Author Contributions: The use of equipment manufacturer and software developer names in this article are provided solely for informational purposes and does not imply endorsement by the authors or the organizations they represent.

References

1. Pavelis, G.A. Economic survey of farm drainage. In *Farm Drainage in the United States: History, Status, and Prospects, Miscellaneous Publication Number 1455*; Pavelis, G.A., Ed.; U.S. Dept. of Agriculture, Economic Research Service: Washington, DC, USA, 2007; pp. 110–136.
2. Allred, B.; Eash, N.; Freeland, R.; Martinez, L.; Wishart, D. Effective and efficient agricultural drainage pipe mapping with UAS thermal infrared imagery: A case study. *Agric. Water Manag.* **2018**, *197*, 132–137. [CrossRef]

3. Chow, T.L.; Rees, H.W. Identification of subsurface drain locations with ground-penetrating radar. *Can. J. Soil Sci.* **1989**, *69*, 223–234. [CrossRef]
4. Allred, B.J.; Fausey, N.R.; Peters, L., Jr.; Chen, C.; Daniels, J.J.; Youn, H. Detection of buried agricultural drainage pipe with geophysical methods. *Appl. Eng. Agric.* **2004**, *20*, 307–318. [CrossRef]
5. Allred, B.J.; Daniels, J.J.; Fausey, N.R.; Chen, C.; Peters, L., Jr.; Youn, H. Important considerations for locating buried agricultural drainage pipe using ground penetrating radar. *Appl. Eng. Agric.* **2005**, *21*, 71–87. [CrossRef]
6. Allred, B.J.; Redman, D. Agricultural drainage pipe assessment using ground penetrating radar: Impact of pipe condition, shallow hydrology, and antenna characteristics. *J. Environ. Eng. Geophys.* **2010**, *15*, 119–134. [CrossRef]
7. Allred, B.J. A GPR agricultural drainage pipe detection case study: Effects of antenna orientation relative to drainage pipe directional trend. *J. Environ. Eng. Geophys.* **2013**, *18*, 55–69. [CrossRef]
8. ODOT's VRS RTK Network. Available online: http://www.dot.state.oh.us/Divisions/Engineering/CaddMapping/Survey/Pages/VRSRTK-.aspx (accessed on 9 September 2018).
9. Topp, G.C.; Davis, J.L.; Annan, A.P. Electromagnetic determination of soil water content—Measurements in coaxial transmission lines. *Water Resour. Res.* **1980**, *16*, 574–582. [CrossRef]
10. Conyers, L.B. *Ground-Penetrating Radar for Archaeology*; AltaMira Press: Walnut Creek, CA, USA, 2004; pp. 23–80.
11. Sharma, P.V. *Environmental and Engineering Geophysics*; Cambridge University Press: Cambridge, UK, 1997; pp. 309–328.
12. Brown, L.A.; Zucker, L.A. *Agricultural Drainage: Water Quality Impacts and Subsurface Drainage Studies in the Midwest*; Extension Bulletin 871; The Ohio State University: Columbus, OH, USA, 1998; pp. 8–9.
13. Allred, B.J.; Brown, L.C.; Fausey, N.R.; Cooper, R.L.; Clevenger, W.B.; Prill, G.L.; La Barge, G.A.; Thornton, C.; Riethman, D.T.; Chester, P.W.; et al. A water table management approach to enhance crop yields in a wetland reservoir subirrigation system. *Appl. Eng. Agric.* **2003**, *19*, 407–421. [CrossRef]
14. Allred, B.J.; Gamble, D.L.; Clevenger, W.B.; LaBarge, G.A.; Prill, G.L.; Czartoski, B.J.; Fausey, N.R.; Brown, L.C. Crop yield summary for three wetland reservoir subirrigation systems in northwest Ohio. *Appl. Eng. Agric.* **2014**, *30*, 889–903.

14

Thermal Imaging Reliability for Estimating Grain Yield and Carbon Isotope Discrimination in Wheat Genotypes: Importance of the Environmental Conditions

Sebastián Romero-Bravo [1,2,*], Ana María Méndez-Espinoza [2], Miguel Garriga [2], Félix Estrada [2], Alejandro Escobar [2], Luis González-Martinez [2], Carlos Poblete-Echeverría [3], Daniel Sepulveda [4], Ivan Matus [5], Dalma Castillo [5], Alejandro del Pozo [2] and Gustavo A. Lobos [2,*]

1 Department of Agricultural Sciences, Universidad Católica del Maule, Curicó P.O. Box 684, Chile
2 Plant Breeding and Phenomic Center, Faculty of Agricultural Sciences, Universidad de Talca, Talca P.O. Box 747, Chile; anmendez@utalca.cl (A.M.M.-E.); mgarriga@utalca.cl (M.G.); festrada@alumnos.utalca.cl (F.E.); escobar.opazo@gmail.com (A.E.); l.gonzalez.m@ieee.org (L.G.-M.); adelpozo@utalca.cl (A.d.P.)
3 Department of Viticulture and Oenology, Stellenbosch University, Matieland 7602, South Africa; CPE@sun.ac.za
4 Centro de Investigación y Transferencia en Riego y Agroclimatología (CITRA), Talca P.O. Box 747, Chile; dsepulveda18@gmail.com
5 Centro Regional Investigación Quilamapu, Instituto de Investigaciones Agropecuarias, Chillán P.O. Box 426, Chile; imatus@inia.cl (I.M.); dalma.castillo@inia.cl (D.C.)
* Correspondence: sromero@ucm.cl (S.R.-B.); globosp@utalca.cl (G.A.L.)

Abstract: Canopy temperature (Tc) by thermal imaging is a useful tool to study plant water status and estimate other crop traits. This work seeks to estimate grain yield (GY) and carbon discrimination ($\Delta^{13}C$) from stress degree day (SDD = Tc − air temperature, Ta), considering the effect of a number of environmental variables such as the averages of the maximum vapor pressure deficit (VPDmax) and the ambient temperature (Tmax), and the soil water content (SWC). For this, a set of 384 and a subset of 16 genotypes of spring bread wheat were evaluated in two Mediterranean-climate sites under water stress (WS) and full irrigation (FI) conditions, in 2011 and 2012, and 2014 and 2015, respectively. The relationship between the GY of the 384 wheat genotypes and SDD was negative and highly significant in 2011 (r^2 = 0.52 to 0.68), but not significant in 2012 (r^2 = 0.03 to 0.12). Under WS, the average GY, $\Delta^{13}C$, and SDD of wheat genotypes growing in ten environments were more associated with changes in VPDmax and Tmax than with the SWC. Therefore, the amount of water available to the plant is not enough information to assume that a particular genotype is experiencing a stress condition.

Keywords: remote sensing; phenotype; phenotyping; phenomics; *Triticum aestivum*; water deficit; stress; infrared

---

1. Introduction

Since the 1960s, crop temperature has been recognized as an indicator of water status [1]. When the plant is facing a water deficit, the stomata begin to close, reducing the transpiratory capacity (evaporative cooling) [2] and this results in increases in canopy temperature [3–7].

The development of infrared sensors/cameras has allowed a faster characterization of canopy temperatures [8]. At the same time, through computational analysis, it is possible to split the different parts of the image (e.g., soil, air, leaves, weeds) focusing only on the fraction(s) of interest [4,9,10]. Although thermal imaging does not directly measure stomatal conductance, the stomatal response is the main cause of changes in canopy temperature [10], so it is a useful tool to indirectly study spatial and temporal heterogeneity of leaf/canopy transpiration and the photosynthetic rate [10–12]. Indeed, in bread wheat grown in hot environments in Mexico under irrigation, a high correlation has been reported between temperature depression (TD = Ta − Tc) and leaf stomatal conductance ($r = 0.76 - 0.85$) and grain yield (GY; up to $r = 0.84$) [11,13]. Other researchers have used the concept of stress degree day (SDD), defined as the difference between leaf/canopy temperature (Tc) and air temperature (Ta) (SDD = Tc − Ta), which is equivalent to TD (but with positive values), mostly because canopy temperature in rainfed environments is lower than air temperature.

The main problem with the use of thermal assessments to estimate physiological and agronomic traits is that Tc is influenced by several environmental factors, such as air temperature and humidity, wind speed, net radiation, and soil water content [14–17]. Therefore, without detailed information about environmental factors, measurements of Tc are not sufficient to properly perform agronomic or physiological trait estimations.

Unlike irrigated conditions, a good correlation between Tc and GY under water deficit is not always expected [18]. However, it would be very useful for breeding programs to find such associations in stressful environments because the focus is on developing drought-tolerant cultivars with higher GY under water-limiting conditions.

It has been established that measurements of carbon isotope discrimination ($\Delta^{13}C$) in wheat are crucial for the selection of individuals with efficient water-use, mainly because this parameter is positively correlated with GY and negatively correlated with water-use efficiency (WUE) in moderately water-stressed to non-water-stressed environments [19–25]. The determination of $\Delta^{13}C$ is simple and relatively fast but needs expensive equipment or engagement of a paid analysis service; attempts have also been made to estimate $\Delta^{13}C$ by modeling the canopy spectral reflectance [24,26]. Under non-stressed conditions, the stomata remain open and the substomatal cavity is enriched with ^{12}C relative to the air; the heavier isotopic $^{13}CO_2$ has a lower diffusion speed than the lighter $^{12}CO_2$ [20]. Additionally, the ribulose bisphosphate carboxylase/oxygenase (RUBISCO) carboxylation enzyme in C3 plants has a higher affinity to $^{12}CO_2$. On the other hand, when stress forces the stomata to close, the proportion of $^{12}CO_2$ in the substomatal cavity is reduced, thus increasing the amount of fixed $^{13}CO_2$ [20]. Thus, daily conditions throughout the season will be summarized in the $\Delta^{13}C$ of leaves and kernels (calculation details in Section 2.2.1). In this sense, under the expected climate change scenarios predicted for the coming decades [27], the estimation of $\Delta^{13}C$ should be relevant in plant breeding programs oriented to environmental constraints [28–30].

Like all species, the phenotype of wheat plants is controlled by a large number of genes, and the expression of these is modulated, predominantly, in response to the environmental conditions (GxE) [31–33]. Consequently, it was hypothesized that the environmental conditions during and between seasons could interfere with the ability of canopy thermal imaging to estimate GY and $\Delta^{13}C$; in particular, the vapor pressure deficit (VPDmax) and soil water content (SWC), which can have a strong influence on canopy temperatures [34]. Therefore, the aim of this work was to study the reliability for estimating grain yield and carbon isotope discrimination in wheat genotypes growing under water stress (WS) and full irrigation (FI) conditions using thermal images, considering the relevance of the prevailing environmental conditions in estimation of the results.

2. Materials and Methods

2.1. Plant Material and Experimental Conditions

During four growing seasons (2011, 2012, 2014, and 2015), two sets of plant material were evaluated in two Mediterranean environments: (1) Cauquenes (c) (35°58′ S, 72°17′ W; 177 m.a.s.l.) under WS (rainfed) conditions during seasons 2011, 2012, and 2015, and under FI in 2015; and (2) Santa Rosa (sr)

(36°32′ S, 71°55′ W; 217 m.a.s.l.) under WS in 2011 and 2015, and FI conditions in 2011, 2012, 2014, and 2015. Each combination of season (year), location (c or sr), and water condition (WS or FI) was considered as an environment.

A collection of 384 advanced lines and cultivars of spring bread wheat (*Triticum aestivum* L.), were evaluated during 2011 and 2012. Plant material originated from three breeding programs: the Instituto de Investigaciones Agropecuarias in Chile (INIA-Chile) (153 genotypes), INIA-Uruguay (178 genotypes), and the International Wheat and Maize Improvement Centre CIMMYT (53 genotypes). In 2014 and 2105, a subset of 16 genotypes with contrasting tolerance to water deficit was studied.

Each genotype was established in plots of five rows (2.0 × 0.2 m) with a seeding rate of 2 0g m^{-2}. Plots were fertilized with 260 kg ha^{-1} of ammonium phosphate (46% P_2O_5 and 18% N), 90 kg ha^{-1} of potassium chloride (60% K_2O), 200 kg ha^{-1} of Sul-Po-Mag (22% K_2O, 18% MgO, and 22% S), 10 kg ha^{-1} of Boronatrocalcita (11% B), and 3 kg ha^{-1} of zinc sulfate (35% Zn). Fertilizers were incorporated with a cultivator before sowing. During tillering, an extra 153 kg ha^{-1} of N was applied. Weeds were controlled with the application of flufenacet + flurtamone + diflufenican (96 g a.i.) as pre-emergence and a further application of MCPA (525 g a.i.) + metsulfuron-methyl (5 g a.i.) as post-emergent [35]; dates of sowing and the main phenological stages are shown in Table 1. Furrow irrigation was used at Santa Rosa, with the WS trials including one irrigation at the end of tillering (Zadocks stage 21–Z21; [36]) and FI comprising irrigations at the end of tillering (Z21), the flag leaf stage (Z37), heading (Z50), and early grain filling (Z71). At Cauquenes, the WS trials were purely rainfed and the FI trial during 2015 received sprinkler irrigation at Z37, Z50, and Z71. Approximately 50 mm was applied during each furrow/sprinkler irrigation application.

At each location, soil volumetric content (m^3 m^{-3}) was monitored periodically using 10HS sensors (Decagon Devices, Pullman, WA, USA), scanning the first 50 cm depth every 4 h. In order to generate the soil water content (SWC; mm), the volumetric values were multiplied by the soil depth (500 mm). Precipitation (mm), ambient temperature (°C), and relative humidity (%) were monitored hourly by autonomous weather stations (AWSs) belonging to the Red Agroclimática Nacional (National Agroclimatic Network, available at: www.agromet.inia.cl). Vapor pressure deficit (VPD; kPa) was determined hourly by the use of ambient temperature and relative humidity, according to Reference [37]. For analysis purposes, each environmental variable was studied as follows from sowing to harvest: (1) precipitation: daily summation; (2) ambient temperature: average of the daily maximum temperatures (Tmax); and (3) VPDmax: estimated at the highest ambient temperature and the corresponding relative humidity of each day, and then the average of the daily maximum VPDs (VPDmax) was calculated. Because water deficit in Mediterranean environments is present, primarily, between anthesis to grain filling, SWC was considered as the average of the daily mean values between anthesis and grain maturity.

Tmax, VPDmax, and SWC are summarized in Table 1 and Figure S1, and rainfall in Figure S2.

2.2. *Evaluations*

2.2.1. Grain Yield and Carbon Isotope Discrimination

Grain yield was evaluated by harvesting the whole plot (2 m^2) and was expressed as t ha^{-1}. Carbon isotope composition ($\delta^{13}C$) was determined in mature kernels using an elemental analyzer (ANCA-SL, PDZ Europa, UK) coupled with an isotope ratio mass spectrometer, at the Laboratory of Applied Physical Chemistry at Ghent University (Belgium): $\delta^{13}C$ (‰) = $(^{13}C/^{12}C)_{sample}/(^{13}C/^{12}C)_{standard} - 1$ [20], where the $^{13}C/^{12}C$ ratio of the sample refers to plant material and the $^{13}C/^{12}C$ ratio of the standard is calibrated against the international standards from Iso-Analytical (Crewe, Cheshire, UK). The carbon isotope discrimination ($\Delta^{13}C$) of kernels was calculated as: $\Delta^{13}C$ (‰) = $(\delta^{13}C_a - \delta^{13}C_p)/[1 + (\delta^{13}C_p)/1000]$, where a and p refer to air and the plant, respectively [20]. $\delta^{13}C_a$ from the air was taken as −8.0‰.

2.2.2. Thermography

Thermal infrared images were taken using a portable infrared camera (i40, FLIR Systems, Sweden), at the soft dough (Z85) phenological stage. This camera provides images of 120 × 120 pixels (every pixel shows a temperature value) and has an uncooled infrared detector (microbolometer) in the spectral range from 7.5 to 13 microns. Infrared images were taken at ±2 h from the zenith (12:00 to 16:00 h), at a position of 45° from the horizontal, 0.5 m above the plant canopy, and a 3 m distance from the plot. Images were filtered using a process of interactive segmentation to exclude foreign matter from the picture (i.e., soil, weeds, neighboring plots, and air) using a custom MATLAB code [38]. To avoid surrounding plot noise, only the center of the image (30 × 30 pixels) was analyzed with a temperature frequency histogram (percentile level). The hottest and coldest pixels were eliminated, taking as a threshold the percentiles 1 downwards and 97.5 upwards, respectively. The remaining pixels were used to calculate the average canopy temperature (Tc), while the air temperature (Ta) was recorded from the AWS at the precise time the image was taken. Finally, Tc and Ta were used to calculate the SDD (°C) [39,40].

2.3. *Statistical Design and Data Analysis*

The experimental design for the trials at Cauquenes and Santa Rosa in seasons 2011 and 2012 was an alpha-lattice with two replicates; for this study, just one replicate (n = 384 genotypes) was assessed by thermography in each trial. For seasons 2014 and 2015, the experimental design was a random block with four replicates (16 genotypes; n = 64).

Correlations (x versus y) were performed through regression analysis: (1) genotype values: SDD versus GY and Δ^{13}C; (2) environmental values: SDD, Tmax, VPDmax, and SWC versus GY and Δ^{13}C; and (3) environmental values: VPDmax, Tmax, and SWC versus SDD.

Using the environmental (Tmax, VPDmax, and SWC), phenological (days between stages), physiological (Δ^{13}C and SDD) and productive (GY) information (Table 1), a clustering analysis was performed to verify whether the two water regimes evaluated (i.e., FI and WS) were grouped together, within and between seasons and locations, which is important in modeling and validation procedures. This consisted of a series of steps necessary to achieve a correct execution of the analysis methodology. For this study, clustering and hierarchical clustering were used, with the purpose of grouping the different environments studied. A group was defined as the set of elements that have a greater degree of similarity between the objects that belong to the same set [41]. The steps performed in the analysis were the following: obtaining the data, eliminating the columns that do not provide information to the grouping model, normalizing the data, then applying a method of hierarchical clustering using the "ward.D2" method [42] as a grouping form, and plotting the Euclidean distance between elements as a dendrogram. For clustering of groupings, a tree cluster was considered, which uses the Euclidean distance to identify the closeness of the nodes (environmental data points). In addition, this algorithm applies the principal component analysis (PCA) method to show the results with greater clarity [41]. The "ward.D2" was set to find two and three main data groups.

All statistical analysis was performed using R 3.0.0 [43].

3. Results

3.1. *Environmental Conditions, Grain Yield, Carbon Isotope Discrimination, and Stress Degree Days*

In general terms, the environmental conditions (Tmax, VPDmax, and SWC) varied according to the seasons, both within and between FI and WS conditions (Table 1, Figures S1 and S2). Under each water supply condition, minimum and maximum values from sowing to harvest were (Table 1): Tmax: 19.1 and 23.5 °C (FI) and 19.1 and 25.4 °C (WS); VPDmax: 1.35 and 1.92 kPa (FI) and 1.35 and 2.39 kPa (WS); and SWC: 198.3–542.7 mm (FI) and 180.4–418.8 mm (WS).

Table 1. Dates of sowing, anthesis, grain filling, and harvest, and mean values of grain yield (GY), carbon isotope discrimination in kernels ($\Delta^{13}C$), and stress degree day (SDD = Tc − Ta), determined at the soft dough stage (Z85), for wheat genotypes grown under full irrigation (FI) and water stress (WS) conditions, at Cauquenes (c) and Santa Rosa (sr) in 2011, 2012, 2014, and 2015. Each trial code is a combination of water regime, site, and season. Also, mean values (from sowing to harvest) of daily maximum temperature (Tmax), maximum vapor pressure deficit (VPDmax), and in the case of soil water content (SWC), the average of the daily mean values (from anthesis to mature grain) between 0 and 50 cm depth, are presented.

Trial Code	*n*	Dates					Means					
		Sowing	Anthesis	Grain filling	Harvesting		GY (t ha^{-1})	$\Delta^{13}C$ (‰)	SDD (°C)	Tmax (°C)	VPDmax (kPa)	SWC (mm)
FIsr 2011	384	31 Aug. 2011	24 Nov. 2011	22 Dec. 2011	11 Jan. 2012		8.03	18.0	1.81	23.4	1.76	542.7
FIsr 2012	384	07 Aug. 2012	05 Nov. 2012	19 Nov. 2012	28 Jan. 2013		9.83	18. 8	−1.70	21.5	1.52	404.8
FIsr 2014	64	27 Aug. 2014	24 Nov. 2014	17 Dec. 2014	22 Jan. 2015		9.90	-	1.60	23.5	1.92	212.3
FIsr 2015	64	29 Jul. 2015	20 Nov. 2015	08 Dec. 2015	25 Jan. 2016		9.38	18.8	1.38	21.5	1.63	198.3
FIc 2015	64	18 May 2015	23 Oct. 2015	12 Nov. 2015	23 Dec. 2015		8.46	16.9	1.12	19.1	1.35	246.2
					FI	Average	9.12	18.1	0.84	21.8	1.64	320.9
					FI	SD	0.84	0.90	1.44	1.79	0.22	148.9
					FI	Min.	8.03	16.9	−1.70	19.1	1.35	198.3
					FI	Max.	9.90	18.8	1.81	23.5	1.92	542.7
WSsr 2011	384	31 Aug. 2011	24 Nov. 2011	22 Dec. 2011	11 Jan. 2012		4.81	16.5	6.44	23.4	1.76	418.8
WSc 2011	384	07 Sep. 2011	29 Nov. 2011	13 Dec. 2011	05 Jan. 2012		1.68	14.2	12.29	25.4	2.39	320.9
WSc 2012	384	23 May 2012	11 Sep. 2012	25 Oct. 2012	23 Dec. 2012		3.18	15.0	2.17	20.6	1.51	225.5
WSsr 2015	64	29 Jul. 2015	20 Nov. 2015	08 Dec. 2015	25 Jan. 2016		7.40	18.5	3.31	21.5	1.63	180.4
WSc 2015	64	18 May 2015	23 Oct. 2015	12 Nov. 2015	23 Dec. 2015		8.13	16.9	2.16	19.1	1.35	283.8
					WS	Average	5.04	16.2	5.27	22.0	1.74	285.9
					WS	SD	2.45	1.49	4.29	2.18	0.36	91.8
					WS	Min.	1.68	14.2	2.16	19.1	1.35	180.4
					WS	Max.	8.13	18.5	12.3	25.4	2.39	418.8

Grain yield under WS conditions was 45% lower than under FI (Table 1). Also, the range of variation among seasons was much greater under WS (1.68–8.13 t ha^{-1}) compared to FI (8.03–9.9 t ha^{-1}). The $\Delta^{13}C$ data showed lower values (10.5%) and higher variability under WS conditions compared to FI conditions (Table 1). The average SDD was much higher (5.3 fold) under WS and had greater variability compared to FI conditions (Table 1).

3.2. Relationships between Grain Yield and Canopy and Ambient Temperatures in 384 Wheat Genotypes

The relationship between GY and SDD of the 384 genotypes was negative and highly significant in 2011 (r^2 = 0.52–0.68; $p < 0.001$) (Figure 1A). However, when SSD was compared with $\Delta^{13}C$, the determination coefficients (r^2) were significant only in FIsr and WSsr (0.22 and 0.32, respectively), but not in WSc (Figure 1C). During the second season, r^2 values were much lower and not significant for both GY (r^2 = 0.03–0.12; $p > 0.05$) (Figure 1B) and $\Delta^{13}C$ (r^2 = 0.0002–0.04; $p > 0.05$) (Figure 1D). In terms of environmental conditions, both seasons showed important differences; Tmax (°C) values were higher in 2011 (WSc = 25.4, FIsr and WSsr = 23.4) than in 2012 (WSc = 20.6 and FIsr = 21.5). Consequently, VPDmax (kPa) in 2011 (WSc = 2.4, FIsr and WSsr = 1.8) was higher than in 2012 (WSc and FIsr = 1.5). In the case of SWC (mm), the values in 2011 (WSc = 381.7, FIsr = 550.7 and WSsr = 507.8) exceeded the values recorded in 2012 (WSc = 256.9 and FIsr = 399.0). Also, SDD (°C) was different between seasons, being higher in 2011 (WSc = 12.3, FIsr = 1.8, and WSsr = 6.4) than in 2012 (WSc = 2.2 and FIsr = −1.7).

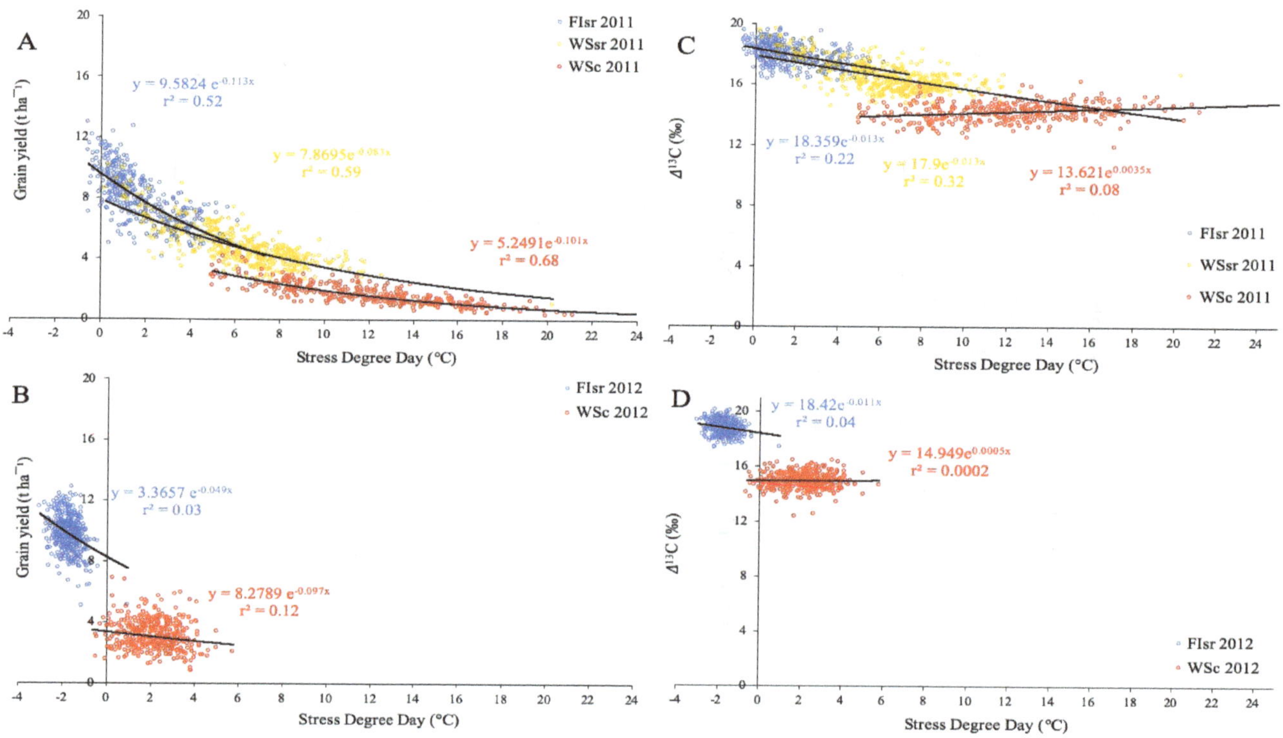

Figure 1. Relationship between stress degree day (SDD = Tc − Ta; where Tc is crop temperature and Ta air temperature, both measured at the soft dough stage (Z85) versus grain yield and carbon isotope discrimination in kernels for 384 spring bread wheat genotypes grown under two water regimes (full irrigation (FI) and water stress (WS)), in two locations (Santa Rosa (sr) and Cauquenes (c)), during the 2011 ((**A**,**C**) respectively) and 2012 seasons ((**B**,**D**), respectively). Regression lines and equations are presented for each water regime and location (determination coefficients are also included).

3.3. Environmental Effects on Grain Yield, Carbon Isotope Discrimination, and Stress Degree Day

The average GY of wheat genotypes under FI and WS conditions indicated different responses to environmental variables (Figure 2). Under WS conditions, GY decreased exponentially as SDD, Tmax, and VPDmax increased, which was not the case under FI conditions (Figure 2A–C). Similarly, $\Delta^{13}C$

also decreased incrementally in SDD, Tmax, and VPDmax (Figure 2E–G). No significant relationships were found between SWC and GY or $\Delta^{13}C$ (Figure 2D,H).

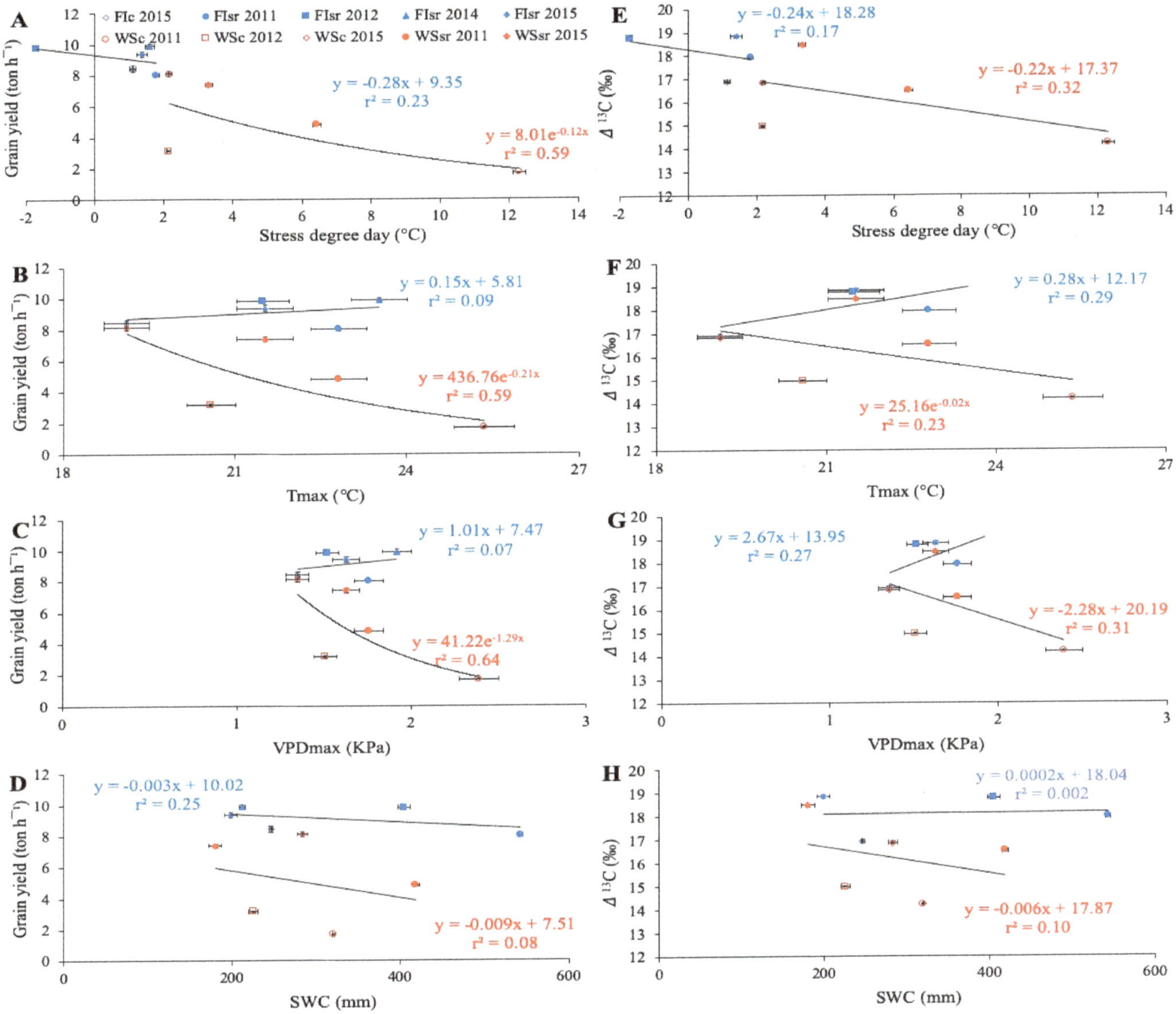

Figure 2. Average grain yield (**A–D**) and carbon isotope discrimination in kernels ($\Delta^{13}C$; **E–H**) of wheat genotypes growing in ten environments, in relation to the stress degree day (SDD = Tc − Ta; Tc is crop temperature and Ta air temperature, both measured at the soft dough stage Z85); **A** and **E**), the seasonal averages of daily maximum temperature (Tmax; **B** and **F**) and maximum vapor pressure deficit (VPDmax; **C** and **G**) and the soil water content between 0 and 50 cm depth (SWC; **D** and **H**). The environments corresponded to the water regime applied (full irrigation (FI) and water stress (WS)), the trial location (Santa Rosa (sr) and Cauquenes (c)), and growing seasons (2011, 2012, 2014, and 2015); the trial code is a combination of these factors. Regression lines and equations are presented for each water regime (determination coefficients are also included).

In relation to the environmental conditions during the study, when all the environments were combined (Table 2), SDD was only correlated with Tmax ($r = 0.64$; $p < 0.05$) and VPDmax ($r = 0.80$; $p < 0.01$). Under each water regime, close and significant relationships were found between SDD and Tmax and VPDmax, but only in plants growing in WS conditions (Figure 3). The relationship between SDD and SWC was not significant under either WS or FI conditions (Figure 3C).

Pearson correlation analysis (Table 2) showed that mean values of GY in the ten environments were highly correlated with SDD ($r = -0.81$; $p < 0.01$) and $\Delta^{13}C$ ($r = 0.92$; $p < 0.01$). Also, $\Delta^{13}C$ was negatively correlated with SDD ($r = -0.71$; $p < 0.05$). In concordance with this, GY and $\Delta^{13}C$ were primarily affected by Tmax and VPDmax but not by SWC (Figure 2B–F,C–G,D–H, respectively).

Table 2. Pearson's correlation matrix for stress degree day (SDD = Tc − Ta; Tc is crop temperature and Ta air temperature, both measured at the soft dough stage Z85), grain yield (GY), carbon isotope discrimination in kernels ($\Delta^{13}C$), and seasonal averages of daily maximum temperature (Tmax), maximum vapor pressure deficit (VPDmax), and soil water content between 0 and 50 cm depth (SWC; between anthesis and mature grain). Data from the water regime applied (full irrigation and water stress), the trial location (Santa Rosa and Cauquenes), and the evaluated season (2011, 2012, 2014, and 2015).

	SDD	**GY**	**$\Delta^{13}C$**	**Tmax**	**VPDmax**
GY	−0.81 **	-	-	-	-
$\Delta^{13}C$	−0.71 *	0.92 **	-	-	-
Tmax	0.64 *	−0.38	−0.16	-	-
VPDmax	0.80 **	−0.52	−0.35	0.95 **	-
SWC	0.06	−0.06	−0.05	0.29	0.16

* Statistically significant ($p < 0.05$); ** Highly statistically significant ($p < 0.01$).

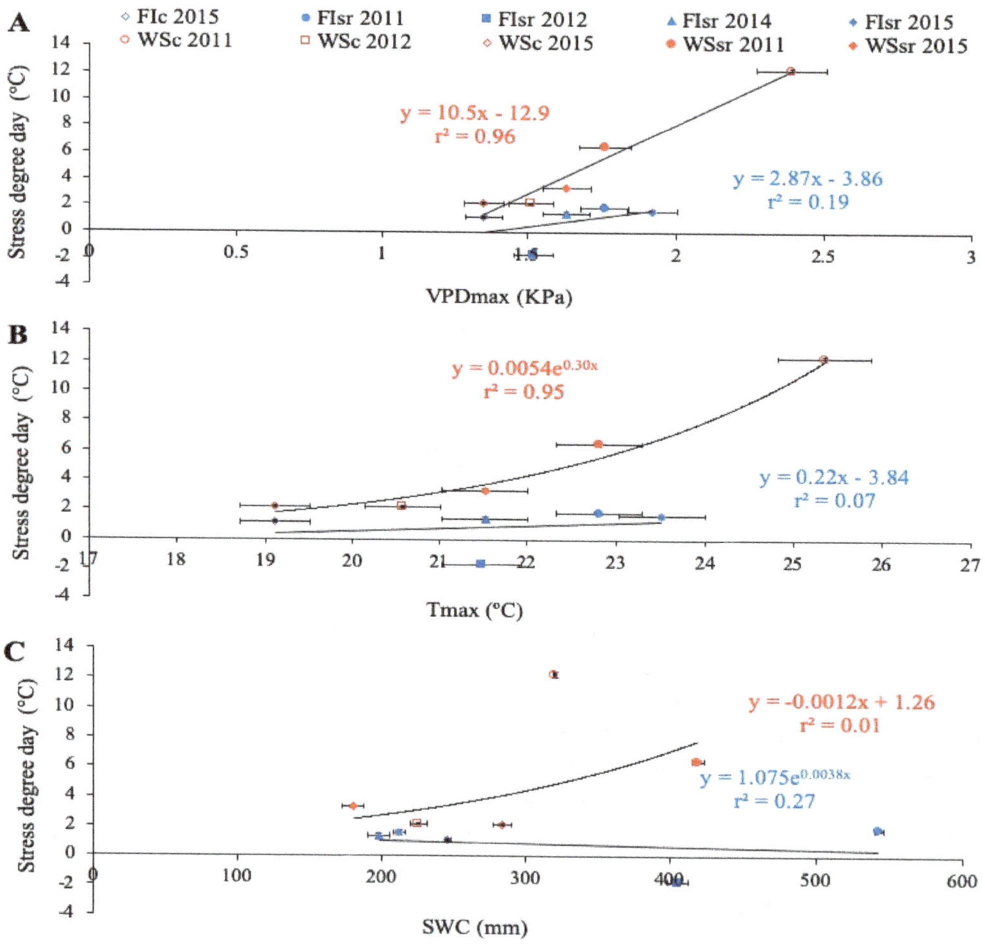

Figure 3. Relationships between the stress degree day (SDD = Tc − Ta; Tc is crop temperature and Ta air temperature, both measured at the soft dough stage Z85), and the seasonal averages of daily maximum vapor pressure deficit (VPDmax) (**A**), maximum temperature (Tmax; **B**) and soil water content between 0 and 50 cm depth (SWC; (**C**)). Mean values were the average of all genotypes growing in the particular environment according to the water regime (full irrigation (FI) and water stress (WS)), the trial location (Santa Rosa (sr) and Cauquenes (c)), and growing seasons (2011, 2012, 2014, and 2015). Regression lines and equations are presented for each water regime (determination coefficients are also included).

Finally, when the environmental, phenological, physiological, and productive information (Table 1) was included to generate the two- and three-group cluster dendrograms and plots (Figure 4A,B and Figure 4C,D, respectively), the main difference was found in the division of the first branch of the less stressful environment (green lines in Figure 4A) at the height of the first knot of the most stressful environment (origin of the blue and green lines in Figure 4B). Differences between the cluster plots were according to changes in the cluster dendrograms; the three groups were (Figure 4B): (i) lowest environmental limitations: FIsr 2012, FIsr 2014, FIsr 2015, and WSsr 2015; (ii) intermediate environmental limitations: WSc 2012, Fic 2015, and WSc 2015; and (iii) highest environmental limitations: WSc 2011, FIsr 2011, and WSsr 2011. The cluster plot that explains 61.3% of the variance (Figure 4C) shows a clear distance or separation between the groups with the lowest and the highest environmental constraints (green and red colors in Figure 4, respectively). In the three-group cluster plot, two of the groups overlap. However, even though WSc 2011, FIsr 2011, and WSsr 2011 had the highest SWC values, they also presented, on average, the greatest VPDmax, Tmax and SDD but the lowest GY (Table 1).

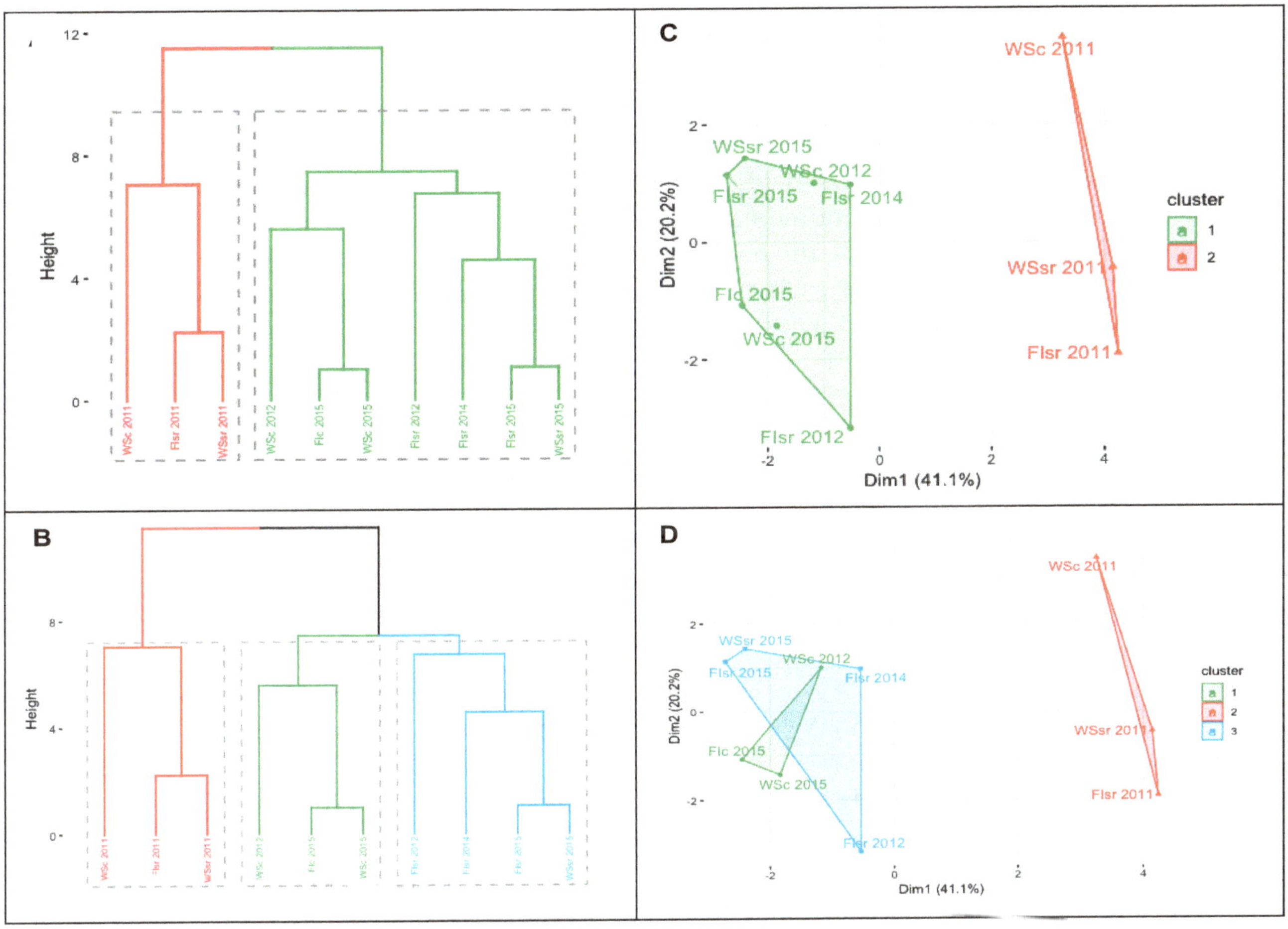

Figure 4. Cluster dendrogram ((**A**) two groups and (**B**) three groups) and plot ((**C**) two groups and (**D**) three groups) for the general characterization of the assessed environments according to the water regime applied (full irrigation (FI) and water stress (WS)), the trial location (Santa Rosa (sr) and Cauquenes (c)), and the evaluated season (2011, 2012, 2014, and 2015); the trial code is a combination of these factors. Data included the phenological (dates), productive (grain yield—GY), physiological (carbon isotope discrimination in kernels (Δ^{13}C) and the stress degree day measured at the soft dough stage Z85 (SDD)), and environmental information (seasonal averages of daily maximum temperature and maximum vapor pressure deficit, and the soil water content between 0 and 50 cm depth). In the case of GY, Δ^{13}C, and SDD, the mean values analyzed were the average of all genotypes growing in the particular environment.

4. Discussion

4.1. Environmental Effects on Grain Yield and Carbon Isotope Discrimination

Tolerance to WS usually implies some improvement or maintenance of metabolic processes that enables the plant to regulate cell water status and maintain leaf turgor under stressful conditions. One of the first mechanisms involved in reducing water loss by transpiration is stomatal control, which partially closes the stomata, thus affecting carbon assimilation and storage [44]. This gas exchange limitation between the atmosphere and the substomatal cavity is primarily driven by the surrounding environmental conditions (e.g., water availability, ambient temperature, relative humidity, wind speed, light intensity). To the extent that the diffusion of CO_2 through the stomata is more restrictive, the carbon isotope discrimination ($\Delta^{13}C$) between ^{12}C and ^{13}C will also be reduced, increasing the proportion of ^{13}C [24]. Therefore, in a particular environment, $\Delta^{13}C$ at the grain level provides an integrated assessment of the transpiration efficiency during the whole season [25,45]. As in other cereal studies [24,35,46–49], GY and $\Delta^{13}C$ in the current work had a strong and positive association ($r = 0.92$) (Table 2). Additionally, the evaluated environmental conditions generated similar responses in GY and $\Delta^{13}C$ (Figure 2), reaffirming the strong relationship that existed between these characters.

The SWC has also been used as an indicator of water stress in plants and is positively related to GY in wheat [25,50–52]. Working in the same species, Reference [53] evaluated the effect of water content in different soil profiles, concluding that a soil that was well irrigated throughout the first 50 cm of the profile obtained a greater yield and harvest index than a soil with dry upper layers. In the present work, even though FI environments always showed higher SWC than under WS in the same location and season (Figure S2A–D), the results did not show a significant correlation between SWC and GY (Table 2). Moreover, while the GY and $\Delta^{13}C$ were both higher under FI than WS (Figure 2D,H), neither GY nor $\Delta^{13}C$ were affected by increases in SWC (~200 to 550 mm) under either FI or WS.

Contrastingly, the environmental water demand (VPD; [54]), which is mainly driven by the ambient temperature and relative humidity, proved to influence both GY and $\Delta^{13}C$ under WS but not in FI (Figure 2C,G). The combination of high temperatures and low relative humidity, which is frequently encountered in the late stages of the growing season in Mediterranean climates (e.g., Santa Rosa and Cauquenes), caused an increase in the VPD. References [55,56] have assessed the effect of environmental variables on wheat physiology and GY, proposing that a high VPD environment should vary between 2.5 and 3.9 kPa. Therefore, the average values of VPDmax found in the present study (1.35 to 2.39 kPa; Figure 2C,G) could be considered moderately low to moderately high, although maximum values reached as high as 6.34 kPa in WSc in 2011 (Figure S1B).

Several studies have proven that growing cereals under non-limiting water conditions but with high VPD values leads to reduced GY and $\Delta^{13}C$ [21,55–57]. The present work, considering all measurements performed, shows a non-significant relationship between GY and VPDmax (Table 2). When FI and WS were analyzed separately, it was only the genotypes growing under WS that showed lower GY and $\Delta^{13}C$ as VPDmax increased (Figure 2C,G) and the VPD had a higher association with GY than $\Delta^{13}C$ (Table 2).

Likewise, the Tmax trends were similar to VPDmax (Figure 2B,F). However, despite the Tmax and VPDmax being relatively low (20.6 °C and 1.51 kPa, respectively) in the WSc 2012 trial, the lowest SWC (256.9 mm) was registered, especially after anthesis, and this generated a low GY (3.18 t ha^{-1}) (Table 1 and Figure S2B). On the other hand, the WSc 2011 trial had a relatively adequate SWC (320 mm) between anthesis and grain filling (Figure S2B), but due to the late sowing date (Table 1), the plants were exposed to higher Tmax and VPDmax (25.4 °C and 2.39 kPa, respectively) (Table 1), reaching ~40 °C and ~6 kPa for Tmax and VPDmax, respectively (Figure S1A,B), resulting in this trial having the lowest GY (1.68 t ha^{-1}).

Under high VPD, guard cell turgor may be decreased by direct evaporative losses from the guard cells and/or decreased water supply to the guard cells if the root or shoot hydraulic conductance is limiting [58], and this leads to a detrimental effect on plant production. Reference [56] tested the effect of VPD and ambient temperature on gas exchange and GY in wheat, finding that environments with high VPD (3.9 kPa) and high temperature (36 °C) increased respiration by up to 22% and decreased photosynthetic water-use efficiency by up to 64% compared to environments with high temperature and lower VPD (1.5 kPa). Indeed, environments with high VPD and temperature caused a reduction in leaf area and net assimilation of CO_2; however, in the case of plants under the same conditions but without water restriction, there was no decrease in GY. The same authors showed that GY was reduced by 7% in environments with water stress compared to no stress, which is concordant with the findings of the present study, where at similar VPDmax and Tmax (between 1.5 and 2 kPa, and between 21 and 24 °C, respectively), plants grown under WS showed lower GY than plants under FI (Figure 2B,C). In FI conditions, the plants had a GY that was higher than 8 t ha^{-1}, while in WS environments, the GY never exceeded that threshold; an exception was WSc 2015 (8.1 t ha^{-1}), which was influenced by an abnormally rainy season ("El Niño" phenomenon; Figure S2D).

4.2. *The Potential of Stress Degree Day to Estimate Grain Yield*

Stomatal closure causes a decrease in the transpiration rate, and as a consequence, there is a reduction in the cooling effect, which finally increases leaf/canopy temperature [10]. The reduction in the stomatal conductance could be a consequence of the limitations of the roots to absorb enough water to supply the atmospheric water demand [6]. Numerous studies have confirmed that the temperature of the canopy is associated with crop yield [59–61], as well as a series of physiological characteristics, including stomatal conductance [11], the hydric state of the plant [59], and the presence of deep roots.

In general, the present study establishes a negative and highly significant correlation ($r = -0.81$) between SDD and GY (Table 2). Analysis of the responses according to water regime (FI or WS) indicated that there was no significant relationship between GY and SDD in plants growing under FI conditions, but under WS conditions the correlation was moderately high ($r^2 = 0.59$) (Figure 2A). The $\Delta^{13}C$ followed the same pattern, but with lower determination coefficients (FI = 0.17 and WS = 0.34). When Reference [12] studied the relationship between GY and Tc in wheat genotypes grown with similar water regimes (FI and WS), they also found a stronger association under WS ($r^2 = 0.66$) than FI ($r^2 = 0.58$).

Despite similarities between the studies described above, there are also contradictory results for the relationship between SDD and GY. For example, References [62,63] found no significant relationships, whereas Reference [11] found a high and significant association in irrigated environments. These differences could be explained by the lower VPD registered in the studies of References [62,63] (~2.4 kPa) compared to that of Reference [11] (~5.5 kPa), with the latter case allowing a greater expression of the tolerance of each genotype to the environmental conditions. Although in the present work, there were trials that reached a VPDmax of 6.5 kPa (WSc 2011) (Figure S1), the seasonal averages were ~2.4 kPa.

Therefore, the low SDD values of plants growing under FI is likely due to the ability to meet the water demand of the air (VPD), thus maintaining a high transpiratory rate and allowing the plants to cool down their leaves; under this condition there is more CO_2 fixation, explaining the higher yields in FI. In this kind of environment where soil water availability is enough to compensate for VPD, the plants do not need to express their water deficit tolerance mechanisms, which in this case means there is a lower SDD versus GY data dispersion, implying lower coefficients of determination.

Similar to the GY and $\Delta^{13}C$, SDD was more sensitive to the VPDmax and Tmax than to SWC, with the WS environment having the most effect on plant temperature. Interestingly, an SDD of 2 °C seems to be the threshold between FI and WS environments (Figure 2A,E); SDD averages in FI were lower than 2 °C, while in WS they were greater than 2 °C.

When the relationship between SDD and GY was studied in individual genotypes under contrasting environments (seasons 2011 and 2012), the association (r^2 values) between these two variables depended on the environment. While in 2011 the relationships in FIsr, WSsr, and WSc were negative and moderately high ($r^2 = 0.52$, 0.59, and 0.68, respectively) (Figure 1A), the FIsr and WSc relationships in 2012 were low ($r^2 = 0.03$ and 0.12, respectively; WSsr was not sown in 2012) (Figure 1B). As seen before by the use of the average values per environment, the best determination coefficients were observed in more stressful conditions (FI < WSsr < WSc), likely associated with the higher trait-range during the first season; Fisr 2011, WSsr 2011, and WSc 2011 had a higher SDD data dispersion in relation to FIsr 2012 and WSc 2012 (Figure 1 and Figure S1E).

Except for the WSsr 2011 trial, the minimum SDD values of 2011 corresponded, approximately, to the maximums registered during 2012. As already explained, GY is influenced by SDD (Figure 2A,E, respectively), which in turn depends on the VPDmax and Tmax (Figure 3A,B). In this sense, although there was less SWC during the second season, Tmax and VPDmax were lower too, reaffirming that these last two variables would have a more significant impact on the transpiratory and cooling capacity than even the SWC.

Finally, because the main differences in the dendrograms between the two- and three-environment groups were found in the division of the first branch of the less stressful environment (green lines in Figure 4A) at the height of the first knot of the most stressful environment (origin of blue and green lines in Figure 4B), it is logical to think that there were at least three environmental conditions across the trials. In order to establish the possible differences between the superimposed groups (lower and intermediate environmental constraints; green and blue lines in Figure 4D), each trait average was contrasted (percentage of change) with the respective value in the higher environmental limitations group (2011 data; red line in Figure 4D). Using this as a form of normalization, it was possible to establish which traits had the largest and smallest differences between the superimposed groups; GY varied by 52.4% while SDD, Δ^{13}C, Tmax, VPDmax, and SWC only varied within the range of 9.77 and 15.1% (data not shown), and these were traits that probably shared similar spatial coordinates in the cluster plot (Figure 4B).

For modeling purposes, it would then be desirable to cease grouping collected data according to text code treatments (e.g., FI and WS according to only the amount of applied water) and start associating them with the environmental conditions existing during the season (e.g., VPDmax). For an adequate estimation of GY and carbon isotope discrimination by thermal imaging, we will then need to use more complex models (e.g., tree-based neural networks) that allow us to identify the "type of environment" in which the collected data should be manipulated to generate and apply the appropriate model. Thus, a deeper environmental characterization would allow development of models with better fit and consistency between years.

5. Conclusions and Future Perspectives

The ability to predict GY through the use of thermal images is highly variable and will not only depend on the amount of water stored in the soil profile, but also on other environmental variables such as VPDmax and Tmax. To the extent that better environmental characterization can be achieved, an objective and integral classification of the assessed environment should then be possible. Because the environmental information usually originates from a standard AWS, characterization of the environment at the canopy level or in the first few centimeters above it would also be an important consideration. This would help to generate models with better predictive capacity, thus improving the consistency between seasons.

Supplementary Materials:

Figure S1: (**A**) Canopy temperature—air temperature (SDD; (A) at the early dough stage of grain filling (Zadoks Z83); (**B**) grain yield; (**C**) daily maximum temperature (from sowing to harvest; carbon isotope discrimination (Δ^{13}C) in kernels; and (**D**) daily maximum VPD from sowing to harvest, for genotypes of wheat grown under full irrigation (FI) and water stress (WS) at Santa Rosa (sr) and Cauquenes (c) during four growing seasons (2011, 2012, 2014, and 2015). Box and whiskers show minimum, 25th percentile, median, mean, 75th percentile, and maximum values. Open symbols represent outlier data; Figure S2: Soil water content between 0 and 50 cm depth and precipitation according to the water regime applied (full irrigation (FI) and water stress (WS)), the trial location (Santa Rosa (sr) and Cauquenes (c)), and the evaluated seasons: 2011 (**A**), 2012 (**B**), 2014 (**C**), and 2015 (**D**); the trial code is a combination of these factors. Bars represent the precipitation and the arrows the phenological stages at Santa Rosa (orange) and Cauquenes (black). Solid arrows indicate anthesis and dashed arrows the early dough stage (Z85 from the Zadoks scale).

Author Contributions: S.R.-B., A.d.P., and G.A.L. contributed to the conception and design of the work. S.R.-B., A.M.M.-E., M.G., F.E., A.E., L.G.-M., C.P.-E., D.S., A.d.P. and G.A.L. performed acquisition, analysis, and interpretation of the data. D.C. and I.M. was in charge of the management of the field experiments and evaluation of agronomic traits. S.R.-B., A.d.P. and G.A.L. collaborated to generate and validate the version for publication.

References

1. Tanner, C.B. Plant Temperatures 1. *Agron. J.* **1963**, *55*, 210–211. [CrossRef]
2. Zia, S.; Spohrer, K.; Merkt, N.; Wenyong, D.; He, X.; Müller, J. Non-invasive water status detection in grapevine (*Vitis vinifera* L.) by thermography. *Int. J. Agric. Biol. Eng.* **2009**, *2*, 46–54. [CrossRef]
3. Jones, H.G.; Stoll, M.; Santos, T.; Sousa, C.D.; Chaves, M.M.; Grant, O.M. Use of infrared thermography for monitoring stomatal closure in the field: Application to grapevine. *J. Exp. Bot.* **2002**, *53*, 2249–2260. [CrossRef] [PubMed]
4. Leinonen, I.; Jones, H.G. Combining thermal and visible imagery for estimating canopy temperature and identifying plant stress. *J. Exp. Bot.* **2004**, *55*, 1423–1431. [CrossRef] [PubMed]
5. Möller, M.; Alchanatis, V.; Cohen, Y.; Meron, M.; Tsipris, J.; Naor, A.; Cohen, S. Use of thermal and visible imagery for estimating crop water status of irrigated grapevine. *J. Exp. Bot.* **2007**, *58*, 827–838. [CrossRef] [PubMed]
6. Wang, W.S.; Pan, Y.J.; Zhao, X.Q.; Dwivedi, D.; Zhu, L.H.; Ali, J.; Li, Z.K. Drought-induced site-specific DNA methylation and its association with drought tolerance in rice (*Oryza sativa* L.). *J. Exp. Bot.* **2010**, *62*, 1951–1960. [CrossRef] [PubMed]
7. Zia, S.; Spohrer, K.; Wenyong, D.; Spreer, W.; Romano, G.; Xiongkui, H.; Joachim, M. Monitoring physiological responses to water stress in two maize varieties by infrared thermography. *Int. J. Agric. Biol. Eng.* **2011**, *4*, 7–15. [CrossRef]
8. Jackson, R.D.; Idso, S.B.; Reginato, R.J.; Pinter, P.J., Jr. Canopy temperature as a crop water stress indicator. *Water Resour. Res.* **1981**, *17*, 1133–1138. [CrossRef]
9. Luquet, D.; Bégué, A.; Vidal, A.; Clouvel, P.; Dauzat, J.; Olioso, A.; Tao, Y. Using multidirectional thermography to characterize water status of cotton. *Remote Sens. Environ.* **2003**, *84*, 411–421. [CrossRef]
10. Jones, H.G. Application of thermal imaging and infrared sensing in plant physiology and ecophysiology. *Adv. Bot. Res.* **2004**, *41*, 107–163. [CrossRef]
11. Amani, I.; Fischer, R.A.; Reynolds, M.P. Canopy temperature depression association with yield of irrigated spring wheat cultivars in a hot climate. *J. Agron. Crop Sci.* **1996**, *176*, 119–129. [CrossRef]
12. Cossani, C.M.; Pietragalla, J.; Reynolds, M. Temperatura del dosel vegetal y características de la relación planta-agua. In *Fitomejoramiento fisiológico I: Enfoques interdisciplinarios para mejorar la adaptación del cultivo*; Reynolds, M.P., Pask, A.J.D., Mullan, D.M., Chávez-Dulanto, P.N., Eds.; CIMMYT: Ciudad de Mexico, Mexico, 2013; pp. 60–68. ISBN 978-607-8263-18-9.
13. Fischer, R.A.; Rees, D.; Sayre, K.D.; Lu, Z.M.; Condon, A.G.; Saavedra, A.L. Wheat yield progress associated with higher stomatal conductance and photosynthetic rate, and cooler canopies. *Crop Sci.* **1998**, *38*, 1467–1475. [CrossRef]
14. Leinonen, I.; Grant, O.M.; Tagliavia, C.P.P.; Chaves, M.M.; Jones, H.G. Estimating stomatal conductance with thermal imagery. *Plant Cell Environ.* **2006**, *29*, 1508–1518. [CrossRef] [PubMed]
15. Grant, O.M.; Tronina, Ł.; Jones, H.G.; Chaves, M. Exploring thermal imaging variables for the detection of stress responses in grapevine under different irrigation regimes. *J. Exp. Bot.* **2007**, *58*, 815–825. [CrossRef] [PubMed]
16. Prashar, A.; Jones, H.G. Infra-red thermography as a high-throughput tool for field phenotyping. *Agronomy* **2014**, *4*, 397–417. [CrossRef]

17. Prashar, A.; Jones, H.G. Assessing drought responses using thermal infrared imaging. In *Environmental Responses in Plants. Methods and Protocols, Methods in Molecular Biology*; Duque, P., Ed.; Humana Press: New York, NY, USA, 2016; Volume 1398, pp. 209–219. [CrossRef]

18. Blum, A.; Shpiler, L.; Golan, G.; Mayer, J. Yield stability and canopy temperature of wheat genotypes under drought-stress. *Field Crops Res.* **1989**, *22*, 289–296. [CrossRef]

19. Condon, A.G.; Richards, R.A.; Farquhar, G.D. Carbon isotope discrimination is positively correlated with grain yield and dry matter production in field-grown wheat. *Crop Sci.* **1987**, *27*, 996–1001. [CrossRef]

20. Farquhar, G.D.; Hubick, K.T.; Condon, A.G.; Richards, R.A. Carbon isotope fractionation and plant water-use efficiency. In *Stable Isotopes in Ecological Research*; Rundel, P.W., Ehleringer, J.R., Nagy, K.A., Eds.; Springer: New York, NY, USA, 1989; pp. 21–40. [CrossRef]

21. Condon, A.G.; Richards, R.A.; Farquhar, G.D. The effect of variation in soil water availability, vapour pressure deficit and nitrogen nutrition on carbon isotope discrimination in wheat. *Aust. J. Agric. Res.* **1992**, *43*, 935–947. [CrossRef]

22. Araus, J.L.; Amaro, T.; Casadesus, J.; Asbati, A.; Nachit, M.M. Relationships between ash content, carbon isotope discrimination and yield in durum wheat. *Funct. Plant Biol.* **1998**, *25*, 835–842. [CrossRef]

23. Araus, J.L.; Villegas, D.; Aparicio, N.; Del Moral, L.F.; El Hani, S.; Rharrabti, Y.; Royo, C. Environmental factors determining carbon isotope discrimination and yield in durum wheat under Mediterranean conditions. *Crop Sci.* **2003**, *43*, 170–180. [CrossRef]

24. Lobos, G.A.; Matus, I.; Rodriguez, A.; Romero-Bravo, S.; Araus, J.L.; del Pozo, A. Wheat genotypic variability in grain yield and carbon isotope discrimination under mediterranean conditions assessed by spectral reflectance. *J. Integr. Plant Biol.* **2014**, *56*, 470–479. [CrossRef] [PubMed]

25. Cabrera-Bosquet, L.; Albrizio, R.; Nogués, S.; Araus, J.L. Dual $\Delta^{13}C/\delta^{18}O$ response to water and nitrogen availability and its relationship with yield in field-grown durum wheat. *Plant Cell Environ.* **2010**, *34*, 418–433. [CrossRef] [PubMed]

26. Garriga, M.; Romero-Bravo, S.; Estrada, F.; Escobar, A.; Matus, I.; del Pozo, A.; Astudillo, C.; Lobos, G. Assessing wheat traits by spectral reflectance: Do we really need to focus on predicted trait-values or directly identify the elite genotypes group? *Front. Plant Sci.* **2017**, *8*, 280. [CrossRef] [PubMed]

27. Moretti, C.L.; Mattos, L.M.; Calbo, A.G.; Sargent, S.A. Climate changes and potential impacts on postharvest quality of fruit and vegetable crops: A review. *Food Res. Int.* **2010**, *43*, 1824–1832. [CrossRef]

28. Lobos, G.A.; Hancock, J.F. Breeding blueberries for a changing global environment: A review. *Front. Plant Sci.* **2015**, *6*, 782. [CrossRef] [PubMed]

29. Camargo, A.V.; Lobos, G.A. Latin America: A development pole for phenomics. *Front. Plant Sci.* **2016**, *7*, 1729. [CrossRef] [PubMed]

30. Lobos, G.A.; Camargo, A.V.; del Pozo, A.; Araus, J.L.; Ortiz, R.; Doonan, J.H. Plant Phenotype and phenomics for plant breeding. *Front. Plant Sci.* **2017**, *8*, 2181. [CrossRef]

31. Sadras, V.O.; Reynolds, M.P.; De la Vega, A.J.; Petrie, P.R.; Robinson, R. Phenotypic plasticity of yield and phenology in wheat, sunflower and grapevine. *Field Crops Res.* **2009**, *110*, 242–250. [CrossRef]

32. Mora, F.; Castillo, D.; Lado, B.; Matus, I.; Poland, P.; Belzile, F.; von Zitzewitz, J.; del Pozo, A. Genome-wide association mapping of agronomic traits and carbon discrimination in a worldwide germplasm collection of spring wheat using SNP markers. *Mol. Breed.* **2015**, *35*. [CrossRef]

33. del Pozo, A.; Yáñez, A.; Matus, I.A.; Tapia, G.; Castillo, D.; Sanchez-Jardón, L.; Araus, J.L. Physiological traits associated with wheat yield potential and performance under water-stress in a mediterranean environment. *Front. Plant Sci.* **2016**, *7*, 987. [CrossRef]

34. Zhang, D.; Du, Q.; Zhang, Z.; Jiao, X.; Song, X.; Li, J. Vapour pressure deficit control in relation to water transport and water productivity in greenhouse tomato production during summer. *Sci. Rep.* **2017**, *7*, 43461. [CrossRef] [PubMed]

35. Hernández, J.; Lobos, G.A.; Matus, I.; del Pozo, A.; Silva, P.; Galleguillos, M. Using ridge regression models to estimate grain yield from field spectral data in bread wheat (*Triticum aestivum* L.) grown under three water regimes. *Remote Sens.* **2015**, *7*, 2109–2126. [CrossRef]

36. Zadoks, J.C.; Chang, T.T.; Konzak, C.F. A decimal code for the growth stages of cereals. *Weed Res.* **1974**, *14*, 415–421. [CrossRef]
37. Allen, R.G.; Pereira, L.S.; Raes, D.; Smith, M. Crop Evapotranspiration: Guidelines for Computing Crop Water Requirements. In *FAO Irrigation and Drainage Paper N° 56*; FAO: Rome, Italy, 1998; 300p.
38. Fuentes, S.; De Bei, R.; Pech, J.; Tyerman, S. Computational water stress indices obtained from thermal image analysis of grapevine canopies. *Irrig. Sci.* **2012**, *30*, 523–536. [CrossRef]
39. Idso, S.B.; Jackson, R.D.; Reginato, R.J. Remote sensing of crop yields. *Science* **1977**, *196*, 19–25. [CrossRef]
40. Jackson, R.D.; Reginato, R.J.; Idso, S.B. Wheat canopy temperature: A practical tool for evaluating water requirements. *Water Resour. Res.* **1977**, *13*, 651–656. [CrossRef]
41. Nerurkar, P.; Shirke, A.; Chandane, M.; Bhirud, S. Empirical analysis of data clustering algorithms. *Procedia Comput. Sci.* **2018**, *125*, 770–779. [CrossRef]
42. Murtagh, F.; Legendre, P. Ward's hierarchical agglomerative clustering method: Which algorithms implement ward's criterion? *J. Classif.* **2014**, *31*, 274–295. [CrossRef]
43. R Development Core Team. *R: A Language and Environment for Statistical Computing*; R Foundation for Statistical Computing: Vienna, Austria, 2011.
44. Ryan, A.C.; Dodd, I.C.; Rothwell, S.A.; Jones, R.; Tardieu, F.; Draye, X.; Davies, W.J. Gravimetric phenotyping of whole plant transpiration responses to atmospheric vapour pressure deficit identifies genotypic variation in water use efficiency. *Plant Sci.* **2016**, *251*, 101–109. [CrossRef]
45. Farquhar, G.D.; O'Leary, M.H.; Berry, J.A. On the relationship between carbon isotope discrimination and the intercellular carbon dioxide concentration in leaves. *Funct. Plant Biol.* **1982**, *9*, 121–137. [CrossRef]
46. Acevedo, E.H.; Baginsky, C.G.; Solar, B.R.; Ceccarelli, S. Discriminación isotópica de C13 y su relación con el rendimiento y la eficiencia de transpiracion de genotipos locales y mejorados de cebada bajo diferentes condiciones hídricas. *Inv. Agric.* **1997**, *17*, 41–54.
47. Rebetzke, G.J.; Condon, A.G.; Farquhar, G.D.; Appels, R.; Richards, R.A. Quantitative trait loci for carbon isotope discrimination are repeatable across environments and wheat mapping populations. *Theor. Appl. Genet.* **2008**, *118*, 123–137. [CrossRef] [PubMed]
48. del Pozo, A.; Castillo, D.; Inostroza, L.; Matus, I.; Méndez, A.M.; Morcuende, R. Physiological and yield responses of recombinant chromosome substitution lines of barley to terminal drought in a mediterranean type environment. *Ann. Appl. Biol.* **2012**, *160*, 157–167. [CrossRef]
49. Araus, J.L.; Cabrera-Bosquet, L.; Serret, M.D.; Bort, J.; Nieto-Taladriz, M.T. Comparative performance of $\delta^{13}C$, $\delta^{18}O$ and $\delta^{15}N$ for phenotyping durum wheat adaptation to a dryland environment. *Funct. Plant Biol.* **2013**, *40*, 595–608. [CrossRef]
50. Passioura, J.B. Grain yield, harvest index, and water use of wheat. *J. Aust. Inst. Agric. Sci.* **1977**, *43*, 117–120.
51. French, R.J.; Schultz, J.E. Water use efficiency of wheat in a Mediterranean-type environment. I. The relation between yield, water use and climate. *Aust. J. Agric. Res.* **1984**, *35*, 743–764. [CrossRef]
52. Sun, H.Y.; Liu, C.M.; Zhang, X.Y.; Shen, Y.J.; Zhang, Y.Q. Effects of irrigation on water balance, yield and WUE of winter wheat in the north China plain. *Agric. Water Manag.* **2006**, *85*, 211–218. [CrossRef]
53. Li, F.M.; Liu, X.L.; Li, S.Q. Effects of early soil water distribution on the dry matter partition between roots and shoots of winter wheat. *Agric. Water Manag.* **2001**, *49*, 163–171. [CrossRef]
54. Anderson, D.B. Relative humidity or vapor pressure deficit. *Ecology* **1936**, *17*, 277–282. [CrossRef]
55. Dreccer, M.F.; Fainges, J.; Whish, J.; Ogbonnaya, F.C.; Sadras, V.O. Comparison of sensitive stages of wheat, barley, canola, chickpea and field pea to temperature and water stress across Australia. *Agric. For. Meteorol.* **2018**, *248*, 275–294. [CrossRef]
56. Rashid, M.A.; Andersen, M.N.; Wollenweber, B.; Zhang, X.; Olesen, J.E. Acclimation to higher VPD and temperature minimized negative effects on assimilation and grain yield of wheat. *Agric. For. Meteorol.* **2018**, *248*, 119–129. [CrossRef]
57. Sharifi, M.R.; Rundel, P.W. The effect of vapour pressure deficit on carbon isotope discrimination in the desert shrub *Larrea tridentata* (creosote bush). *J. Exp. Bot.* **1993**, *44*, 481–487. [CrossRef]
58. Franks, P.J. Stomatal control and hydraulic conductance, with special reference to tall trees. *Tree Physiol.* **2004**, *24*, 865–878. [CrossRef] [PubMed]
59. Blum, A.; Mayer, J.; Gozlan, G. Infrared thermal sensing of plant canopies as a screening technique for dehydration avoidance in wheat. *Field Crops Res.* **1982**, *5*, 137–146. [CrossRef]

60. Reynolds, M.P.; Balota, M.; Delgado, M.I.B.; Amani, I.; Fischer, R.A. Physiological and morphological traits associated with spring wheat yield under hot, irrigated conditions. *Funct. Plant Biol.* **1994**, *21*, 717–730. [CrossRef]
61. Olivares-Villegas, J.J.; Reynolds, M.P.; McDonald, G.K. Drought-adaptive attributes in the Seri/Babax hexaploid wheat population. *Funct. Plant Biol.* **2007**, *34*, 189–203. [CrossRef]
62. Idso, S.B.; Reginato, R.J.; Clawson, K.L.; Anderson, M.G. On the stability of non-water-stressed baselines. *Agric. For. Meteorol.* **1984**, *32*, 177–182. [CrossRef]
63. Hatfield, J.L. Measuring plant stress with an infrared thermometer. *HortScience* **1990**, *25*, 1535–1538.

Permissions

All chapters in this book were first published in MDPI; hereby published with permission under the Creative Commons Attribution License or equivalent. Every chapter published in this book has been scrutinized by our experts. Their significance has been extensively debated. The topics covered herein carry significant findings which will fuel the growth of the discipline. They may even be implemented as practical applications or may be referred to as a beginning point for another development.

The contributors of this book come from diverse backgrounds, making this book a truly international effort. This book will bring forth new frontiers with its revolutionizing research information and detailed analysis of the nascent developments around the world.

We would like to thank all the contributing authors for lending their expertise to make the book truly unique. They have played a crucial role in the development of this book. Without their invaluable contributions this book wouldn't have been possible. They have made vital efforts to compile up to date information on the varied aspects of this subject to make this book a valuable addition to the collection of many professionals and students.

This book was conceptualized with the vision of imparting up-to-date information and advanced data in this field. To ensure the same, a matchless editorial board was set up. Every individual on the board went through rigorous rounds of assessment to prove their worth. After which they invested a large part of their time researching and compiling the most relevant data for our readers.

The editorial board has been involved in producing this book since its inception. They have spent rigorous hours researching and exploring the diverse topics which have resulted in the successful publishing of this book. They have passed on their knowledge of decades through this book. To expedite this challenging task, the publisher supported the team at every step. A small team of assistant editors was also appointed to further simplify the editing procedure and attain best results for the readers.

Apart from the editorial board, the designing team has also invested a significant amount of their time in understanding the subject and creating the most relevant covers. They scrutinized every image to scout for the most suitable representation of the subject and create an appropriate cover for the book.

The publishing team has been an ardent support to the editorial, designing and production team. Their endless efforts to recruit the best for this project, has resulted in the accomplishment of this book. They are a veteran in the field of academics and their pool of knowledge is as vast as their experience in printing. Their expertise and guidance has proved useful at every step. Their uncompromising quality standards have made this book an exceptional effort. Their encouragement from time to time has been an inspiration for everyone.

The publisher and the editorial board hope that this book will prove to be a valuable piece of knowledge for researchers, students, practitioners and scholars across the globe.

List of Contributors

Ariel Freidenreich, Gabriel Barraza, Krishnaswamy Jayachandran and Amir Ali Khoddamzadeh
Department of Earth and Environment, Florida International University, 11200 SW 8th St, Miami, FL 33199, USA

Gerassimos G. Peteinatos and Roland Gerhards
Institute of Phytomedicine, University Hohenheim, Otto-Sander-Straße 5, Stuttgart 70599, Germany

Audun Korsaetha and Therese W. Berge
Norwegian Institute of Bioeconomy Research (NIBIO), N-1431 Ås, Norway

Daniele Vidal and Rui Pitarma
Unit for Inland Development, Polytechnic of Guarda, Avenida Doutor Francisco Sá Carneiro N°50, 6300-559 Guarda, Portugal

Eric T. Psota, Mateusz Mittek and Lance C. Pérez
Department of Electrical and Computer Engineering, University of Nebraska–Lincoln, Lincoln, NE 68505, USA

Ty Schmidt and Benny Mote
Department of Animal Science, University of Nebraska–Lincoln, Lincoln, NE 68588, USA

Yuxuan Wang, Shamaila Zia-Khan, Sebastian Owusu-Adu and Joachim Müller
Tropics and Subtropics Group (440e), Institute of Agricultural Engineering, Universität Hohenheim, Garbenstraße 9, 70593 Stuttgart, Germany

Thomas Miedaner
State Plant Breeding Institute (720), Universität Hohenheim, Fruwirthstr. 21, 70593 Stuttgart, Germany

Sigfredo Fuentes, Gabriela Chacon, Andrea Zarate and Claudia Gonzalez Viejo
School of Agriculture and Food, Faculty of Veterinary and Agricultural Sciences, University of Melbourne, Melbourne, VIC 3010, Australia

Damir D. Torrico
School of Agriculture and Food, Faculty of Veterinary and Agricultural Sciences, University of Melbourne, Melbourne, VIC 3010, Australia
Department of Wine, Food and Molecular Biosciences, Faculty of Agriculture and Life Sciences, Lincoln University, Lincoln 7647, New Zealand

Luca Stevanato, Cristiano Lino Fontana, Marcello Lunardon, Sandra Moretto and Luca Morselli
Department of Physics and Astronomy, University of Padova, Via Marzolo 8, 35131 Padova, Italy

Gabriele Baroni
Department of Agricultural and Food Sciences, University of Bologna, Viale Fanin 50, 40127 Bologna, Italy

Yafit Cohen
Department of Sensing, Information and Mechanization Engineering, Agricultural Research Organization (ARO), Volcani Center, Rishon Lezion 7505101, Israel

Simone Gatto and Francesco Marinello
Department of Land, Environment, Agriculture and Forestry, University of Padova, Viale dell'Università 16, 35020 Legnaro, Italy

Sergio Castro-Garcia, Fernando Aragon-Rodriguez, Rafael R. Sola-Guirado and Jesús A. Gil-Ribes
Department of Rural Engineering, Universidad de Cordoba, 14004 Cordoba, Spain

Antonio J. Serrano and Emilio Soria-Olivas
IDAL, Intelligent Data Analysis Laboratory, Universidad de Valencia, 46100 Valencia, Spain

Dele Raheem
Northern Institute of Environmental and Minority Law, Arctic Centre, University of Lapland, 96101 Rovaniemi, Finland

Maxim Shishaev
Department of Information Systems and Technologies, Murmansk Arctic State University, 183038 Murmansk, Russia
Institute for Informatics and Mathematical Modelling, Kola Science Center of RAS, 184209 Apatity, Russia

Vladimir Dikovitsky
Institute for Informatics and Mathematical Modelling, Kola Science Center of RAS, 184209 Apatity, Russia

Chris Hacking and Nitesh Poona
Department of Geography and Environmental Studies, Stellenbosch University, Private Bag X1, Matieland 7602, South Africa

Nicola Manzan
Dipartimento di Scienze AgroAlimentari, Ambientali e Animali, University of Udine, Via delle Scienze 208, 33100 Udine, Italy

Carlos Poblete-Echeverría
Department of Viticulture and Oenology, Faculty of AgriSciences, Stellenbosch University, Private Bag X1, Matieland 7602, South Africa

Maggie Mulley and Lammert Kooistra
Laboratory of Geo-Information Science and Remote Sensing, Wageningen University and Research, 6700 AA Wageningen, The Netherlands

Laurens Bierens
TEC-IB B.V., Oude Veiling 29, 2635 GK Den Hoorn (ZH), The Netherlands

Elie Zemmour, Polina Kurtser and Yael Edan
Department of Industrial Engineering and Management, Ben-Gurion University of the Negev, Beer-Sheva 8410501, Israel

Barry Allred and Luis Martinez
U.S. Dept. of Agriculture, Agricultural Res. Serv., Soil Drainage Res. Unit, Columbus, OH 43210, USA

DeBonne Wishart
C.J. International Center for Water Resources Management, Central State University, Wilberforce, OH 45384, USA

Harry Schomberg, Steven Mirsky and George Meyers
U.S. Dept. of Agriculture, Agricultural Res. Serv., Beltsville Agric. Res. Cntr., Beltsville, MD 20705, USA

John Elliott and Christine Charyton
Veselka Farms, Pataskala, OH 43062, USA

Sebastián Romero-Bravo
Department of Agricultural Sciences, Universidad Católica del Maule, Curicó, Chile
Plant Breeding and Phenomic Center, Faculty of Agricultural Sciences, Universidad de Talca, Talca, Chile

Ana María Méndez-Espinoza, Miguel Garriga, Félix Estrada, Alejandro Escobar, Luis González-Martinez, Alejandro del Pozo and Gustavo A. Lobos
Plant Breeding and Phenomic Center, Faculty of Agricultural Sciences, Universidad de Talca, Talca, Chile

Daniel Sepulveda
Centro de Investigación y Transferencia en Riego y Agroclimatología (CITRA), Talca, Chile

Ivan Matus and Dalma Castillo
Centro Regional Investigación Quilamapu, Instituto de Investigaciones Agropecuarias, Chillán, Chile

Index

www.ingramcontent.com/pod-product-compliance
Lightning Source LLC
LaVergne TN
LVHW060512210926
848485LV00005B/1071